Springer Series in Computational Physics

Springer
Berlin
Heidelberg
New York
Barcelona
Budapest
Hong Kong
London
Milan
Paris
Santa Clara
Singapore
Tokyo

Springer Series in Computational Physics

C. A. J. Fletcher

Computational Techniques for Fluid Dynamics 1

Fundamental and General Techniques

Second Edition
With 138 Figures

 Springer

Professor Clive A. J. Fletcher

The University of New South Wales
Centre for Advanced Numerical Computation
in Engineering and Science – CANCES
Sydney 2052, Australia

2nd Edition 1991 · 6th Printing 2005

ISSN 0172-5726

ISBN 3-540-53058-4 2nd Edition
Springer-Verlag Berlin Heidelberg New York

ISBN 3-540-18151-2 1st Edition
Springer-Verlag Berlin Heidelberg New York

Library of Congress Cataloging-in-Publication Data. Fletcher, C. A. J. Computational techniques for fluid dynamics / C. A. J. Fletcher. - 2nd ed. p. cm. - (Springer series in computional physics) Includes bibliographical references and index. Contents: 1.Fundamental and general techniques. ISBN 3-540-53058-4 (Springer-Verlag Berlin, Heidelberg, New York). ISBN 0-387-53058-4 (Springer-Verlag New-York, Berlin, Heidelberg) 1. Fluid dynamics - Mathematics. 2. Fluid dynamics - Data processing. 3. Numercial analysis. ITitle. IISeries. QC151.F58 1991 532'.05'0151 dc20 90-22257

©Springer-Verlag Berlin Heidelberg 1998, 1991
Printed in Germany

Cover design: *design & production* GmbH, Heidelberg
Typesettig: Macmillan India Ltd., India
55/3111 – 5   – Printed on acid-free paper   SPIN: 11428404

Preface to the Second Edition

The purpose and organisation of this book are described in the preface to the first edition (1988). In preparing this edition minor changes have been made, particularly to Chap. 1 to keep it reasonably current. However, the rest of the book has required only minor modification to clarify the presentation and to modify or replace individual problems to make them more effective. The answers to the problems are available in *Solutions Manual for Computational Techniques for Fluid Dynamics* by C. A. J. Fletcher and K. Srinivas, published by Springer-Verlag, Heidelberg, 1991. The computer programs have also been reviewed and tidied up. These are available on an IBM-compatible floppy disc direct from the author.

I would like to take this opportunity to thank the many readers for their usually generous comments about the first edition and particularly those readers who went to the trouble of drawing specific errors to my attention. In this revised edition considerable effort has been made to remove a number of minor errors that had found their way into the original. I express the hope that no errors remain but welcome communication that will help me improve future editions.

In preparing this revised edition I have received considerable help from Dr. K. Srinivas, Nam-Hyo Cho, Zili Zhu and Susan Gonzales at the University of Sydney and from Professor W. Beiglböck and his colleagues at Springer-Verlag. I am very grateful to all of them.

Sydney, November 1990

C. A. J. Fletcher

Preface to the First Edition

The purpose of this two-volume textbook is to provide students of engineering, science and applied mathematics with the specific techniques, and the framework to develop skill in using them, that have proven effective in the various branches of computational fluid dynamics (CFD). Volume 1 describes both fundamental and general techniques that are relevant to all branches of fluid flow. Volume 2 provides specific techniques, applicable to the different categories of engineering flow behaviour, many of which are also appropriate to convective heat transfer.

An underlying theme of the text is that the competing formulations which are suitable for computational fluid dynamics, e.g. the finite difference, finite element, finite volume and spectral methods, are closely related and can be interpreted as part of a unified structure. Classroom experience indicates that this approach assists, considerably, the student in acquiring a deeper understanding of the strengths and weaknesses of the alternative computational methods.

Through the provision of 24 computer programs and associated examples and problems, the present text is also suitable for established research workers and practitioners who wish to acquire computational skills without the benefit of formal instruction. The text includes the most up-to-date techniques and is supported by more than 300 figures and 500 references.

For the conventional student the contents of Vol. 1 are suitable for introductory CFD courses at the final-year undergraduate or beginning graduate level. The contents of Vol. 2 are applicable to specialised graduate courses in the engineering CFD area. For the established research worker and practitioner it is recommended that Vol. 1 is read and the problems systematically solved before the individual's CFD project is started, if possible. The contents of Vol. 2 are of greater value after the individual has gained some CFD experience with his own project.

It is assumed that the reader is familiar with basic computational processes such as the solution of systems of linear algebraic equations, non-linear equations and ordinary differential equations. Such material is provided by Dahlquist, Bjorck and Anderson in *Numerical Methods;* by Forsythe, Malcolm and Moler in *Computer Methods for Mathematical Computation;* and by Carnaghan, Luther and Wilkes in *Applied Numerical Analysis.* It is also assumed that the reader has some knowledge of fluid dynamics. Such knowledge can be obtained from *Fluid Mechanics* by Streeter and Wylie; from *An Indroduction of Fluid Dynamics* by Batchelor; or from *Incompressible Flow* by Panton, amongst others.

Computer programs are provided in the present text for guidance and to make it easier for the reader to write his own programs, either by using equivalent constructions, or by modifying the programs provided. In the sense that the CFD practitioner is as likely to inherit an existing code as to write his own from scratch, some practice in modifying existing, but simple, programs is desirable. An IBM-compatible floppy disk containing the computer programs may be obtained from the author.

The contents of Vol. 1 are arranged in the following way. Chapter 1 contains an introduction to computational fluid dynamics, designed to give the reader an appreciation of why CFD is so important, the sort of problems it is capable of solving and an overview of how CFD is implemented. The equations governing fluid flow are usually expressed as partial differential equations. Chapter 2 describes the different classes of partial differential equations and appropriate boundary conditions and briefly reviews traditional methods of solution.

Obtaining computational solutions consists of two stages: the reduction of the partial differential equations to algebraic equations and the solution of the algebraic equations. The first stage, called discretisation, is examined in Chap. 3 with special emphasis on the accuracy. Chapter 4 provides sufficient theoretical background to ensure that computational solutions can be related properly to the usually unknown "exact" solution. Weighted residual methods are introduced in Chap. 5 as a vehicle for investigating and comparing the finite element, finite volume and spectral methods as alternative means of discretisation. Specific techniques to solve the algebraic equations resulting from discretisation are described in Chap. 6. Chapters 3 – 6 provide essential background information.

The one-dimensional diffusion equation, considered in Chap. 7, provides the simplest model for highly dissipative fluid flows. This equation is used to contrast explicit and implicit methods and to discuss the computational representation of derivative boundary conditions. If two or more spatial dimensions are present, splitting techniques are usually required to obtain computational solutions efficiently. Splitting techniques are described in Chap. 8. Convective (or advective) aspects of fluid flow, and their effective computational prediction, are examined in Chap. 9. The convective terms are usually nonlinear. The additional difficulties that this introduces are considered in Chap. 10. The general techniques, developed in Chaps. 7 – 10, are utilised in constructing specific techniques for the different categories of flow behaviour, as is demonstrated in Chaps. 14 – 18 of Vol. 2.

In preparing this textbook I have been assisted by many people. In particular I would like to thank Dr. K. Srinivas, Nam-Hyo Cho and Zili Zhu for having read the text and made many helpful suggestions. I am grateful to June Jeffery for producing illustrations of a very high standard. Special thanks are due to Susan Gonzales, Lyn Kennedy, Marichu Agudo and Shane Gorton for typing the manuscript and revisions with commendable accuracy, speed and equilibrium while coping with both an arbitrary author and recalcitrant word processors.

It is a pleasure to acknowledge the thoughtful assistance and professional competence provided by Professor W. Beiglböck, Ms. Christine Pendl, Mr. R. Michels and colleagues at Springer-Verlag in the production of this textbook. Finally I express deep gratitude to my wife, Mary, who has been unfailingly supportive while accepting the role of book-widow with her customary good grace.

Sydney, October 1987   *C. A. J. Fletcher*

Contents

1. Computational Fluid Dynamics: An Introduction

This chapter provides an overview of computational fluid dynamics (CFD) with emphasis on its cost-effectiveness in design. Some representative applications are described to indicate what CFD is capable of. The typical structure of the equations governing fluid dynamics is highlighted and the way in which these equations are converted into computer-executable algorithms is illustrated. Finally attention is drawn to some of the important sources of further information.

1.1 Advantages of Computational Fluid Dynamics

The establishment of the science of fluid dynamics and the practical application of that science has been under way since the time of Newton. The theoretical development of fluid dynamics focuses on the construction and solution of the governing equations for the different categories of fluid dynamics and the study of various approximations to those equations.

The governing equations for Newtonian fluid dynamics, the unsteady Navier-Stokes equations, have been known for 150 years or more. However, the development of reduced forms of these equations (Chap. 16) is still an active area of research as is the turbulent closure problem for the Reynolds-averaged Navier-Stokes equations (Sect. 11.5.2). For non-Newtonian fluid dynamics, chemically reacting flows and two-phase flows the theoretical development is at a less advanced stage.

Experimental fluid dynamics has played an important role in validating and delineating the limits of the various approximations to the governing equations. The wind tunnel, as a piece of experimental equipment, provides an effective means of simulating real flows. Traditionally this has provided a cost-effective alternative to full-scale measurement. In the design of equipment that depends critically on the flow behaviour, e.g. aircraft design, full-scale measurement as part of the design process is economically unavailable.

The steady improvement in the speed of computers and the memory size since the 1950s has led to the emergence of computational fluid dynamics (CFD). This branch of fluid dynamics complements experimental and theoretical fluid dynamics by providing an alternative cost-effective means of simulating real flows. As such it offers the means of testing theoretical advances for conditions unavailable experimentally. For example wind tunnel experiments are limited to a certain range of Reynolds numbers, typically one or two orders of magnitude less than full scale.

Computational fluid dynamics also provides the convenience of being able to switch off specific terms in the governing equations. This permits the testing of theoretical models and, inverting the connection, suggests new paths for theoretical exploration.

The development of more efficient computers has generated the interest in CFD and, in turn, this has produced a dramatic improvement in the efficiency of the computational techniques. Consequently CFD is now the preferred means of testing alternative designs in many branches of the aircraft, flow machinery and, to a lesser extent, automobile industries.

Following Chapman et al. (1975), Chapman (1979, 1981), Green (1982), Rubbert (1986) and Jameson (1989) CFD provides five major advantages compared with experimental fluid dynamics:

(i) Lead time in design and development is significantly reduced.
(ii) CFD can simulate flow conditions not reproducible in experimental model tests.
(iii) CFD provides more detailed and comprehensive information.
(iv) CFD is increasingly more cost-effective than wind-tunnel testing.
(v) CFD produces a lower energy consumption.

Traditionally, large lead times have been caused by the necessary sequence of design, model construction, wind-tunnel testing and redesign. Model construction is often the slowest component. Using a well-developed CFD code allows alternative designs (different geometric configurations) to be run over a range of parameter values, e.g. Reynolds number, Mach number, flow orientation. Each case may require 15 min runs on a supercomputer, e.g. CRAY Y-MP. The design optimisation process is essentially limited by the ability of the designer to absorb and assess the computational results. In practice CFD is very effective in the early elimination of competing design configurations. Final design choices are still confirmed by wind-tunnel testing.

Rubbert (1986) draws attention to the speed with which CFD can be used to redesign minor components, if the CFD packages have been thoroughly validated. Rubbert cites the example of the redesign of the external contour of the Boeing 757 cab to accommodate the same cockpit components as the Boeing 767 to minimise pilot conversion time. Rubbert indicates that CFD provided the external shape which was incorporated into the production schedule before any wind-tunnel verification was undertaken.

Wind-tunnel testing is typically limited in the Reynolds number it can achieve, usually short of full scale. Very high temperatures associated with coupled heat transfer fluid flow problems are beyond the scope of many experimental facilities. This is particularly true of combustion problems where the changing chemical composition adds another level of complexity. Some categories of unsteady flow motion cannot be properly modelled experimentally, particularly where geometric unsteadiness occurs as in certain categories of biological fluid dynamics. Many geophysical fluid dynamic problems are too big or too remote in space or time to simulate experimentally. Thus oil reservoir flows are generally inaccessible to detailed experimental measurement. Problems of astrophysical fluid dynamics are too remote spatially and weather patterns must be predicted *before* they occur. All of these categories of fluid motion are amenable to the computational approach.

Experimental facilities, such as wind tunnels, are very effective for obtaining global information, such as the complete lift and drag on a body and the surface pressure distributions at key locations. However, to obtain detailed velocity and pressure distributions throughout the region surrounding a body would be prohibitively expensive and very time consuming. CFD provides this detailed information at no additional cost and consequently permits a more precise understanding of the flow processes to be obtained.

Perhaps the most important reason for the growth of CFD is that for much mainstream flow simulation, CFD is significantly cheaper than wind-tunnel testing and will become even more so in the future. Improvements in computer hardware performance have occurred hand in hand with a decreasing hardware cost. Consequently for a given numerical algorithm and flow problem the relative cost of a computational simulation has decreased significantly historically (Fig. 1.1). Paralleling the improvement in computer hardware has been the improvement in the efficiency of computational algorithms for a given problem. Current improvements in hardware cost and computational algorithm efficiency show no obvious sign of reaching a limit. Consequently these two factors combine to make CFD increasingly cost-effective. In contrast the cost of performing experiments continues to increase.

The improvement in computer hardware and numerical algorithms has also brought about a reduction in energy consumption to obtain computational flow simulations. Conversely, the need to simulate more extreme physical conditions, higher Reynolds number, higher Mach number, higher temperature, has brought about an increase in energy consumption associated with experimental testing.

The chronological development of computers over the last thirty years has been towards faster machines with larger memories. A modern supercomputer such as

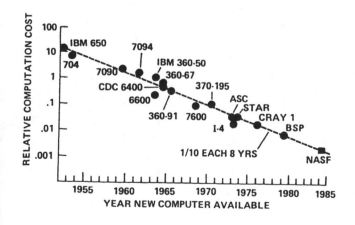

Fig. 1.1. Relative cost of computation for a given algorithm and flow (after Chapman, 1979; reprinted with permission of AIAA)

the CRAY Y-MP is capable of operating at more than 2000 Megaflops (Dongarra 1989). A Megaflop is one million floating-point arithmetic operations per second. More recent supercomputers, e.g. the NEC SX3, are capable of theoretical speeds of 20 000 Megaflops. The speed comes partly from a short machine cycle time, that is the time required for each cycle of logic operations. The CRAY Y-MP has a cycle time of 6 nanoseconds (6×10^{-9} s) whereas the NEC SX3 has a cycle time of 2.9 ns.

A specific operation, e.g. a floating point addition, can be broken up into a number of logic operations each one of which requires one machine cycle to execute. If the same operation, e.g. floating point addition, is to be applied sequentially to a large number of elements in a vector, it is desirable to treat each logic operation sequentially but to permit different logic operations associated with each vector element to be executed concurrently. Thus there is a considerable overlap and a considerable speed-up in the overall execution time if the computational algorithm can exploit such a pipeline arrangement.

Modern supercomputers have special vector processors that utilise the pipeline format. However vector processors have an effective "start-up" time that makes them slower than scalar processors for very short vectors. One can define a break-even vector length, N_b, for which the vector processor has the same speed as a scalar processor. For very long vectors ($N = \infty$) the theoretical vector processor speed is achieved.

To compare the efficiency of different vector-processing computers it is (almost) standard practice to consider $N_{1/2}$ (after Hockney and Jesshope 1981), which is the vector length for which half the asymptotic peak vector processing performance ($N = \infty$) is achieved. The actual $N_{1/2}$ is dependent on the specific operations being performed as well as the hardware. For a SAXPY operation ($S = AX + Y$), $N_{1/2} = 37$ for a CRAY X-MP and $N_{1/2} = 238$ for a CYBER 205. For most modern supercomputers, $30 \leq N_{1/2} \leq 100$.

The speed-up due to vectorisation is quantifiable by considering Amdahl's law which can be expressed as (Gentzsch and Neves 1988)

$$G = [(1 - P) + P/R]^{-1} \quad \text{and} \quad R = V(N)/S \tag{1.1}$$

where G is the overall gain in speed of the process (overall speed-up ratio)
$V(N)$ is the vector processor speed for an N component vector process
S is the scalar processor speed for a single component process
P is the proportion of the process that is vectorized and
R is the vector processor speed-up ratio.

As is indicated in Fig. 1.2 a vector processor with a theoretical ($N = \infty$) vector speed-up ratio, $R = 10$, must achieve a high percentage vectorisation, say $P > 0.75$, to produce a significant overall speed-up ratio, G. But at this level $\partial G/\partial P \gg \partial G/\partial R$. Thus modification of the computer program to increase P will provide a much bigger increase in G than modifying the hardware to increase V and hence R. In addition unless a large proportion of the computer program can be written so that vector lengths are significantly greater than $N_{1/2}$, the overall speed-up ratio, G, will not be very great.

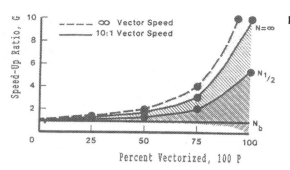

Fig. 1.2. Amdahl's Law

The ability to increase the overall execution speed to the limit set by the hardware depends partly on the ability of the operating system and compiler to vectorise the computational algorithm and partly on choosing computational algorithms that are inherently vectorisable (Ortega and Voigt 1985). The structuring of computational algorithms to permit vectorisation is an important research topic but is beyond the scope of this book (see Gentzsch and Neves 1988). The long term trend would appear to be towards making the operating system and compiler take care of the vectorisation with less emphasis on the user having to manipulate the basic algorithm.

With a pipeline architecture, an efficient vector instruction set and as small a cycle time as possible the major means of further increasing the processing speed is to introduce multiple processors operating in parallel. Supercomputers are typically being designed with up to sixteen processors in parallel. Theoretically this should provide up to a factor of sixteen improvement in speed. Experiments by Grassl and Schwarzmeier (1990) with an eight-processor CRAY Y-MP indicate that 84% of the theoretical improvement can be achieved for a typical CFD code such as ARC3D (Vol. 2, Sect. 18.4.1).

The concept of an array of processors each operating on an element of a vector has been an important feature in the development of more efficient computer architecture (Hockney and Jesshope 1981). The Illiac IV had 64 parallel processors and achieved an overall processing speed comparable to the CRAY-1 and CYBER-205 even though the cycle time was only 80 ns. However Amdahl's law, (1.1), also applies to parallel processors if R is replaced by N_P, the number of parallel processors, and P is the proportion of the process that is parallelisable. The relative merits of pipeline and parallel processing are discussed in general terms by Levine (1982), Ortega and Voigt (1985) and in more detail by Hockney and Jesshope (1981) and Gentzsch and Neves (1988).

The development of bigger and cheaper memory modules is being driven by the substantial commercial interest in data storage and manipulation. For CFD applications it is important that the complete program, both instructions and variable storage, should reside in main memory. This is because the speed of data transfer from secondary (disc) storage to main memory is much slower than data transfer rates between the main memory and the processing units. In the past the main memory size has typically limited the complexity of the CFD problems under investigation.

The chronological trend of increasing memory capacity for supercomputers is impressive. The CDC-7600 (1970 technology) had a capacity of 4×10^5 64-bit words. The CYBER-205 (1980 technology) has a capacity of 3×10^7 64-bit words and the CRAY-2 (1990 technology) has a capacity of 10^9 64-bit words.

Significant developments in minicomputers in the 1970s and microcomputers in the 1980s have provided many alternative paths to cost-effective CFD. The relative cheapness of random access memory implies that large problems can be handled efficiently on micro- and minicomputers. The primary difference between microcomputers and mainframes is the significantly slower cycle time of a microcomputer and the simpler, less efficient architecture. However the blurring of the distinction between microcomputers and personal workstations, such as the SUN Sparcstation, and the appearance of minisupercomputers has produced a *price/performance continuum* (Gentzsch and Neves 1988).

The coupling of many, relatively low power, parallel processors is seen as a very efficient way of solving complex CFD problems. Each processor can use fairly standard microcomputer components; hence the potentially low cost. A typical system, QCDPAX, is described by Hoshino (1989). This system has from 100 to 1000 processing units, each based on the L64132 floating point processor. Thus a system of 400 processing units is expected to deliver about 2000 Megaflops when operating on a representative CFD code.

To a certain extent the relative slowness of microcomputer-based systems can be compensated for by allowing longer running times. Although 15 mins on a

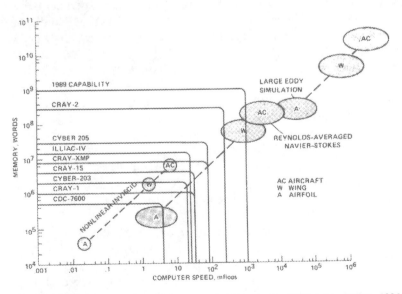

Fig. 1.3. Computer speed and memory requirements for CFD (after Bailey, 1986; reprinted with permission of Japan Society of Computational Fluid Dynamics)

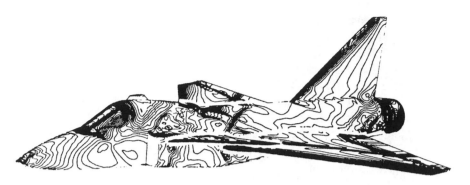

Fig. 1.4. Surface pressure distribution on a typical military aircraft. Surface pressure contours, $\Delta C_p = 0.02$ (after Arlinger, 1986; reprinted with permission of Japan Society of Computational Fluid Dynamics)

supercomputer appears to be the accepted norm (Bailey 1986) for routine design work, running times of a few hours on a microcomputer may well be acceptable in the research and development area. This has the advantage of allowing the CFD research worker adequate time to interpret the results and to prepare additional cases.

The future trends for computer speed and memory capacity are encouraging. Predictions by Simon (1989) indicate that by 2000 one may expect sustained computer speeds up to 10^6 Megaflops and main memory capacities of 50 000 Megawords. This is expected to be adequate (Fig. 1.3) for predictions of steady viscous (turbulent) compressible flow around complete aircraft and to allow global design optimisation to be considered seriously.

1.2 Typical Practical Problems

Computational fluid dynamics, particularly in engineering, is still at the stage of development where "problems involving complex geometries can be treated with simple physics and those involving simple geometry can be treated with complex physics" (Bailey 1986). What is changing is the accepted norm for simplicity and complexity. Representative examples are provided below.

1.2.1 Complex Geometry, Simple Physics

The surface pressure distribution on a typical supersonic military aircraft is shown in Fig. 1.4. The freestream Mach number is 1.8 and the angle of attack is 8°. The aircraft consists of a fuselage, canopy, engine inlets, fin, main delta wing and forward (canard) wings. In addition control surfaces at the trailing edge of the delta wing are deflected upwards 10°. Approximately 19 000 grid points are required in each cross-section plane at each downstream location. The complexity of the geometry places a considerable demand on the grid generating procedure. Arlinger (1986) uses an algebraic grid generation technique based on transfinite interpolation (Sect. 13.3.4).

The flow is assumed inviscid and everywhere supersonic so that an explicit marching scheme in the freestream direction can be employed. This is equivalent to the procedure described in Sect. 14.2.4. The explicit marching scheme is particularly efficient with the complete flowfield requiring 15 minutes on a CRAY-1. The finite volume method (Sect. 5.2) is used to discretise the governing equations. Arlinger stresses that the key element in obtaining the results efficiently is the versatile grid generation technique.

1.2.2 Simpler Geometry, More Complex Physics

The limiting particle paths on the upper surface of a three-dimensional wing for increasing freestream Mach number, M_∞, are shown in Fig. 1.5. The limiting particle paths correspond to the surface oil-flow patterns that would be obtained experimentally. The results shown in Fig. 1.5 come from computations (Holst et al. 1986) of the transonic viscous flow past a wing at $2°$ angle of attack, with an aspect ratio of 3 and a chord Reynolds number of 8×10^6.

For these conditions a shock wave forms above the wing and interacts with the upper surface boundary layer causing massive separation. The region of separation changes and grows as M_∞ increases. The influence of the flow past the wingtip makes the separation pattern very three-dimensional. The terminology, spiral node, etc., indicated in Fig. 1.5 is appropriate to the classification of three-dimensional separation (Tobak and Peake 1982).

The solutions require a three-dimensional grid of approximately 170 000 points separated into four partially overlapping zones. The two zones immediately above and below the wing have a fine grid in the normal direction to accurately predict the severe velocity gradients that occur. In these two zones the thin layer Navier-Stokes equations (Sect. 18.1.3) are solved. These equations include viscous terms only associated with the normal direction. They are an example of reduced Navier-Stokes equations (Chap. 16). In the two zones away from the wing the flow is assumed inviscid and governed by the Euler equations (Sect. 11.6.1).

The grid point solutions in all zones are solved by marching a pseudo-transient form (Sect. 6.4) of the governing equations in time until the solution no longer changes. To do this an implicit procedure is used similar to that described in Sect. 14.2.8. The zones are connected by locally interpolating the overlap region, typically two cells. Holst indicates that stable solutions are obtained even though severe gradients cross zonal boundaries.

By including viscous effects the current problem incorporates significantly more complicated flow behaviour, and requires a more sophisticated computational algorithm, than the problem considered in Sect. 1.2.1. However, the shape of the computational domain is considerably simpler. In addition the computational grid is generated on a zonal basis which provides better control over the grid point locations.

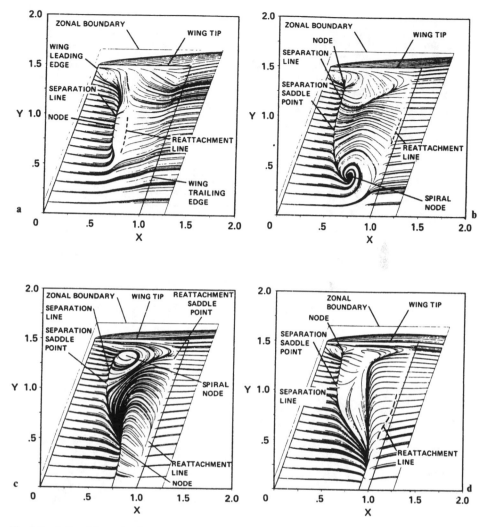

Fig. 1.5a–d. Particle paths for upper wing surface flow. (a) $M_\infty = 0.80$ (b) $M_\infty = 0.85$ (c) $M_\infty = 0.90$ (d) $M_\infty = 0.95$ (after Holst et al., 1986; reprinted with permission of Japan Society of Computational Fluid Dynamics)

1.2.3 Simple Geometry, Complex Physics

To illustrate this category a meteorological example is used instead of an engineering example. Figure 1.6 shows a four-day forecast (b) of the surface pressure compared with measurements (a). This particular weather pattern was associated with a severe storm on January 29, 1990 which caused substantial property damage in the southern part of England. The computations predict the developing weather pattern quite closely.

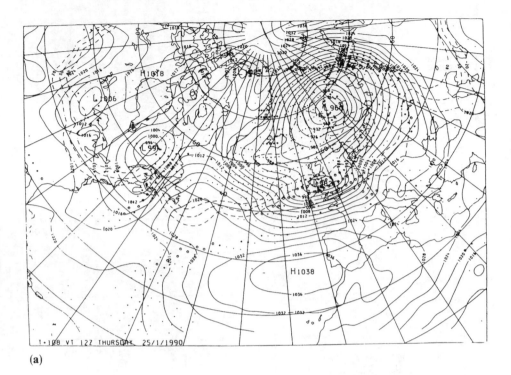

(a)

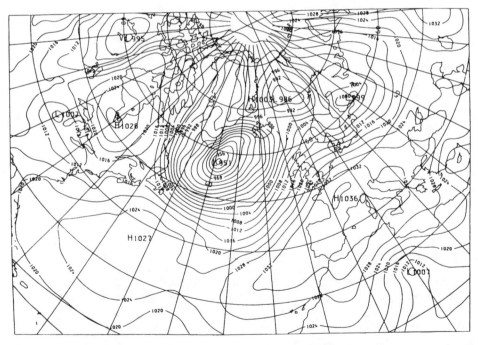

(b)

Fig. 1.6a, b. Surface pressure comparison. (a) Measurements; (b) Predictions (after Cullen, 1990; reprinted with permission of the Meteorological Office, U.K.)

The governing equations (Cullen 1983) are essentially inviscid but account for wind, temperature, pressure, humidity, surface stresses over land and sea, heating effect, precipitation and other effects (Haltiner and Williams 1980). The equations are typically written in spherical polar coordinates parallel to the earth's surface and in a normalised pressure coordinate perpendicular to the earth's surface. Consequently difficulties associated with an irregular computational boundary and grid generation are minimal.

Cullen (1990) indicates that the results shown in Fig. 1.6 were obtained on a $192 \times 120 \times 15$ grid and used a split explicit finite difference scheme to advance the solution in time. This permits the complete grid to be retained in main memory. 432 time steps are used for a $4\frac{1}{2}$ day forecast and require 20 minutes processing time on a CYBER 205.

Cullen (1983) reports that the major problem in extending accurate large-scale predictions beyond 3 to 4 days is obtaining initial data of sufficient quality. For more refined local predictions further difficulties arise in preventing boundary disturbances from contaminating the interior solution and in accurately representing the severe local gradients associated with fronts.

For global circulation modelling and particularly for long-term predictions the spectral method (Sect. 5.6) is well suited to spherical polar geometry. Spectral methods are generally more economical than finite difference or finite element methods for comparable accuracy, at least for global predictions. The application of spectral methods to weather forecasting is discussed briefly by Fletcher (1984) and in greater detail by Bourke et al. (1977). Chervin (1989) provides a recent indication of the capability of CFD for climate modelling.

The above examples are indicative of the current status of CFD. For the future Bailey (1986) states that "more powerful computers with more memory capacity are required to solve problems involving both complex geometries and complex physics". The growth in human expectations will probably keep this statement current for a long time to come.

1.3 Equation Structure

A connecting feature of the categories of fluid dynamics considered in this book is that the fluid can be interpreted as a continuous medium. As a result the behaviour of the fluid can be described in terms of the velocity and thermodynamic properties as continuous functions of time and space.

Application of the principles of conservation of mass, momentum and energy produces systems of partial differential equations (Vol. 2, Chap. 11) for the velocity and thermodynamic variables as functions of time and position. With boundary and initial conditions appropriate to the given flow and type of partial differential equation the mathematical description of the problem is established.

Many flow problems involve the developing interaction between convection and diffusion. A simple example is indicated in Fig. 1.7, which shows the temperature distribution of fluid in a pipe at different times. It is assumed that the fluid

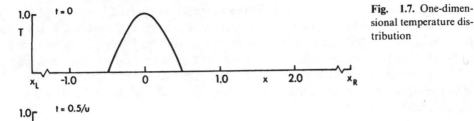

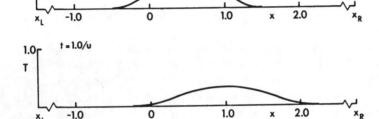

Fig. 1.7. One-dimensional temperature distribution

is moving to the right with constant velocity u and that the temperature is constant across the pipe.

The temperature as a function of x and t is governed by the equation

$$\frac{\partial T}{\partial t} + u\frac{\partial T}{\partial x} - \alpha\frac{\partial^2 T}{\partial x^2} = 0 \quad \text{for } x_L \leq x \leq x_R \quad \text{and} \quad t > 0 . \tag{1.2}$$

With a suitable nondimensionalisation, appropriate boundary and initial conditions are

$$T(x_L, t) = T(x_R, t) = 0 \quad \text{and} \tag{1.3}$$

$$T(x, 0) = \cos \pi x, \quad -0.5 \leq x \leq 0.5$$
$$= 0 , \qquad x < -0.5 \quad \text{and} \quad x > 0.5 . \tag{1.4}$$

Equations (1.2–4) provide a mathematical description of the problem. The term $\alpha\partial^2 T/\partial x^2$ is the diffusion term and α is the thermal diffusivity. This term is responsible for the spread of the nonzero temperature both to the right and to the left; if α is small the spread is small. Computational techniques for dealing with equations containing such terms are dealt with in Chaps. 7 and 8.

The term $u\partial T/\partial x$ is the convection term and is responsible for the temperature distribution being swept bodily to the right with the known velocity u. The treatment of this term and the complete transport equation (1.2) are considered in Chap. 9. In more than one dimension convective and diffusive terms appear associated with each direction (Sect. 9.5).

Since u is known, (1.2) is linear in T. However, when solving for the velocity field it is necessary to consider equations with nonlinear convective terms. A prototype for such a nonlinearity is given by Burgers' equation (Sect. 10.1)

$$\frac{\partial u}{\partial t} + u\frac{\partial u}{\partial x} - \alpha\frac{\partial^2 u}{\partial x^2} = 0 \ . \tag{1.5}$$

The nonlinear convective term, $u\partial u/\partial x$, permits very steep gradients in u to develop if α is very small. Steep gradients require finer grids and the presence of the nonlinearity often necessitates an additional level of iteration in the computational algorithm.

Some flow and heat transfer problems are governed by Laplace's equation,

$$\frac{\partial^2 \phi}{\partial x^2} + \frac{\partial^2 \phi}{\partial y^2} = 0 \ . \tag{1.6}$$

This is the case for a flow which is inviscid, incompressible and irrotational. In that case ϕ is the velocity potential (Sect. 11.3). Laplace's equation is typical of the type of equation that governs equilibrium or steady problems (Chap. 6). Laplace's equation also has the special property of possessing simple exact solutions which can be added together (superposed) since it is linear. These properties are exploited in the techniques described in Sect. 14.1.

For many flow problems more than one dependent variable will be involved and it is necessary to consider systems of equations. Thus one-dimensional unsteady inviscid compressible flow is governed by (Sect. 10.2)

$$\frac{\partial \varrho}{\partial t} + \frac{\partial(\varrho u)}{\partial x} = 0 \ , \tag{1.7a}$$

$$\frac{\partial(\varrho u)}{\partial t} + \frac{\partial}{\partial x}(\varrho u^2 + p) = 0 \ , \tag{1.7b}$$

$$\frac{\partial E}{\partial t} + \frac{\partial}{\partial x}[u(p + E)] = 0 \ , \tag{1.7c}$$

where p is the pressure and E is the total energy per unit volume given by

$$E = \frac{p}{\gamma - 1} + 0.5\,\varrho u^2 \ , \tag{1.8}$$

and γ is the ratio of specific heats. Although equations (1.7) are nonlinear the structure is similar to (1.5) without the diffusive terms. The broad strategy of the computational techniques developed for scalar equations will also be applicable to systems of equations.

For flow problems where the average properties of the turbulence need to be included the conceptual equation structure could be written as follows

$$\frac{\partial u}{\partial t} + u\frac{\partial u}{\partial x} - \frac{\partial}{\partial x}\left(\alpha\frac{\partial u}{\partial x}\right) = S \ , \tag{1.9}$$

where "α" is now a function of the dependent variable u, and S is a source term containing additional turbulent contributions. However, it should be made clear (Sects. 11.4.2 and 11.5.2) that turbulent flows are at least two-dimensional and often three-dimensional and that a system of equations is required to describe the flow.

1.4 Overview of Computational Fluid Dynamics

The total process of determining practical information about problems involving fluid motion can be represented schematically as in Fig. 1.8.

The governing equations (Chap. 11) for flows of practical interest are usually so complicated that an exact solution is unavailable and it is necessary to seek a computational solution. Computational techniques replace the governing partial differential equations with systems of algebraic equations, so that a computer can be used to obtain the solution. This book will be concerned with the computational techniques for obtaining and solving the systems of algebraic equations.

For local methods, like the finite difference, finite element and finite volume methods, the algebraic equations link together values of the dependent variables at adjacent grid points. For this situation it is understood that a grid of discrete points is distributed throughout the computational domain, in time and space. Consequently one refers to the process of converting the continuous governing equations

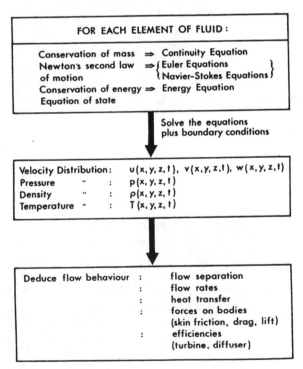

Fig. 1.8. Overview of computational fluid dynamics

to a system of algebraic equations as discretisation (Chap. 3). For a global method, like the spectral method, the dependent variables are replaced with amplitudes associated with different frequencies, typically.

The algebraic equations produced by discretisation could arise as follows. A typical finite difference representation of (1.2) would be

$$\frac{T_j^{n+1} - T_j^n}{\Delta t} + \frac{u(T_{j+1}^n - T_{j-1}^n)}{2\Delta x} = \frac{\alpha(T_{j-1}^n - 2T_j^n + T_{j+1}^n)}{\Delta x^2} , \tag{1.10}$$

where $x = j\,\Delta x$ and $t = n\,\Delta t$.

If the solution is known at all grid points x_j at time level n, (1.10) can be used to provide an algorithm for T_j^{n+1}, i.e.

$$T_j^{n+1} = T_j^n - \left(\frac{u\Delta t}{2\Delta x}\right)(T_{j+1}^n - T_{j-1}^n) + \left(\frac{\alpha\Delta t}{\Delta x^2}\right)(T_{j-1}^n - 2T_j^n + T_{j+1}^n) . \tag{1.11}$$

Repeated use of (1.11) generates the solution at all interior grid points, x_j, at time level $n+1$. Incrementing n and substituting the values T^{n+1} into the right-hand side of (1.11) allows the discrete solution to be marched forward in time.

For a local method, e.g. the finite difference method, the required number of grid points for an accurate solution typically depends on the dimensionality, the geometric complexity and severity of the gradients of the dependent variables. For the flow about a complete aircraft a grid of ten million points might be required. At each grid point each dependent variable and certain auxiliary variables must be stored. For turbulent compressible three-dimensional flow this may require anywhere between five and thirty dependent variables per grid point. For efficient computation all of these variables must be stored in main memory.

Since the governing equations for most classes of fluid dynamics are nonlinear the computational solution usually proceeds iteratively. That is, the solution for each dependent variable at each grid point is sequentially corrected using the discretised equations. The iterative process is often equivalent to advancing the solution over a small time step (Chap. 6). The number of iterations or time steps might vary from a few hundred to several thousand.

The discretisation process introduces an error that can be reduced, in principle, by refining the grid as long as the discrete equations, e.g. (1.10), are faithful representations of the governing equations (Sect. 4.2). If the numerical algorithm that performs the iteration or advances in time is also stable (Sect. 4.3), then the computational solution can be made arbitrarily close to the true solution of the governing equations, by refining the grid, if sufficient computer resources are available.

Although the solution is often sought in terms of discrete nodal values some methods, e.g., the finite element and spectral methods, do explicitly introduce a continuous representation for the computational solution. Where the underlying physical problem is smooth such methods often provide greater accuracy per unknown in the discretised equations. Such methods are discussed briefly in Chap. 5.

1.5 Further Reading

The purpose of the present text is to provide an introduction to the computational techniques that are appropriate for solving flow problems. More specific information is available in other books, review articles, journal articles and conference proceedings.

Richtmyer and Morton (1967) construct a general theoretical framework for analysing computational techniques relevant to fluid dynamics and discuss specific finite difference techniques for inviscid compressible flow. Roache (1976) examines viscous separated flow for both incompressible and compressible conditions but concentrates on finite difference techniques. More recently, Peyret and Taylor (1983) have considered computational techniques for the various branches of fluid dynamics with more emphasis on finite difference and spectral methods. Holt (1984) describes very powerful techniques for boundary layer flow and inviscid compressible flow. Book (1981) considers finite difference techniques for both engineering and geophysical fluid dynamics where the diffusive mechanisms are absent or very small.

Thomasset (1981), Baker (1983) and Glowinski (1984) examine computational techniques based on the finite element method and Fletcher (1984) provides techniques for the finite element and spectral methods. Canuto et al. (1987) analyse computational techniques based on spectral methods. Haltiner and Williams (1980) discuss computational techniques for geophysical fluid dynamics.

The review articles by Chapman (1975, 1979, 1981), Green (1982), Krause (1985), Kutler (1985) and Jameson (1989) indicate what engineering CFD is currently capable of and what will be possible in the future. These articles have a strong aeronautical leaning. A more general review is provided by Turkel (1982). Cullen (1983) and Chervin (1989) review the current status of meteorological CFD. Review papers on specific branches of computational fluid dynamics appear in Annual Reviews of Fluid Dynamics, in the lecture series of the von Karman Institute and in the monograph series of Pineridge Press. More advanced computational techniques which exploit vector and parallel computers will not be covered in this book. However Ortega and Voigt (1985) and Gentzsch and Neves (1988) provide a comprehensive survey of this area.

Relevant journal articles appear in AIAA Journal, Journal of Computational Physics, International Journal of Numerical Methods in Fluids, Computer Methods in Applied Mechanics and Engineering, Computers and Fluids, Applied Mathematical Modelling, Comm nications in Applied Numerical Methods, Theoretical and Computational Fluid Dynamics, Numerical Heat Transfer, Journal of Applied Mechanics and Journal of Fluids Engineering. Important conferences are the International Conference series on Numerical Methods in Fluid Dynamics, International Symposium series on Computational Fluid Dynamics, the AIAA CFD conference series, the GAMM conference series, Finite Elements in Flow Problems conference series, the Numerical Methods in Laminar and Turbulent Flow conference series and many other specialist conferences.

2. Partial Differential Equations

In this chapter, procedures will be developed for classifying partial differential equations as elliptic, parabolic or hyperbolic. The different types of partial differential equations will be examined from both a mathematical and a physical viewpoint to indicate their key features and the flow categories for which they occur.

The governing equations for fluid dynamics (Vol. 2, Chap. 11) are partial differential equations containing first and second derivatives in the spatial coordinates and first derivatives only in time. The time derivatives appear linearly but the spatial derivatives often appear nonlinearly. Also, except for the special case of potential flow, systems of governing equations occur rather than a single equation.

2.1 Background

For linear partial differential equations of second-order in two independent variables a simple classification (Garabedian 1964, p. 57) is possible. Thus for the partial differential equation (PDE)

$$A\frac{\partial^2 u}{\partial x^2} + B\frac{\partial^2 u}{\partial x \partial y} + C\frac{\partial^2 u}{\partial y^2} + D\frac{\partial u}{\partial x} + E\frac{\partial u}{\partial y} + Fu + G = 0 \ , \tag{2.1}$$

where A to G are constant coefficients, three categories of partial differential equation can be distinguished. These are

$$\begin{aligned}
&\text{elliptic PDE:} \quad B^2 - 4AC < 0 \ , \\
&\text{parabolic PDE:} \quad B^2 - 4AC = 0 \ , \\
&\text{hyperbolic PDE:} \quad B^2 - 4AC > 0 \ .
\end{aligned} \tag{2.2}$$

It is apparent that the classification depends only on the highest-order derivatives in each independent variable.

For two-dimensional steady compressible potential flow about a slender body the governing equation, similar to (11.109), is

$$(1 - M_\infty^2)\frac{\partial^2 \phi}{\partial x^2} + \frac{\partial^2 \phi}{\partial y^2} = 0 \ . \tag{2.3}$$

Applying the criteria (2.2) indicates that (2.3) is elliptic for subsonic flow ($M_\infty < 1$) and hyperbolic for supersonic flow ($M_\infty > 1$).

If the coefficients, A to G in (2.1), are functions of x, y, u, $\partial u/\partial x$ or $\partial u/\partial y$, (2.2) can still be used if A, B and C are given a local interpretation. This implies that the classification of the governing equations can change in different parts of the computational domain.

The governing equation for steady, compressible, potential flow, (11.103), can be written in two-dimensional natural coordinates as

$$(1 - M^2)\frac{\partial^2 \phi}{\partial s^2} + \frac{\partial^2 \phi}{\partial n^2} = 0 , \tag{2.4}$$

where s and n are parallel and perpendicular to the local streamline direction, and M is the local Mach number. Applying conditions (2.2) on a local basis indicates that (2.4) is elliptic, parabolic or hyperbolic as $M < 1$, $M = 1$ or $M > 1$. A typical distribution of local Mach number, M, for the flow about an aerofoil or turbine blade, is shown in Fig. 11.15. The feature that the governing equation can change its type in different parts of the computational domain is one of the major complicating factors in computing transonic flow (Sect. 14.3).

The introduction of simpler flow categories (Sect. 11.2.6) may introduce a change in the equation type. The governing equations for two-dimensional steady, incompressible viscous flow, (11.82–84) without the $\partial u/\partial t$ and $\partial v/\partial t$ terms, are elliptic. However, introduction of the boundary layer approximation produces a parabolic system of PDEs, that is (11.60 and 61).

For equations that can be cast in the form of (2.1) the classification of the PDE can be determined by inspection, using (2.2). When this is not possible, e.g. systems of PDEs, it is usually necessary to examine the characteristics (Sect. 2.1.3) to determine the correct classification.

The different categories of PDEs can be associated, broadly, with different types of flow problems. Generally time-dependent problems lead to either parabolic or hyperbolic PDEs. Parabolic PDEs govern flows containing dissipative mechanisms, e.g. significant viscous stresses or thermal conduction. In this case the solution will be smooth and gradients will reduce for increasing time if the boundary conditions are not time-dependent. If there are no dissipative mechanisms present, the solution will remain of constant amplitude if the PDE is linear and may even grow if the PDE is nonlinear. This solution is typical of flows governed by hyperbolic PDEs. Elliptic PDEs usually govern steady-state or equilibrium problems. However, some steady-state flows lead to parabolic PDEs (steady boundary layer flow) and to hyperbolic PDEs (steady inviscid supersonic flow).

2.1.1 Nature of a Well-Posed Problem

Before proceeding further with the formal classification of partial differential equations it is worthwhile embedding the problem formulation and algorithm construction in the framework of a well-posed problem. The governing equations and auxiliary (initial and boundary) conditions are well-posed mathematically if the following three conditions are met:

i) the solution exists,
ii) the solution is unique,
iii) the solution depends continuously on the auxiliary data.

The question of existence does not usually create any difficulty. An exception occurs in introducing exact solutions of Laplace's equation (Sect. 11.3) where the solution may not exist at isolated points. Thus it does not exist at the location of the source, $r = r_s$ in (11.53). In practice this problem is often avoided by placing the source outside the computational domain, e.g. inside the body in Fig. 11.7.

The usual cause of non-uniqueness is a failure to properly match the auxiliary conditions to the type of governing PDE. For the potential equation governing inviscid, irrotational flows, and for the boundary layer equations, the appropriate initial and boundary conditions are well established. For the Navier-Stokes equations the proper boundary conditions at a solid surface are well known but there is some flexibility in making the correct choice for farfield boundary conditions. In general an underprescription of boundary conditions leads to non-uniqueness and an overprescription to unphysical solutions adjacent to the boundary in question.

There are some flow problems for which multiple solutions may be expected on physical grounds. These problems would fail the above criteria of mathematical well-posedness. This situation often arises for flows undergoing transition from laminar to turbulent motion. However, the broad understanding of fluid dynamics will usually identify such classes of flows for which the computation may be complicated by concern about the well-posedness of the mathematical formulation.

The third criterion above requires that a small change in the initial or boundary conditions should cause only a small change in the solution. The auxiliary conditions are often introduced approximately in a typical computational algorithm. Consequently if the third condition is not met the errors in the auxiliary data will propagate into the interior causing the solution to grow rapidly, particularly for hyperbolic PDEs.

The above criteria are usually attributed to Hadamard (Garabedian 1964, p. 109). In addition we could take a simple parallel and require that for a well-posed computation:

i) the computational solution exists,
ii) the computational solution is unique,
iii) the computational solution depends continuously on the approximate auxiliary data.

The process of obtaining the computational solution can be represented schematically as in Fig. 2.1. Here the specified data are the approximate implementation of the initial and boundary conditions. If boundary conditions are placed on derivatives of u an error will be introduced in approximating the boundary conditions. The computational algorithm is typically constructed from

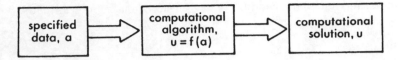

Fig. 2.1. Computational procedure

the governing PDE (Sect. 3.1) and must be stable (Sect. 4.3) in order for the above three conditions to be met.

Therefore for a well-posed computation it is necessary that not only should both the underlying PDE and auxiliary conditions be well-posed but that the algorithm should be well-posed (stable) also. It is implicit here that the approximate solution produced by a well-posed computation will be close, in some sense, to the exact solution of the well-posed problem. This question will be pursued in Sect. 4.1.

2.1.2 Boundary and Initial Conditions

It is clear from the discussion of well-posed problems and well-posed computations in Sect. 2.1.1 that the auxiliary data are, in a sense, the starting point for obtaining the interior solution, particularly for propagation problems. If we don't distinguish between time and space as independent variables then the auxiliary data specified on ∂R, Fig. 2.2, is "extrapolated" by the computational algorithm (based on the PDE) to provide the solution in the interior, R.

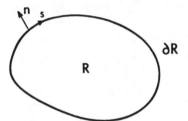

Fig. 2.2. Computational domain R

Auxiliary conditions are specified in three ways:

i) Dirichlet condition, e.g. $u=f$ on ∂R.
ii) Neumann (derivative) condition, e.g. $\partial u/\partial n=f$ or $\partial u/\partial s=g$ on ∂R,
iii) mixed or Robin condition, e.g. $\partial u/\partial n+ku=f$, $k>0$ on ∂R.

In auxiliary conditions ii) and iii), $\partial/\partial n$ denotes the outward normal derivative.

For most flows, which require the solution of the Navier-Stokes equations in primitive variables $(u, v, p, \text{etc.})$, at least one velocity component is given on an inflow boundary. This provides a Dirichlet boundary condition on the velocity. For the velocity potential equation governing inviscid compressible flow, the condition that $\partial\phi/\partial n=0$ at the body surface is a Neumann boundary condition. Mixed conditions are rare in fluid mechanics but occur in convective heat transfer. Computationally, Dirichlet auxiliary conditions can be applied exactly as long as f is analytic. However, errors are introduced in representing Neumann or mixed conditions (Sect. 7.3).

2.1.3 Classification by Characteristics

For partial differential equations that are functions of two independent variables the classification into elliptic, parabolic or hyperbolic type can be achieved by first seeking characteristic directions along which the governing equations only involve total differentials.

For a single first-order PDE in two independent variables,

$$A \frac{\partial u}{\partial t} + B \frac{\partial u}{\partial x} = C \ , \tag{2.5}$$

a single real characteristic exists through every point and the characteristic direction is defined by (Fig. 2.5)

$$\frac{dx}{dt} = \frac{B}{A} \ . \tag{2.6}$$

Along the characteristic direction, (2.5) reduces to

$$\frac{du}{dt} = \frac{C}{A} \quad \text{and} \quad \frac{du}{dx} = \frac{C}{B} \ . \tag{2.7}$$

Equation (2.5) is a hyperbolic PDE and it is possible to integrate (2.7) as ordinary differential equations along a grid defined by (2.6), as long as the initial data are given on a non-characteristic line.

The same concept of characteristic directions can be used in conjunction with PDEs of second order in two independent variables, e.g. (2.1). Since it is found from (2.1 and 2) that only the coefficients of the highest derivatives determine the category of PDE, it is convenient to write (2.1) as

$$A \frac{\partial^2 u}{\partial x^2} + B \frac{\partial^2 u}{\partial x \partial y} + C \frac{\partial^2 u}{\partial y^2} + H = 0 \ , \tag{2.8}$$

where H contains all the first derivative terms etc. in (2.1) and A, B and C may be functions of x, y. It is possible to obtain, for each point in the domain, two directions along which the integration of (2.8) involves only total differentials. The existence of these (characteristic) directions relates directly to the category of PDE.

For ease of presentation the following notation is introduced:

$$P = \frac{\partial u}{\partial x} \ , \quad Q = \frac{\partial u}{\partial y} \ , \quad R = \frac{\partial^2 u}{\partial x^2} \ , \quad S = \frac{\partial^2 u}{\partial x \partial y} \ , \quad T = \frac{\partial^2 u}{\partial y^2} \ . \tag{2.9}$$

A curve K is introduced in the interior of the domain on which P, Q, R, S, T and u satisfy (2.8). Along a tangent to K the differentials of P and Q satisfy

$$dP = R \, dx + S \, dy \ , \tag{2.10}$$

$$dQ = S \, dx + T \, dy \ , \tag{2.11}$$

and (2.8) can be written as

$$AR + BS + CT + H = 0 \ . \tag{2.12}$$

In (2.10 and 11), dy/dx defines the slope of the tangent to K. Using (2.10 and 11), R and T can be eliminated from (2.12) to give

$$S\left[A\left(\frac{dy}{dx}\right)^2 - B\left(\frac{dy}{dx}\right) + C\right] - \left\{\left[A\left(\frac{dP}{dx}\right) + H\right]\frac{dy}{dx} + C\frac{dQ}{dx}\right\} = 0 \ . \tag{2.13}$$

If dy/dx is chosen such that

$$A\left(\frac{dy}{dx}\right)^2 - B\left(\frac{dy}{dx}\right) + C = 0 \ , \tag{2.14}$$

(2.13) reduces to the simpler relationship between dP/dx and dQ/dx,

$$\left[A\left(\frac{dP}{dx}\right) + H\right]\frac{dy}{dx} + C\frac{dQ}{dx} = 0 \ . \tag{2.15}$$

The two solutions to (2.14) define the characteristic directions for which (2.15) holds. Comparing (2.14) with (2.2) it is clear that if (2.8) is:

i) a hyperbolic PDE, two real characteristics exist,
ii) a parabolic PDE, one real characteristic exists,
iii) an elliptic PDE, the characteristics are complex.

Thus a consideration of the discriminant, $B^2 - 4AC$, determines both the type of PDE and the nature of the characteristics.

The classification of the partial differential equation type has been undertaken in Cartesian coordinates, so far. An important question is whether a coordinate transformation, such as will be described in Chap. 12, can alter the type of the partial differential equation.

Thus new independent variables (ξ, η) are introduced in place of (x, y) and it is assumed that the transformation, $\xi = \xi(x, y)$ and $\eta = \eta(x, y)$ is known. Derivatives are transformed as (Sect. 12.1)

$$\frac{\partial u}{\partial x} = \xi_x \frac{\partial u}{\partial \xi} + \eta_x \frac{\partial u}{\partial \eta} \ , \tag{2.16}$$

where $\xi_x \equiv \partial\xi/\partial x$, etc. After some manipulation, (2.8) becomes

$$A'\frac{\partial^2 u}{\partial \xi^2} + B'\frac{\partial^2 u}{\partial \xi \partial \eta} + C'\frac{\partial^2 u}{\partial \eta^2} + H' = 0 \ , \tag{2.17}$$

where

$$A' = A\xi_x^2 + B\xi_x\xi_y + C\xi_y^2 ,$$

$$B' = 2A\xi_x\eta_x + B(\xi_x\eta_y + \xi_y\eta_x) + 2C\xi_y\eta_y , \quad \text{and} \qquad (2.18)$$

$$C' = A\eta_x^2 + B\eta_x\eta_y + C\eta_y^2 .$$

The discriminant, $(B')^2 - 4A'C'$, then becomes

$$(B')^2 - 4A'C' = J^2[B^2 - 4AC] , \qquad (2.19)$$

where the Jacobian of the transform is $J = \xi_x\,\eta_y - \xi_y\,\eta_x$. Equation (2.19) gives the important result that the classification of the PDE is precisely the same whether it is determined in Cartesian coordinates from (2.8) or in (ξ, η) coordinates from (2.17 and 18). Thus, introducing a coordinate transformation does not change the type of PDE.

To extend the examination of characteristics beyond two independent variables is less useful. In m dimensions $(m-1)$ dimensional surfaces must be considered. However, an examination of the coefficients multiplying the highest-order derivatives can, in principle, furnish useful information. For example, in three dimensions (2.8) would be replaced by

$$A\frac{\partial^2 u}{\partial x^2} + B\frac{\partial^2 u}{\partial x \partial y} + C\frac{\partial^2 u}{\partial y^2} + D\frac{\partial^2 u}{\partial x \partial z} + E\frac{\partial^2 u}{\partial z^2} + F\frac{\partial^2 u}{\partial y \partial z} + H = 0 . \qquad (2.20)$$

It is necessary to obtain a transformation, $\xi = \xi(x, y, z), \eta = \eta(x, y, z), \zeta = \zeta(x, y, z)$ such that all cross derivatives in (ξ, η, ζ) coordinates disappear. This approach will fail for more than three independent variables, in which case it is convenient to replace (2.20) with

$$\sum_{j=1}^{N} \sum_{k=1}^{N} a_{jk} \frac{\partial^2 u}{\partial x_j \partial x_k} + H = 0 , \qquad (2.21)$$

where N is the number of independent variables and the coefficients a_{jk} replace A to F in (2.20). The previously mentioned transformation to remove cross derivatives is equivalent to finding the eigenvalues λ of the matrix $\underline{A}$ with elements a_{jk} (see footnote).

The following classification, following Chester (1971, p. 134), can be given:

i) If any of the eigenvalues λ is zero, (2.21) is parabolic.
ii) If all eigenvalues are non-zero and of the same sign, (2.21) is elliptic.
iii) If all eigenvalues are non-zero and all but one are of the same sign, (2.21) is hyperbolic.

For three independent variables Hellwig (1964, p. 60) provides an equivalent

---

Underlined bold type denotes matrix or tensor

classification in terms of the coefficients multiplying the derivatives in the transformed equations.

In more than two independent variables useful information can often be determined about the behaviour of the partial differential equation by considering two-dimensional surfaces, i.e. by choosing particular coordinate values. Thus for (2.20) the character of the equation can be established in the plane $x = $ constant by temporarily freezing all terms involving x derivatives and treating the resulting equation as though it were a function of two independent variables.

2.1.4 Systems of Equations

A consideration of Chap. 11 indicates that the governing equations for fluid dynamics often form a system, rather than being a single equation. A two-component system of first-order PDEs, in two independent variables, could be written

$$A_{11}\frac{\partial u}{\partial x} + B_{11}\frac{\partial u}{\partial y} + A_{12}\frac{\partial v}{\partial x} + B_{12}\frac{\partial v}{\partial y} = E_1 \ , \tag{2.22}$$

$$A_{21}\frac{\partial u}{\partial x} + B_{21}\frac{\partial u}{\partial y} + A_{22}\frac{\partial v}{\partial x} + B_{22}\frac{\partial v}{\partial y} = E_2 \ . \tag{2.23}$$

Since both u and v are functions of x and y the following relationships hold:

$$du = \left(\frac{\partial u}{\partial x}\right)dx + \left(\frac{\partial u}{\partial y}\right)dy \ , \tag{2.24}$$

$$dv = \left(\frac{\partial v}{\partial x}\right)dx + \left(\frac{\partial v}{\partial y}\right)dy \ . \tag{2.25}$$

For the problem shown in Fig. 2.3 it is assumed that the solution has already been determined in the region ACPDB. As before, two directions, dy/dx, through P are

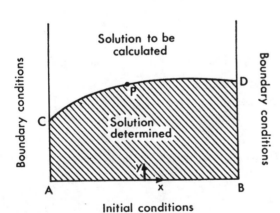

Fig. 2.3. Schematic representation of the computational domain for a propagation problem

sought along which only total differentials, du and dv, appear. For the system of equations (2.22, 23) this is equivalent to seeking multipliers, L_1 and L_2, such that

$$L_1 \times (2.22) + L_2 \times (2.23) \Rightarrow m_1\, du + m_2\, dv = (L_1 E_1 + L_2 E_2) \ . \tag{2.26}$$

Expansion of the terms making up (2.26) establishes the relationships

$$L_1 A_{11} + L_2 A_{21} = m_1 dx \ , \qquad L_1 B_{11} + L_2 B_{21} = m_1 dy \ ,$$

$$L_1 A_{12} + L_2 A_{22} = m_2 dx \ , \qquad L_1 B_{12} + L_2 B_{22} = m_2 dy \ . \tag{2.27}$$

Eliminating m_1 and m_2 and rearranging gives

$$\begin{bmatrix} (A_{11}dy - B_{11}dx) & (A_{21}dy - B_{21}dx) \\ (A_{12}dy - B_{12}dx) & (A_{22}dy - B_{22}dx) \end{bmatrix} \begin{bmatrix} L_1 \\ L_2 \end{bmatrix} = 0 \ . \tag{2.28}$$

Since this system is homogeneous in L_i it is necessary that

$$\det[\underline{A}\, dy - \underline{B}\, dx] = 0 \ , \tag{2.29}$$

for a nontrivial solution. For the above example, (2.29) takes the form

$$(A_{11}A_{22} - A_{21}A_{12})\left(\frac{dy}{dx}\right)^2 - (A_{11}B_{22} - A_{21}B_{12} + B_{11}A_{22}$$

$$- B_{21}A_{12})\frac{dy}{dx} + (B_{11}B_{22} - B_{21}B_{12}) = 0 \ . \tag{2.30}$$

Equation (2.30) has two solutions and the nature of the solutions depends on the discriminant

$$\text{DIS} = (A_{11}B_{22} - A_{21}B_{12} + A_{22}B_{11} - A_{12}B_{21})^2$$

$$- 4(A_{11}A_{22} - A_{21}A_{12})(B_{11}B_{22} - B_{21}B_{12}) \ , \tag{2.31}$$

which also determines the classification of the system (2.22, 23). The different possibilities are indicated in Table 2.1.

On characteristics the derivatives $\partial u/\partial x$, $\partial u/\partial y$, $\partial v/\partial x$ and $\partial v/\partial y$ are not defined uniquely. In fact, in crossing the characteristics discontinuities in the normal derivatives can occur whereas tangential derivatives are continuous.

An example using the above classification can be developed as follows. The governing equations for two-dimensional compressible potential flow, (11.103), can

Table 2.1. Classification of (2.22, 23)

DIS	Roots of (2.30)	Classification of the system (2.22, 23)
positive	2 real	hyperbolic
zero	1 real	parabolic
negative	2 complex	elliptic

be recast in terms of the velocity components, i.e.

$$\left(\frac{u^2}{a^2}-1\right)\frac{\partial u}{\partial x}+\left(\frac{uv}{a^2}\right)\frac{\partial u}{\partial y}+\left(\frac{uv}{a^2}\right)\frac{\partial v}{\partial x}+\left(\frac{v^2}{a^2}-1\right)\frac{\partial v}{\partial y}=0 \tag{2.32}$$

and

$$-\frac{\partial u}{\partial y}+\frac{\partial v}{\partial x}=0 \ . \tag{2.33}$$

Equations (2.32, 33) have the same structure as (2.22, 23). The evaluation of (2.31) gives

$$\text{DIS}=4(M^2-1), \quad \text{where} \quad M^2=(u^2+v^2)/a^2 \ ,$$

and indicates that the system (2.32, 33) is hyperbolic if $M>1$. This is the same result as was found in considering the compressible potential equation (2.4). This is to be expected since, although the equations are different, they govern the same physical situation.

The construction used to derive (2.28 and 29) can be generalised to a system of n first-order equations (Whitham 1974, p. 116). Equation (2.28) is replaced by

$$\left[\underline{A}\left(\frac{dy}{dx}\right)^{(k)}-\underline{B}\right]L^{(k)}=0 \ , \quad k=1,\dots,n \ . \tag{2.34}$$

The character of the system (Hellwig 1964, p. 70) depends on the solution of (2.29).

i) If n real roots are obtained the system is hyperbolic.
ii) If v real roots, $1\leq v\leq n-1$, and no complex roots are obtained, the system is parabolic.
iii) If no real roots are obtained the system is elliptic.

For large systems some roots may be complex and some may be real; this gives a mixed system. The most important division is between elliptic and non-elliptic partial differential equations since elliptic partial differential equations preclude time-like behaviour. Therefore the system of equations will be assumed to be elliptic if any complex roots occur.

The above classification extends to systems of second-order equations in two independent variables since auxiliary variables can be introduced to generate an even larger system of first-order equations. However, there is a risk that both $\underline{A}$ and $\underline{B}$ are singular so that it may be necessary to consider combinations of the equations to avoid this degenerate behaviour (Whitham 1974, p. 115).

For systems of more than two independent variables (2.29) can be partially generalised as follows. A system of first-order equations in three independent variables could be written

$$\underline{A}\frac{\partial q}{\partial x}+\underline{B}\frac{\partial q}{\partial y}+\underline{C}\frac{\partial q}{\partial z}=D \ , \tag{2.35}$$

where q is the vector of n dependent variables. Equation (2.35) leads to the nth order characteristic polynomial (Chester 1971, p. 272)

$$\det[\underline{A}\lambda_x + \underline{B}\lambda_y + \underline{C}\lambda_z] = 0 \ , \tag{2.36}$$

where λ_x, λ_y, λ_z define a normal direction to a surface at (x, y, z). Equation (2.36) generalises (2.29) and gives the condition that the surface is a characteristic surface. Clearly, for a real characteristic surface (2.36) must have real roots. If n real roots are obtained the system is hyperbolic.

It is possible to ask what the character of the partial differential equation is with respect to particular directions. For example setting $\lambda_x = \lambda_z = 1$ and solving for λ_y indicates that (2.35) is elliptic with respect to the y direction if any imaginary roots occur. Clearly each direction can be examined in turn.

Here we provide a simple example of a system of equations based on the steady incompressible Navier Stokes equations in two dimensions. In nondimensional form these are

$$u_x + v_y = 0 \ , \tag{2.37a}$$

$$uu_x + vu_y + p_x - \frac{1}{Re}(u_{xx} + u_{yy}) = 0 \ , \tag{2.37b}$$

$$uv_x + vv_y + p_y - \frac{1}{Re}(v_{xx} + v_{yy}) = 0 \ , \tag{2.37c}$$

where $u_x \equiv \partial u/\partial x$, etc., Re is the Reynolds number and u, v, p are the dependent variables. Equations (2.37) are reduced to a first-order system by introducing auxiliary variables $R = v_x$, $S = v_y$ and $T = u_y$. Thus (2.37) can be replaced with

$$
\begin{aligned}
u_y & = T \ , \\
u_x + v_y & = 0 \ , \\
-R_y + S_x & = 0 \ , \\
S_y + T_x & = 0 \ , \\
S_x/Re - T_y/Re + p_x & = uS - vT \ , \\
-R_x/Re - S_y/Re + p_y & = uR - vS \ .
\end{aligned}
\tag{2.38}
$$

The particular choices for (2.38) are made to avoid the equivalent of $\underline{A}$ and $\underline{B}$ in (2.35) being singular. The character of the above set of equations can be determined by replacing $\partial/\partial x$ with λ_x and $\partial/\partial y$ with λ_y and setting the determinant to zero, as in (2.36). The result is

$$(1/Re)\lambda_y^2(\lambda_x^2 + \lambda_y^2)^2 = 0 \ . \tag{2.39}$$

Setting $\lambda_y = 1$ indicates that λ_x is imaginary. Setting $\lambda_x = 1$ indicates that imaginary roots exist for λ_y. Therefore it is concluded that the system (2.37) is elliptic.

The general problem of classifying partial differential equations may be pursued in Garabedian (1964), Hellwig (1964), Courant and Hilbert (1962) and Chester (1971).

2.1.5 Classification by Fourier Analysis

The classification of partial differential equations by characteristics (Sects. 2.1.3 and 2.1.4) leads to the interpretation of the roots of a characteristic polynomial, e.g. (2.36). The roots determine the characteristic directions (or surfaces in more than two independent variables).

However, the same characteristic polynomial can be obtained from a Fourier analysis of the partial differential equation. In this case the roots have a different physical interpretation, although the classification of the partial differential equation in relation to the nature of the roots remains the same. The Fourier analysis approach is useful for systems of equations where higher than first-order derivatives appear, since it avoids the construction of an intermediate, but enlarged, first-order system. The Fourier analysis approach also indicates the expected form of the solution, e.g. oscillatory, exponential growth, etc. This feature is exploited in Chap. 16 in determining whether stable computational solutions of reduced forms of the Navier-Stokes equations can be obtained in a single spatial march.

Suppose a solution of the homogeneous second-order scalar equation

$$A\frac{\partial^2 u}{\partial x^2} + B\frac{\partial^2 u}{\partial x \partial y} + C\frac{\partial^2 u}{\partial y^2} = 0 \tag{2.40}$$

is sought of the form

$$u(x, y) = \frac{1}{4\pi^2} \sum_{j=-\infty}^{\infty} \sum_{k=-\infty}^{\infty} \hat{u}_{jk} \exp[i(\sigma_x)_j x] \exp[i(\sigma_y)_k y] \ . \tag{2.41}$$

The amplitudes of the various modes are determined by the boundary conditions. However, the nature of the solution will depend on the $(\sigma_x)_j$ and $(\sigma_y)_k$ coefficients, which may be complex. If A, B and C are not functions of u the relationship between σ_x and σ_y is the same for all modes so that only one mode need be considered in (2.41). Substituting into (2.40) gives

$$-A\sigma_x^2 - B\sigma_x\sigma_y - C\sigma_y^2 = 0 \qquad \text{or}$$

$$A(\sigma_x/\sigma_y)^2 + B(\sigma_x/\sigma_y) + C = 0 \ . \tag{2.42}$$

This is a characteristic polynomial for σ_x/σ_y equivalent to (2.29). The nature of the partial differential equation (2.40) depends on the nature of the roots, and hence on A, B and C as indicated by (2.2).

The Fourier analysis approach produces the same characteristic polynomial from the *principal part* of the governing equation as does the characteristic analysis. However, if σ_y is assumed real, the form of the solution is wavelike in the y direction.

Then the solution of the characteristic polynomial (2.42) formed from the *complete* equation indicates the form of the solution in the x direction.

An examination of (2.41) indicates the similarity with the Fourier transform definition (Lighthill 1958, p. 8),

$$\hat{u}(\sigma_x, \sigma_y) = \int_{-\infty}^{\infty} \int_{-\infty}^{\infty} u(x, y)\exp(-i\sigma_x x)\exp(-i\sigma_y y)dx\, dy \ , \tag{2.43}$$

or, notationally, $\hat{u} = Fu$.

To analyse the character of partial differential equations, use is made of the following results:

$$i\sigma_x \hat{u} = F\frac{\partial u}{\partial x}\ , \qquad i\sigma_y \hat{u} = F\frac{\partial u}{\partial y}\ . \tag{2.44}$$

Thus the characteristic polynomial is obtained by taking the Fourier transform of the governing equation. As an example (2.40) is transformed to

$$[A(i\sigma_x)^2 + B(i\sigma_x i\sigma_y) + C(i\sigma_y)^2]\hat{u} = 0 \ , \tag{2.45}$$

and (2.42) follows directly. The characteristic polynomial derived via the Fourier transform is often called the *symbol* of the partial differential equation.

The Fourier transform approach to obtaining the characteristic polynomial is applicable if A, B or C are functions of the independent variables. If A, B or C are functions of the dependent variables it is necessary to freeze them at their local values *before* introducing the Fourier transform.

The application of the Fourier analysis approach to systems of equations can be illustrated by considering (2.37). Freezing the coefficients u and v in (2.37b, c) and taking Fourier transforms of u, v and p produces the following homogeneous system of algebraic equations:

$$\begin{bmatrix} i\sigma_x & i\sigma_y & 0 \\ i(u\sigma_x + v\sigma_y) + \dfrac{1}{\mathrm{Re}}(\sigma_x^2 + \sigma_y^2) & 0 & i\sigma_x \\ 0 & i(u\sigma_x + v\sigma_y) + \dfrac{1}{\mathrm{Re}}(\sigma_x^2 + \sigma_y^2) & i\sigma_y \end{bmatrix} \begin{bmatrix} \hat{u} \\ \hat{v} \\ \hat{p} \end{bmatrix} = 0 \ , \tag{2.46}$$

which leads to the characteristic polynomial, $\det[\quad] = 0$, i.e.

$$(\sigma_x^2 + \sigma_y^2)\, [i(u\sigma_x + v\sigma_y) + (1/\mathrm{Re})\, (\sigma_x^2 + \sigma_y^2)] = 0 \ . \tag{2.47}$$

However, (2.47) contains the group $i(u\sigma_x + v\sigma_y)$, which corresponds to first derivatives of u and v. But the character of the system (2.37) is determined by the principal part, which explicitly excludes all but the highest derivatives. In this case (2.47) coincides with (2.39) and leads to the conclusion that (2.37) is an elliptic system. It is clear in comparing (2.46) with (2.38) that the Fourier analysis approach avoids the problem of constructing an equivalent first-order system and the possibility that it may be singular.

The roots of the characteristic polynomial produced by the Fourier analysis are interpreted here in the same way as in the characteristic method to determine the partial differential equation type. An alternative classification based on the magnitude of the largest root of the characteristic polynomial is described by Gelfand and Shilov (1967).

The Fourier analysis approach is made use of in Sect. 16.1 to determine the character of the solution produced by a single downstream march. In that situation all terms in the governing equations, not just the principal part, are retained in the equivalent of (2.47).

2.2 Hyperbolic Partial Differential Equations

The simplest example of a hyperbolic PDE is the wave equation,

$$\frac{\partial^2 u}{\partial t^2} - \frac{\partial^2 u}{\partial x^2} = 0 \ . \tag{2.48}$$

For initial conditions, $u(x, 0) = \sin \pi x$, $\partial u/\partial t(x, 0) = 0$, and boundary conditions, $u(0, t) = u(1, t) = 0$, (2.48) has the exact solution

$$u(x, t) = \sin \pi x \cos \pi t \ . \tag{2.49}$$

The lack of attenuation is a feature of linear hyperbolic PDEs.

The convection equation, considered in Sect. 9.1, is a linear hyperbolic PDE. The equations governing unsteady inviscid flow are hyperbolic, but nonlinear, as are the equations governing steady supersonic inviscid flow (Sect. 14.2).

2.2.1 Interpretation by Characteristics

Hyperbolic PDEs produce real characteristics. For the wave equation (2.48) the characteristic directions are given by $dx/dt = \pm 1$. In the (x, t) plane, the characteristics through a point P are shown in Fig. 2.4.

For the system of equations (2.32, 33) there are two characteristics, given by

$$\frac{dy}{dx} = \left[\frac{uv}{a^2} \pm \left(\frac{u^2 + v^2}{a^2} - 1 \right)^{1/2} \right] \left[\left(\frac{u}{a} \right)^2 - 1 \right]^{-1} \ . \tag{2.50}$$

Clearly the characteristics depend on the local solution and will, in general, be curved (Courant and Friedrichs, 1948).

For the first-order hyperbolic PDE (2.5) a single characteristic, $dt/dx = A/B$, passes through every point (Fig. 2.5). If A and B are constant the characteristics are straight lines. If A and B are functions of u, x or t, they are curved. For the linear

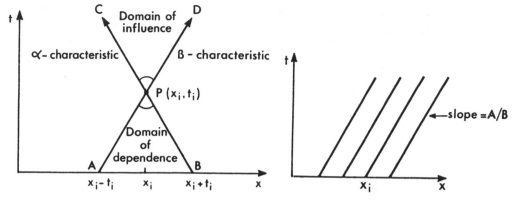

Fig. 2.4. Characteristics for the wave equation

Fig. 2.5. Characteristics for a first-order hyperbolic PDE, (2.5)

convection equation considered in Sect. 9.1 the ratio B/A is just the velocity. For three-dimensional steady inviscid supersonic flow the characteristic surface forms a cone about the local flow direction as indicated in Fig. 11.14.

For hyperbolic PDEs it is possible to use the characteristic directions to develop a computational grid on which the compatibility conditions, for example (2.15), hold. This is the strategy behind the method of characteristics, Sect. 2.5.1. For reasons to be discussed in Sect. 14.2.1, this method is now mainly of historic interest. However it is useful for determining far-field boundary conditions (Sect. 14.2.8).

2.2.2 Interpretation on a Physical Basis

As noted above, hyperbolic PDEs are associated with propagation problems when no dissipation is present. The occurrence of real characteristics, as in Fig. 2.4, implies that a disturbance to the solution u at P can only influence the rest of the solution in the domain CPD. Conversely the solution at P is influenced by disturbances in the domain APB only.

In addition, if initial conditions are specified at $t=0$, i.e. on AB in Fig. 2.4, these are sufficient to determine the solution at P, uniquely. This can be demonstrated, for (2.48) as follows.

New independent variables (ξ, η) are introduced as

$$\xi = x + t, \quad \eta = x - t ,$$ (2.51)

so that (2.48) reduces to

$$\frac{\partial^2 u}{\partial \xi \partial \eta} = 0 ,$$ (2.52)

which has the general solution

$$u(\xi, \eta) = f(\xi) + g(\eta) ,$$ (2.53)

where f and g are arbitrary twice-differentiable functions. If (2.48) is solved as a pure initial value problem it is appropriate to introduce the initial conditions

$$u(x, 0) = S(x) \ , \qquad \frac{\partial u}{\partial t}(x, 0) = T(x) \ . \tag{2.54}$$

It can be shown (Ames 1969, p. 165) that, for $t = 0$,

$$f(x) = 0.5 \left[S(x) + \int_0^x T(r)dr \right] + C \ , \tag{2.55}$$

$$g(x) = 0.5 \left[S(x) - \int_0^x T(r)dr \right] + D \ ,$$

where C and D are integration constants. It then follows from (2.53) that the general solution of (2.48) with initial conditions given by (2.54) is

$$u(x, t) = 0.5 \left[S(x+t) + S(x-t) + \int_{x-t}^{x+t} T(r)dr \right] . \tag{2.56}$$

In particular, if the point P has coordinates (x_i, t_i), the solution at P is

$$u(x_i, t_i) = 0.5 \left[S(x_i + t_i) + S(x_i - t_i) + \int_{x_i - t_i}^{x_i + t_i} T(r)dr \right] , \tag{2.57}$$

i.e. the solution at P is determined uniquely by the initial conditions on AB (Fig. 2.4).

For hyperbolic equations there is no dissipative (or smoothing) mechanism present. This implies that if the initial data (or boundary data) contain discontinuities they will be transmitted into the interior along characteristics, without attenuation of the discontinuity for linear equations. This is consistent with the result indicated in Sect. 2.1.3 that discontinuities in the normal derivatives can occur in crossing characteristics.

It should be emphasised here that in considering the equations that govern supersonic inviscid flow, which are hyperbolic, the discontinuities must be small to be consistent with isentropic flow. For supersonic inviscid isentropic flow the governing equations (2.32, 33) produce characteristic directions given by (2.50). If the solution is such that the characteristics run together a non-unique solution would result (Whitham 1974, p. 24); in practice a shock-wave occurs. However, there is a change in entropy across the shock-wave and this invalidates the assumption of isentropic flow on which (2.32 and 33) are based. Therefore the shock-wave forms a boundary (internal or external) of the domain in which (2.32 and 33) are valid.

2.2.3 Appropriate Boundary (and Initial) Conditions

It has already been indicated (Sect. 2.2.2) that for the wave equation (2.48) the initial conditions (2.54) are suitable, and, depending on the extent of AB, will determine the solution, uniquely, in the domain APB (Fig. 2.4). It is also possible to specify boundary conditions (Sect. 2.1.2), for example as on CD and EF in Fig. 2.8.

Here we reconsider the equations (2.22, 23), since these are directly applicable to supersonic inviscid flow (with particular choices of A_{11}, etc.), and ask what are appropriate choices of the auxiliary conditions so that a unique solution to (2.22 and 23) is possible. The characteristic directions arising from the equivalent of (2.50) will be labelled α and β characteristics. Three cases (shown in Fig. 2.6) are considered initially.

The case shown in Fig. 2.6a is equivalent to that shown in Fig. 2.4. That is, data for both u and v on a non-characteristic curve, AB, uniquely determine the solution

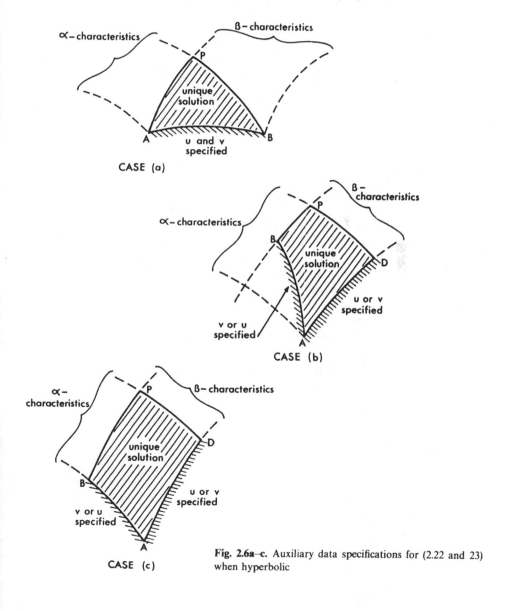

Fig. 2.6a–c. Auxiliary data specifications for (2.22 and 23) when hyperbolic

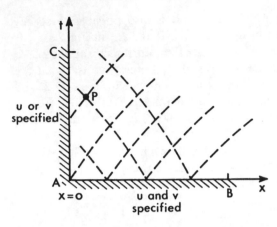

Fig. 2.7. Boundary conditions for the unsteady interpretation of (2.22 and 23)

up to P. For the case shown in Fig. 2.6b AB is a non-characteristic curve but AD is a β characteristic. For this case u or v should be given on one curve matched to v or u on the other. Thus both u and v are known at A. A similar situation occurs for the case shown in Fig. 2.6c except that both AB and AD are characteristic curves.

Equations (2.22 and 23) may be interpreted as unsteady equations by replacing y with t. A consideration (Fig. 2.7) of the computational domain $x \geqq 0$ and $t \geqq 0$ indicates that a point P close to the boundary $x = 0$ is partly determined by boundary conditions on AC and partly by initial conditions on AB, assuming that the governing PDEs are hyperbolic. Appropriate auxiliary conditions for this case are u and v specified on AB and v or u specified on AC.

These examples, Figs. 2.6 and 2.7, illustrate the general rule for hyperbolic PDEs that the number of auxiliary conditions is equal to the number of characteristics pointing into the domain (Whitham 1974, p. 127). The direction along the characteristic needs to be chosen consistently. For time-dependent problems the positive direction will be in the direction of increasing time. For multidimensional steady hyperbolic spatial problems in primitive variables one characteristic ("associated" with the continuity equation) coincides with the local streamline. Thus through a boundary point this characteristic defines the positive direction and indicates the positive direction for the other characteristics through the same point.

2.3 Parabolic Partial Differential Equations

Parabolic PDEs occur when propagation problems include dissipative mechanisms, such as viscous shear or heat conduction. The classical example of a parabolic PDE is the diffusion or heat conduction equation

$$\frac{\partial u}{\partial t} = \frac{\partial^2 u}{\partial x^2} .$$

$$(2.58)$$

Equation (2.58) will be used to introduce different computational techniques in Chap. 7.

For initial conditions $u = \sin \pi x$ and boundary conditions $u(0, t) = u(1, t) = 0$, (2.58) has the exact solution

$$u(x, t) = \sin \pi x \exp(-\pi^2 t) \ . \tag{2.59}$$

The exponential decay in time shown by (2.59) may be contrasted with the oscillatory solution (2.49) of the wave equation (2.48).

The transport equation (Sects. 9.4 and 9.5) is a linear parabolic PDE, and Burgers' equation, considered in Sect. 10.1, is a nonlinear parabolic PDE. However, the Cole–Hopf transformation (Fletcher 1983) permits Burgers' equation to be converted into the diffusion equation (2.58). The unsteady Navier-Stokes equations are parabolic. These equations are used both for unsteady problems and when a pseudo-transient formulation (Sect. 6.4) is introduced to solve a steady problem. For purely steady flow, boundary layers (Chap. 15) and shear layers are typically governed by parabolic PDEs, with the flow direction having a time-like role. Many of the reduced forms of the Navier-Stokes equations (Chap. 16) are governed by parabolic PDEs.

2.3.1 Interpretation by Characteristics

Interpretation of (2.58) as (2.8) with $y = t$ indicates that $A = 1$, $B = C = 0$ so that (2.58) is parabolic. Solution of (2.14) indicates that there is a single characteristic direction defined by $dt/dx = 0$. A typical computational domain for (2.58) is indicated in Fig. 2.8. In contrast to the situation for hyperbolic equations, derivatives of u are always continuous in crossing the $t = t_i$ line. Characteristics do not play such a significant role as for hyperbolic PDEs. There is no equivalent to the method of characteristics for parabolic PDEs. Clearly, laying out a computational grid to follow the local characteristics would never advance the solution in time.

2.3.2 Interpretation on a Physical Basis

Parabolic problems are typified by solutions which march forward in time but diffuse in space. Thus a disturbance to the solution introduced at P (in Fig. 2.8) can

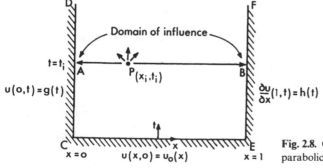

Fig. 2.8. Computational domain for a parabolic PDE

influence any part of the computational domain for $t \geq t_i$. However, the magnitude of the disturbance quickly attenuates in moving away from P. For steady two-dimensional boundary layer flow (Chap. 15) the characteristics are normal to the flow direction and imply no upstream influence.

The incorporation of a dissipative mechanism also implies that even if the initial conditions include a discontinuity, the solution in the interior will always be continuous. Partial differential equations in more than one spatial direction that are parabolic in time become elliptic in the steady state (if a steady-state solution exists).

2.3.3 Appropriate Boundary (and Initial) Conditions

For (2.58) it is necessary to specify Dirichlet initial conditions, e.g.

$$u(x, 0) = u_0(x) \quad \text{for} \quad 0 \leq x \leq 1 . \tag{2.60}$$

Appropriate boundary conditions would be

$$u(0, t) = g(t) \quad \text{and} \quad \frac{\partial u}{\partial x}(1, t) = h(t) . \tag{2.61}$$

For the boundaries CD and EF (Fig. 2.8) any combination of Dirichlet, Neumann or mixed boundary conditions (Sect. 2.1.2) is acceptable. However, it is desirable, in specifying Dirichlet boundary conditions, to ensure continuity with the initial conditions at C and E. Failure to do so will produce a solution with severe gradients adjacent to C and E, which may create difficulties for the computational algorithm. For systems of parabolic PDEs, initial conditions on CE and boundary conditions on CD and EF are necessary for all dependent variables.

2.4 Elliptic Partial Differential Equations

For fluid dynamics, elliptic PDEs are associated with steady-state problems. The simplest example of an elliptic PDE is Laplace's equation,

$$\frac{\partial^2 \phi}{\partial x^2} + \frac{\partial^2 \phi}{\partial y^2} = 0 , \tag{2.62}$$

which governs incompressible, potential flow. For boundary conditions

$$\phi(x, 0) = \sin \pi x , \quad \phi(x, 1) = \sin \pi x \exp(-\pi) , \quad \phi(0, y) = \phi(1, y) = 0 ,$$

(2.62) has the solution

$$\phi(x, y) = \sin \pi x \exp(-\pi y) \tag{2.63}$$

in the domain $0 \leq x \leq 1, 0 \leq y \leq 1$.

The Poisson equation for the stream function, (11.88), in two-dimensional rotational flow is an elliptic PDE. As noted above, the steady Navier-Stokes equations and the steady energy equation are also elliptic.

For second-order elliptic PDEs of the form (2.1), an important maximum principle exists (Garabedian 1964, p. 232). Namely, both the maximum and minimum values of ϕ must occur on the boundary ∂R, except for the trivial case that ϕ is a constant. The maximum principle is useful in testing that computational solutions of elliptic PDEs are behaving properly.

2.4.1 Interpretation by Characteristics

For the general second-order PDE (2.1), which is known to be elliptic, i.e. $4AC < B^2$, the characteristics are complex and cannot be displayed in the (real) computational domain. For elliptic problems in fluid dynamics, identification of characteristic directions serves no useful purpose.

2.4.2 Interpretation on a Physical Basis

The most important feature concerning elliptic PDEs is that a disturbance introduced at an interior point P, as in Fig. 2.9, influences all other points in the computational domain, although away from P the influence will be small. This implies that in seeking computational solutions to elliptic problems it is necessary to consider the global domain. In contrast, parabolic and hyperbolic PDEs can be solved by marching progressively from the initial conditions. Discontinuities in boundary conditions for elliptic PDEs are smoothed out in the interior.

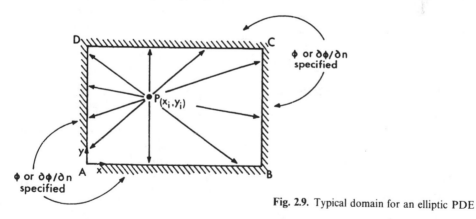

Fig. 2.9. Typical domain for an elliptic PDE

2.4.3 Appropriate Boundary Conditions

The ability to influence all other points in the domain from an interior point implies that boundary conditions are required on all boundaries (Fig. 2.9). The boundary conditions can be any combination of Dirichlet, Neumann or mixed (Sect. 2.1.2) boundary conditions. However, if a Neumann condition, $\partial\phi/\partial n = f(s)$, is applied on all boundaries, where n is the outward normal and s is measured along the boundary contour, care must be taken that the specification is consistent with the governing equation. From Green's theorem,

$$\oint_{\partial R} f \, ds = - \int_R \nabla^2 \phi \, dV \ . \tag{2.64}$$

Clearly, if the governing equation is the Laplace or Poisson equation, (2.64) implies an additional global constraint on the Neumann boundary condition specification. When (2.62) represents steady, incompressible, potential flow and ϕ is the velocity potential, f is just the normal velocity. Thus for steady, incompressible, potential flow, (2.64) coincides with the conservation of mass, (11.7). The computational implementation of (2.64) is discussed in Sect. 16.2.2. For systems of elliptic PDEs boundary conditions are required on all boundaries for all dependent variables.

For parabolic and hyperbolic PDEs it is always possible to obtain the local solution immediately adjacent to a boundary by a series expansion. Attempts to do the same with an elliptic PDE typically produce an infinite solution, due to the fact that elliptic PDEs are not well-posed for the case where boundary conditions are not specified on a closed boundary.

2.5 Traditional Solution Methods

In this section we briefly describe three techniques that may be considered pre-computer methods, requiring only hand or primitive machine calculation. These methods work well for simple model problems but are less effective for the more complicated equations governing fluid flow. However, they are sometimes useful in suggesting a method of solution or obtaining an approximate or local solution.

2.5.1 The Method of Characteristics

This method is only applicable to hyperbolic PDEs. It is described here for a second-order PDE in two independent variables, which was considered previously in Sect. 2.1.3,

$$A \frac{\partial^2 u}{\partial x^2} + B \frac{\partial^2 u}{\partial x \partial y} + C \frac{\partial^2 u}{\partial y^2} + H = 0 \ . \tag{2.65}$$

Solution of (2.14) will furnish two roots,

$$\frac{dy}{dx} = F \quad \text{and} \quad \frac{dy}{dx} = G \ . \tag{2.66}$$

For two adjacent points on the characteristics defined by (2.66) the compatibility equation (2.15) can be approximated by

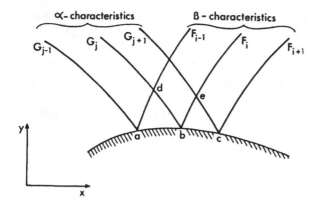

Fig. 2.10. Method of characteristics

$$A \frac{dy}{dx} \Delta P + C\Delta Q + H\Delta y = 0 \ . \tag{2.67}$$

It may be recalled from Sect. 2.1.3 that $P = \partial u/\partial x$ and $Q = \partial u/\partial y$ so that, for the same two adjacent points,

$$\Delta u = P\Delta x + Q\Delta y \ . \tag{2.68}$$

It will be assumed that u, P and Q are known along some non-characteristic boundary (Fig. 2.10). Initially both the solution and the locations for interior points, like d and e, are unknown. Two equations can be obtained from (2.66) to provide the location of d. These are

$$y_d - y_a = \bar{F}_{ad}(x_d - x_a) \quad \text{and} \tag{2.69}$$

$$y_d - y_b = \bar{G}_{bd}(x_d - x_b) \ , \quad \text{where} \tag{2.70}$$

$$\bar{F}_{ad} = 0.5(F_a + F_d) \quad \text{and} \quad \bar{G}_{bd} = 0.5(G_b + G_d) \ .$$

Effectively, the curved lines ad and bd have been replaced by straight lines determined by averaging the slope at the end points.

If x_d and y_d were known it would be possible to obtain P_d and Q_d from (2.67) in the form

$$\bar{A}_{ad}\bar{F}_{ad}(P_d - P_a) + \bar{C}_{ad}(Q_d - Q_a) + \bar{H}_{ad}(y_d - y_a) = 0 \tag{2.71}$$

$$\bar{A}_{bd}\bar{G}_{bd}(P_d - P_b) + \bar{C}_{bd}(Q_d - Q_b) + \bar{H}_{bd}(y_d - y_b) = 0 \ . \tag{2.72}$$

If P_d and Q_d were known (2.68) could be used, as follows, to obtain u_d:

$$u_d - u_a = \bar{P}_{ad}(x_d - x_a) + \bar{Q}_{ad}(y_d - y_a) \ . \tag{2.73}$$

In practice (2.69–73) must be solved iteratively to obtain x_d, y_d, P_d, Q_d and u_d. For the first step of the iteration the following approximations are used:

$$\bar{F}_{ad} = F_a \ , \quad \bar{G}_{bd} = G_b \ , \quad \text{etc.} \tag{2.74}$$

Typically two or three iterations are required as long as d is not too far from a and b. The method progresses by marching along the grid defined by the local characteristics which are determined as part of the solution. The above formulation is described in a fluid dynamic context by Belotserkovskii and Chushkin (1965).

The method of characteristics has been widely used in one-dimensional unsteady gas dynamics and for steady two-dimensional supersonic inviscid flow. However, the method is rather cumbersome when extended to three or four independent variables, or if internal shocks occur. For supersonic inviscid flow the method of characteristics is useful for determining the number and form of appropriate far-field boundary conditions.

2.5.2 Separation of Variables

This method is applicable to PDEs of any classification. It will be illustrated here for the diffusion equation

$$\frac{\partial u}{\partial t} = \frac{\partial^2 u}{\partial x^2} \tag{2.75}$$

in the domain shown in Fig. 2.11. The initial and boundary conditions are also shown in Fig. 2.11. The method introduces a general separable solution

$$u(x, t) = X(x) T(t) . \tag{2.76}$$

Substitution into (2.75) gives

$$\frac{d^2 X}{dx^2} + \lambda X = 0 \quad \text{and} \tag{2.77}$$

$$\frac{dT}{dt} + \lambda t = 0 , \tag{2.78}$$

where λ is an arbitrary constant. Equation (2.77) has an infinite number of solutions of the form

$$X_k(x) = A_k \sin kx , \tag{2.79}$$

where $\lambda_k = k^2, k = 1, 2, 3 \ldots$ and A_k are constants to be determined by the boundary and initial conditions. Consequently (2.78) also has an infinite number of solutions

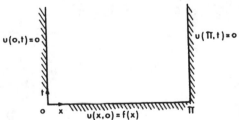

Fig. 2.11. Auxiliary conditions for (2.75)

of the form

$$T_k(t) = B_k \exp(-k^2 t) ,\tag{2.80}$$

where B_k are constants to be determined by the initial and boundary conditions. Substituting (2.79 and 80) into (2.76) implies the general solution

$$u(x,t) = \sum_{k=1}^{\infty} C_k \sin kx \exp(-k^2 t) .\tag{2.81}$$

Equation (2.81) satisfies the boundary conditions of the problem. The constants C_k are obtained from satisfying the initial conditions

$$\sum_{k=1}^{\infty} C_k \sin kx = u(x,0) = f(x) , \text{ i.e.}\tag{2.82}$$

$$C_k = \left(\frac{2}{\pi}\right) \int_0^{\pi} f(s) \sin ks \, ds .\tag{2.83}$$

For this particular problem the contribution from $\exp(-k^2 t)$ for large k is very small for $t > 0$. Therefore it is usually only necessary to retain 10 or 20 terms in the expansion (2.81).

The separation of variables method relies on the availability of a coordinate system for which ∂R coincides with coordinate lines. It also implies that the operators in the PDE will separate. Consequently, although the method is effective on model problems it does not find much direct use for the rather complicated equations governing fluid motion, often in irregular domains. However, an interesting discussion of the method is provided by Gustafson (1980, pp. 115–138).

2.5.3 Green's Function Method

For a PDE written in the general manner

$$Lu = f ,\tag{2.84}$$

a solution can be constructed, in principle, by "inverting" the operator L. The solution is expressed in integral form as

$$u(p) = \int_R G(p,q) f(q) \, dV_q ,\tag{2.85}$$

where $G(p,q)$ is the Green's function. In general $G(p,q)$ contains information equivalent to the operator L, the boundary conditions and the domain. Consequently the major difficulty in using the Green's function method is in determining what the Green's function should be to suit the particular problem. The subsequent evaluation of (2.85) is usually straightforward.

Green's functions can be obtained for relatively simple linear equations like Laplace's equation and the Poisson equation. For example, a point source of unit strength in three dimensions has the Green's function

$$S(p,q)=1/4\pi r_{pq} , \tag{2.86}$$

where r_{pq} is just the distance between p and q. This formula is effectively equivalent to the two-dimensional velocity potential given by (11.53) with $m=1$. Carrying out the required differentiation indicates that

$$-\nabla_q^2 S(p,q)=\delta(p,q) , \tag{2.87}$$

where $\delta(p,q)$ is the Dirac delta centred at p and ∇_q^2 is the Laplacian evaluated at q. A property of the Dirac delta function is that

$$\int_R w(q)\delta(p,q)dV_q=w(p) \quad \text{and} \quad \delta(p,q)=0 \quad \text{if} \quad p\neq q . \tag{2.88}$$

In (2.88) $w(q)$ is an arbitrary smooth function.

A solution procedure can be established by invoking Green's second identity,

$$\int_R (u\nabla^2 v-v\nabla^2 u)dV+\int_{\partial R}\left(u\frac{\partial v}{\partial n}-v\frac{\partial u}{\partial n}\right)dS=0 . \tag{2.89}$$

In the present situation, u in (2.89) is identified with the solution of the Poisson equation

$$\nabla^2 u=-f \text{ on } R, \quad u=0 \quad \text{on} \quad \partial R \tag{2.90}$$

and

$$v=G(p,q)=S(p,q)+g(p,q) \quad \text{with} \quad v=0 \quad \text{on} \quad \partial R . \tag{2.91}$$

Consequently (2.89) reduces to

$$\int_R vfdV=-\int_R u\nabla^2 vdV . \tag{2.92}$$

The function $g(p,q)$ is chosen so that $\nabla_q^2 g=0$ in R and $G(p,q)=0$ when q is on ∂R. As a result, (2.88 and 92) give the solution

$$u(p)=\int_R G(p,q)f(q)dV_q . \tag{2.93}$$

The Green's function method is implicit in the panel method (Sect. 14.1) and is used almost directly in the boundary element method (Sect. 14.1.3).

For some elliptic PDEs it is possible to construct an equivalent variational principle and to use a Rayleigh-Ritz procedure (Gustafson 1980, p. 161). Although such a technique is standard for structural applications of the finite element method, the elliptic PDEs that occur in fluid dynamics do not usually possess an equivalent variational form.

2.6 Closure

In this chapter we have examined the classification of PDEs into hyperbolic, parabolic and elliptic type. All three types occur for various simplifications of the fluid dynamic governing equations (Chap. 11). However, systems of equations may also be of mixed type. Hyperbolic PDEs are usually associated with propagation problems without dissipation (wave-like motion remains unattenuated) and parabolic PDEs are usually associated with propagation problems with dissipation. In fluid dynamics the dissipation usually comes from the viscous or heat conduction terms or eddy-viscosity type turbulence modelling. Elliptic PDEs are associated with steady-state problems.

Each type of PDE requires different boundary (and initial) conditions and may lend themselves to particular solution techniques. For example the method of characteristics is 'natural' for hyperbolic PDEs in two independent variables. For the nonlinear equations governing fluid dynamics the classification of the PDE can change locally. Consequently boundary conditions should be chosen to suit the classification of the PDE adjacent to the boundary.

The changing classification of the governing PDEs in different parts of the domain can be illustrated by considering supersonic viscous flow past a two-dimensional wing. For this example the governing equations are the Navier-Stokes equations which, due to the appearance of the second derivatives, are strictly elliptic when interpreted according to Sect. 2.1.2. However, such a classification takes no account of the magnitude of the relevant terms. In fact the viscous terms are only significant close to the surface where the streamwise viscous dissipation is an order-of-magnitude smaller than the cross-stream viscous dissipation; and the governing equations are mixed parabolic/hyperbolic. Away from the body all the viscous terms are small and the equation system is effectively hyperbolic. When shock waves occur the severe gradients away from the body cause the viscous (and heat conduction) terms to be significant so that the governing equations are locally elliptic (within the thickness of the shock-wave). This is sufficient to replace the discontinuous solution (in the inviscid approximation) with a severe, but continuous, gradient.

Clearly the strict mathematical classification of the governing PDEs should be tempered by a knowledge of the physical processes involved to ensure that correct auxiliary conditions are specified and appropriate computational techniques are used.

2.7 Problems

Background (Sect. 2.1)

2.1   a) Transform Laplace's equation, $\partial^2\phi/\partial x^2 + \partial^2\phi/\partial y^2 = 0$, into generalised coordinates $\xi = \xi(x, y)$, $\eta = \eta(x, y)$ and show that the resulting equation is elliptic.

b) Transform the wave equation, $\partial^2\phi/\partial t^2 - \partial^2\phi/\partial x^2 = 0$, into generalised coordinates $\xi = \xi(t, x)$, $\eta = \eta(t, x)$ and show that the resulting equation is hyperbolic.

2.2 Convert the Kortweg-de Vries equation (Jeffrey and Taniuti 1984 and (9.27)),

$$\frac{\partial u}{\partial t}+u\frac{\partial u}{\partial x}+\frac{\partial^3 u}{\partial x^3}=0$$

into an equivalent system of first-order equations by introducing auxiliary variables $p=\partial u/\partial x$, etc. Deduce that the resulting system of equations is parabolic.

2.3 The nondimensional equations governing steady inviscid incompressible flow are

$$\frac{\partial u}{\partial x}+\frac{\partial v}{\partial y}=0\ ,$$

$$u\frac{\partial u}{\partial x}+v\frac{\partial u}{\partial y}+\frac{\partial p}{\partial x}=0\ ,$$

$$u\frac{\partial v}{\partial x}+v\frac{\partial v}{\partial y}+\frac{\partial p}{\partial y}=0\ .$$

Determine the type of the system of partial differential equations.

Hyperbolic PDEs (Sect. 2.2)

2.4 Show by inspection that the second-order PDE $\partial^2 u/\partial x\partial t=0$ is hyperbolic. Consider the equivalent system

$$\frac{\partial u}{\partial t}-v=0\quad\text{and}$$

$$\frac{\partial v}{\partial x}=0\ .$$

Deduce that this system is hyperbolic and that the x and t axes are characteristics.

2.5 Consider the modified wave equation

$$\frac{\partial^2 u}{\partial t^2}-\beta\frac{\partial^2 u}{\partial x^2}+u=0\ . \tag{2.94}$$

Show, by inspection that this equation is hyperbolic. Consider the related system of equations

$$\frac{\partial u}{\partial t}-w=0\ ,$$

$$\frac{\partial v}{\partial t}-\frac{\partial w}{\partial x}=0\ , \tag{2.95}$$

$$\frac{\partial w}{\partial t}-\beta\frac{\partial v}{\partial x}+u=0\ .$$

Show that this system is hyperbolic and determine the characteristic directions. What is the connection between (2.94) and (2.95)? Does this explain the extra characteristic in (2.95)?

2.6 The governing equations for one-dimensional unsteady isentropic inviscid compressible flow are

$$\frac{\partial \varrho}{\partial t} + \frac{u \partial \varrho}{\partial x} + \frac{\varrho \partial u}{\partial x} = 0 \ ,$$

$$\frac{\partial u}{\partial t} + u \frac{\partial u}{\partial x} + \frac{1}{\varrho} \frac{\partial p}{\partial x} = 0 \ ,$$

where $p = k\varrho^{\gamma}$ and $a^2 = \gamma p/\varrho$. Here a is the speed of sound. Show that this system is hyperbolic and that the characteristics are given by $dx/dt = u \pm a$.

Parabolic PDEs (Sect. 2.3)

2.7 (a) Convert the equation $\partial\phi/\partial t - \alpha \partial^2\phi/\partial x^2 = 0$ to an equivalent system by introducing an auxiliary variable $p = \partial\phi/\partial x$. Show that the system is parabolic.
(b) Analyse $\partial\phi/\partial t - \alpha(\partial^2\phi/\partial x^2 + \partial^2\phi/\partial y^2) = 0$ in a similar way and show that it is parabolic.

2.8 Consider the transport equation $\partial u/\partial t + 2c\partial u/\partial x - d\partial^2 u/\partial x^2 = 0$ with initial conditions $u(x,0) = \exp(cx/d)$ and boundary conditions $u(0,t) = \exp(-c^2t/d)$ and $u(1,t) = (d/c)\partial u/\partial x(1,t)$. Show that the equation is parabolic and determine the solution.

2.9 The equations governing steady incompressible boundary layer flow over a flat plate can be written

$$\frac{\partial u}{\partial x} + \frac{\partial v}{\partial y} = 0 \ ,$$

$$u \frac{\partial u}{\partial x} + v \frac{\partial u}{\partial y} - \frac{1}{Re} \frac{\partial^2 u}{\partial y^2} = 0 \ .$$

Show that this system is parabolic and suggest suitable initial and boundary conditions for u and v.

Elliptic PDEs (Sect. 2.4)

2.10 Consider the equations

$$\frac{\partial u}{\partial x} + \frac{\partial v}{\partial y} = 0 \ , \qquad \frac{\partial u}{\partial y} - \frac{\partial v}{\partial x} = 0 \ . \tag{2.96}$$

Show that this system is elliptic,
(a) directly,
(b) by introducing the variable ϕ, where $u = \partial\phi/\partial x$ and $v = \partial\phi/\partial y$.

2.11 Show that the expressions $u=x/(x^2+y^2)$, $v=y/(x^2+y^2)$ are a solution of (2.96).

2.12 Show that the equations

$$u\frac{\partial u}{\partial x}+v\frac{\partial u}{\partial y}-\frac{1}{Re}\left(\frac{\partial^2 u}{\partial x^2}+\frac{\partial^2 u}{\partial y^2}\right)=0 ,$$

$$u\frac{\partial v}{\partial x}+v\frac{\partial v}{\partial y}-\frac{1}{Re}\left(\frac{\partial^2 v}{\partial x^2}+\frac{\partial^2 v}{\partial y^2}\right)=0$$

form an elliptic system and that they are satisfied by the expressions

$$u=-2[a_1+a_3y+k\{\exp[k(x-x_0)]+\exp[-k(x-x_0)]\}\cos(ky)]/(Re\,D) ,$$

$$v=-2[a_2+a_3x-k\{\exp[k(x-x_0)]+\exp[-k(x-x_0)]\}\sin(ky)]/(Re\,D) ,$$

where

$$D=[a_0+a_1x+a_2y+a_3xy+\{\exp[k(x-x_0)]+\exp[-k(x-x_0)]\}\cos(ky)]$$

and a_0,a_1,a_2,a_3,k and x_0 are arbitrary constants.

Traditional Methods (Sect. 2.5)

2.13 Consider the solution of $\partial^2 T/\partial x^2+\partial^2 T/\partial y^2=0$ on a unit square, with boundary conditions

$$T(0,y)=0 , \qquad T(1,y)=0 , \qquad T(x,0)=T_0,\ldots, \qquad T(x,1)=0 .$$

Apply the separation of variables technique to obtain

$$T(x,y)=\sum_{k=1}^{\infty} A_k\sin(k\pi x)\sinh[k\pi(y-1)] \quad \text{with}$$

$$A_k=\frac{(2T_0/k\pi)[(-1)^k-1]}{\sinh(k\pi)} .$$

2.14 The equation $\partial\phi/\partial t-\alpha\partial^2\phi/\partial x^2=0$ is to be solved in the domain $0\le x\le 1$, $t>0$ with boundary conditions $\phi(0,t)=0$, $\phi(1,t)=\phi_R$ and initial condition $\phi(x,0)=0$. Show, via the separation of variables technique, that the solution is

$$\phi=\phi_R x+\sum_{k=1}^{\infty}\frac{2\phi_R(-1)^k\exp(-k^2\pi^2\alpha t)\sin(k\pi x)}{k\pi} .$$

2.15 Show that the expression

$$\frac{1}{\sqrt{4\pi t}}\exp\left(-\frac{(x-y)^2}{4t}\right)$$

is the Green's function for the heat conduction problem considered in Problem 2.14, by showing that it satisfies (2.75) with y fixed.

3. Preliminary Computational Techniques

In this chapter an examination will be made of some of the basic computational techniques that are required to solve flow problems. For a specific problem the governing equations (Chap. 11) and the appropriate boundary conditions (Chaps. 11 and 2) will be known. Computational techniques are used to obtain an approximate solution of the governing equations and boundary conditions.

For example, for three-dimensional unsteady incompressible flow, velocity and pressure solutions, $u(x, y, z, t)$, $v(x, y, z, t)$, $w(x, y, z, t)$ and $p(x, y, z, t)$, would be computed. The process of obtaining the computational solution consists of two stages that are shown schematically in Fig. 3.1. The first stage converts the continuous partial differential equations and auxiliary (boundary and initial) conditions into a discrete system of algebraic equations. This first stage is called discretisation (Sect. 3.1). The process of discretisation is easily identified if the finite difference method is used (Sect. 3.5) but is slightly less obvious with the finite element, finite volume and spectral methods (Chap. 5).

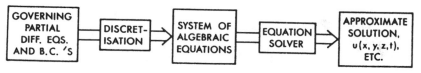

Fig. 3.1. Overview of the computational solution procedure

The replacement of individual differentiated terms in the governing partial differential equations by algebraic expressions connecting nodal values on a finite grid introduces an error. Choosing the algebraic expressions in a way that produces small errors is considered in Sect. 3.2. The achieved accuracy of representing the differentiated terms is examined in Sects. 3.3 and 3.4. Equally important as the error in representing the differentiated terms in the governing equation is the error in the solution. A simple finite difference program is provided in Sect. 3.5 so that the solution error can be examined directly.

In discussing unsteady problems the discretisation process is often identified with the reduction of the governing partial differential equations to a system of ordinary differential equations in time. This is understandable in the sense that techniques for solving ordinary differential equations (Lambert 1973) are so well-known that further discussion may not be required. However, in applying a particular method, the system of ordinary differential equations must be converted to a corresponding system of algebraic equations to obtain the computational solution.

The second stage of the solution process (Fig. 3.1) requires an equation solver to provide the solution of the system of algebraic equations. This stage can also introduce an error but it is usually negligible compared with the error introduced in the discretisation stage, unless the method is unstable (Sect. 4.3). Appropriate methods for solving systems of algebraic equations are discussed in Chap. 6. Systems of algebraic equations typically arise in solving steady flow problems. For unsteady flow problems the use of explicit techniques (e.g. Sect. 7.1.1) may reduce the equation-solving stage to no more than a one-line algorithm.

3.1 Discretisation

To convert the governing partial differential equation(s) to a system of algebraic equations (or ordinary differential equations), a number of choices are available. The most common are the finite difference, finite element, finite volume and spectral methods.

The way the discretisation is performed also depends on whether time derivatives (in time dependent problems) or equations containing only spatial derivatives are being considered. In practice, time derivatives are discretised almost exclusively using the finite difference method. Spatial derivatives are discretised by either the finite difference, finite element, finite volume or spectral method, typically.

3.1.1 Converting Derivatives to Discrete Algebraic Expressions

The discretisation process can be illustrated by considering the equation

$$\frac{\partial \bar{T}}{\partial t} = \alpha \frac{\partial^2 \bar{T}}{\partial x^2} , \tag{3.1}$$

which governs transient heat conduction in one dimension. $\bar{T}$ is the temperature and α is the thermal diffusivity. The overbar ($\bar{\ }$) denotes the exact solution. Typical boundary and initial conditions to suit (3.1) are

$$\bar{T}(0, t) = b , \quad \bar{T}(1, t) = d \quad \text{and} \tag{3.2}$$

$$\bar{T}(x, 0) = T_0(x) , \quad 0 \leq x \leq 1 . \tag{3.3}$$

The most direct means of discretisation is provided by replacing the derivatives by equivalent finite difference expressions. Thus, using (3.21, 25), (3.1) can be replaced by

$$\frac{T_j^{n+1} - T_j^n}{\Delta t} = \frac{\alpha(T_{j-1}^n - 2T_j^n + T_{j+1}^n)}{\Delta x^2} . \tag{3.4}$$

The step sizes Δt, Δx and the meaning of the subscript j and superscript n are indicated in Fig. 3.2. In (3.4) T_j^n is the value of T at the (j, n)th node.

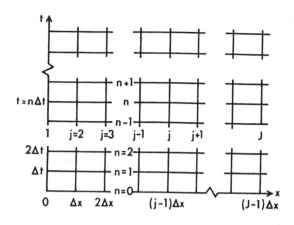

Fig. 3.2. The discrete grid

The process of discretising (3.1) to give (3.4) implies that the problem of finding the exact (continuous) solution $\bar{T}(x, t)$ has been replaced with the problem of finding discrete values T_j^n, i.e. the approximate solution at the (j, n)th node (Fig. 3.2). In turn, two related errors arise, the truncation error and the solution error. The truncation error introduced by the discretisation of (3.1) will be considered in Sects. 3.3 and 3.4. The corresponding (solution) error between the approximate solution and the exact solution will be examined in Sect. 4.1.

The precise value of the approximate solution between the nodal (grid) points is not obvious. Intuitively the solution would be expected to vary smoothly between the nodal points. In principle, the solution at some point (x_r, t_r) that does not coincide with a node can be obtained by interpolating the surrounding nodal point solution. It will be seen (Sect. 5.3) that this interpolation process is automatically built into the finite element method.

It is apparent that, whereas (3.1) is a partial differential equation, (3.4) is an algebraic equation. With reference to Fig. 3.2, (3.4) can be manipulated to give a formula (or algorithm) for the unknown value T_j^{n+1} in terms of the known values T_j^n at the nth time level, i.e.

$$T_j^{n+1} = T_j^n + \frac{\alpha \Delta t}{\Delta x^2} (T_{j-1}^n - 2T_j^n + T_{j+1}^n) \ . \tag{3.5}$$

To provide the complete numerical solution at time level $(n+1)$, (3.5) must be applied for all the nodes $j=2, \ldots, J-1$, assuming that Dirichlet boundary conditions provide the values T_1^{n+1} and T_J^{n+1}.

3.1.2 Spatial Derivatives

It has already been seen how the finite difference method discretises spatial derivatives, e.g. $\partial^2 \bar{T}/\partial x^2$ in (3.1) becomes $(T_{j-1}^n - 2T_j^n + T_{j+1}^n)/\Delta x^2$ in (3.4).

The finite element method (Sect. 5.3) achieves discretisation by first assuming that the local solution for T can be interpolated. Subsequently the local solution is substituted into a suitably weighted integral of the governing equation and the integrals evaluated. A typical result (using linear elements on a uniform mesh) would be

$$\frac{(\Delta x/6)(T_{j-1}^{n+1} - T_{j-1}^n)}{\Delta t} + \frac{(2\Delta x/3)(T_j^{n+1} - T_j^n)}{\Delta t} + \frac{(\Delta x/6)(T_{j+1}^{n+1} - T_{j+1}^n)}{\Delta t}$$

$$= \frac{\alpha(T_{j-1}^n - 2T_j^n + T_{j+1}^n)}{\Delta x} . \tag{3.6}$$

Dividing both sides of (3.6) by Δx produces a result that is similar in structure to (3.4). Equation (3.6) is derived in Sect. 5.5.1.

The spectral method (Sect. 5.6) proceeds in a similar manner to the finite element method except that the assumed solution for T is of the form

$$T = \sum_{j=1}^{J} a_j(t)\phi(x) , \tag{3.7}$$

where $a_j(t)$ are unknown coefficients to be determined as part of the solution and $\phi_j(x)$ are known functions of x (see Sect. 5.6). The final form of the discretised equation using the spectral method can be written

$$\frac{a_j^{n+1} - a_j^n}{\Delta t} = \sum_{j=1}^{J} p_j a_j^n , \tag{3.8}$$

where p_j are known algebraic coefficients.

Whatever method is used to perform the discretisation the subsequent solution of the equations, e.g. using (3.5), is obtained directly from the algebraic equations and is, in a sense, independent of the means of discretisation.

3.1.3 Time Derivatives

The replacement of $\partial \bar{T}/\partial t$ in (3.1) with the one-sided difference formula $(T_j^{n+1} - T_j^n)/\Delta t$ only uses information at time-levels n and $n+1$. Because time only proceeds in the positive direction, information at time-levels $n+2$ and greater is not available. In (3.4) the finite difference representation of the spatial derivative $\partial^2 \bar{T}/\partial x^2$ has been evaluated at time-level n and provides an explicit algorithm for T_j^{n+1}. If the spatial terms were evaluated at time-level $n+1$ the following implicit algorithm would be obtained:

$$-sT_{j-1}^{n+1} + (1 + 2s)T_j^{n+1} - sT_{j+1}^{n+1} = T_j^n , \tag{3.9}$$

where $s = \alpha \Delta t/\Delta x^2$. Equation (3.9) can only be solved as part of a system of equations formed by evaluating it for all nodes $j = 2, \ldots, J-1$ (See Sect. 7.2).

If the centred difference formula $(T_j^{n+1} - T_j^{n-1})/2\Delta t$ were used to replace $\partial \bar{T}/\partial t$ in (3.1) the following explicit algorithm can be constructed for T_j^{n+1}:

$$T_j^{n+1} = T_j^{n-1} + (2\alpha\Delta t/\Delta x^2)(T_{j-1}^n - 2T_j^n + T_{j+1}^n) \; . \tag{3.10}$$

The algorithm (3.10) is more accurate than (3.5) but more complicated since it involves three levels of data, $n-1$, n, $n+1$, rather than two. This particular algorithm is not practical since it is unstable (Sect. 7.1.2). However the use of centred time differencing with other equations, e.g. the convection equation (Sect. 9.1), is stable.

There is an alternative approach to discretising time derivatives which builds on the connection with ordinary differential equations. Equation (3.1) can be written as

$$\frac{\partial \bar{T}}{\partial t} = L\bar{T} \; , \tag{3.11}$$

where L is the differential operator $\alpha \partial^2/\partial x^2$. After spatial discretisation, (3.11) becomes

$$\frac{\partial T_j}{\partial t} = L_a T_j \tag{3.12}$$

where L_a is an algebraic operator resulting from the spatial discretisation. Equation (3.12), repeated for each node, represents a system of ordinary differential equations in time. Consequently, in principle, any of the techniques for integrating ordinary differential equations (Gear 1971) can be applied to (3.12). In general the integration can be written

$$T_j^{n+1} = T_j^n + \int_{t_n}^{t_{n+1}} L_a T_j \, dt' \; . \tag{3.13}$$

The Euler scheme for evaluating (3.13) is

$$T_j^{n+1} = T_j^n + [L_a T_j]^n \Delta t \; , \tag{3.14}$$

which is identical with (3.5) if L_a is the finite difference operator given in (3.4). Because of the errors associated with the spatial discretisation operator L_a, there is usually no advantage in using a very high-order integration formula in (3.13). Some of the more effective algorithms in this category are considered in Sect. 7.4.

3.2 Approximation to Derivatives

In Sect. 3.1 typical algebraic formulae were presented to illustrate the mechanics of discretising derivatives like $\partial^2 T/\partial x^2$. Here such algebraic formulae are constructed, first by inspection of a Taylor series expansion and secondly via a general procedure. In each case an estimate of the error involved in the discretisation process is readily available.

3.2.1 Taylor Series Expansion

As the first step in developing an algorithm to compute values of $\bar{T}$ that appear in (3.1), the space and time derivatives of $\bar{T}$ at the node (j, n) are expressed in terms of the values of $\bar{T}$ at nearby nodes. Taylor series expansions such as

$$\bar{T}_{j+1}^n = \sum_{m=0}^{\infty} \frac{\Delta x^m}{m!} \left[\frac{\partial^m \bar{T}}{\partial x^m} \right]_j^n \tag{3.15}$$

and

$$\bar{T}_j^{n+1} = \sum_{m=0}^{\infty} \frac{\Delta t^m}{m!} \left[\frac{\partial^m \bar{T}}{\partial t^m} \right]_j^n \tag{3.16}$$

are used in the process. These series may be truncated after any number of terms, the resulting (truncation) error being dominated by the next term in the expansion if $\Delta x \ll 1$ in (3.15) or if $\Delta t \ll 1$ in (3.16). Thus we may write

$$\bar{T}_{j+1}^n = \bar{T}_j^n + \Delta x \left[\frac{\partial \bar{T}}{\partial x} \right]_j^n + \frac{\Delta x^2}{2} \left[\frac{\partial^2 \bar{T}}{\partial x^2} \right]_j^n + O(\Delta x^3) \ . \tag{3.17}$$

The term $O(\Delta x^3)$ is interpreted as meaning there exists a positive constant K, depending on $\bar{T}$, such that the difference between $\bar{T}$ at the $(j+1, n)$th node and the first three terms on the right-hand side of (3.17), all evaluated at the (j, n)th node, is numerically less than $K\Delta x^3$ for all sufficiently small Δx. Clearly the error involved in this approximation rapidly reduces in magnitude as the size of Δx decreases.

A consideration of (3.17) suggests that a finite difference expression for $\partial \bar{T}/\partial x$ could be obtained directly. Thus, by rearranging (3.17),

$$\left[\frac{\partial \bar{T}}{\partial x} \right]_j^n = \frac{(\bar{T}_{j+1}^n - \bar{T}_j^n)}{\Delta x} - 0.5\Delta x \left[\frac{\partial^2 \bar{T}}{\partial x^2} \right]_j^n + \cdots . \tag{3.18}$$

It is apparent that using the finite difference replacement

$$\left[\frac{\partial \bar{T}}{\partial x} \right]_j^n \approx \frac{T_{j+1}^n - T_j^n}{\Delta x} \tag{3.19}$$

is accurate to $O(\Delta x)$. The additional terms appearing in (3.18) are referred to as the truncation error. Equation (3.19) is called a forward difference approximation. By expanding T_{j-1}^n as a Taylor series about node (j, n) and rearranging, a backward difference approximation can be constructed:

$$\left[\frac{\partial \bar{T}}{\partial x} \right]_j^n \approx \frac{T_j^n - T_{j-1}^n}{\Delta x} \ . \tag{3.20}$$

This, like (3.19), introduces an error of $O(\Delta x)$. A geometric interpretation of (3.19 and 20) is provided in Fig. 3.3. Equation (3.19) evaluates $[\partial \bar{T}/\partial x]_j^n$ as the slope BC; (3.20) evaluates $[\partial \bar{T}/\partial x]_j^n$ as the slope AB.

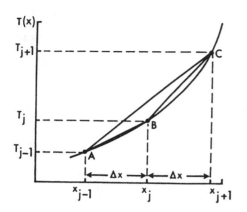

Fig. 3.3 Finite difference representations of $\partial \bar{T}/\partial x$

Equations (3.19 and 20) have been obtained by introducing a Taylor series expansion in space. The Taylor series expansion in time, (3.16), can be manipulated to give the forward difference approximation

$$\left[\frac{\partial \bar{T}}{\partial t}\right]_j^n \approx \frac{T_j^{n+1} - T_j^n}{\Delta t} \; , \tag{3.21}$$

which introduces an error of $O(\Delta t)$, assuming that $\Delta t \ll 1$ and higher-order derivatives are bounded.

3.2.2 General Technique

The finite difference expressions provided in Sect. 3.2.1 have been constructed by a simple manipulation of a single Taylor expansion. A more methodical technique for constructing difference approximations is to start from a general expression, e.g.

$$\left[\frac{\partial \bar{T}}{\partial x}\right]_j^n = a\bar{T}_{j-1}^n + b\bar{T}_j^n + c\bar{T}_{j+1}^n + O(\Delta x^m) \; , \tag{3.22}$$

where a, b and c are to be determined and the term $O(\Delta x^m)$ will indicate the accuracy of the resulting approximation.

Using (3.15) we may write

$$a\bar{T}_{j-1}^n + b\bar{T}_j^n + c\bar{T}_{j+1}^n = (a+b+c)\,\bar{T}_j^n + (-a+c)\Delta x\left[\frac{\partial \bar{T}}{\partial x}\right]_j^n$$

$$+(a+c)\frac{\Delta x^2}{2}\left[\frac{\partial^2 \bar{T}}{\partial x^2}\right]_j^n + (-a+c)\frac{\Delta x^3}{6}\left[\frac{\partial^3 \bar{T}}{\partial x^3}\right]_j^n + \ldots . \tag{3.23}$$

Setting

$$a+b+c=0 \; , \quad (-a+c)\Delta x=1 \quad \text{gives}$$

$$a=c-1/\Delta x \quad \text{and} \quad b=-2c+1/\Delta x \quad \text{for any } c \; .$$

Choosing c so that the third term on the right-hand side of (3.23) disappears produces the most accurate approximation possible with three disposable parameters. That is

$$c = -a = 1/2\Delta x \quad \text{and} \quad b = 0 .$$

Substitution of these values into (3.23) gives

$$\left[\frac{\partial \bar{T}}{\partial x}\right]_j^n = \frac{1}{2\Delta x}(-\bar{T}_{j-1}^n + \bar{T}_{j+1}^n) - \frac{\Delta x^2}{6}\left[\frac{\partial^3 \bar{T}}{\partial x^3}\right]_j^n + \dots$$

Therefore the centred (or central) difference approximation to $[\partial \bar{T}/\partial x]_j^n$, is

$$\left[\frac{\partial \bar{T}}{\partial x}\right]_j^n \approx \frac{T_{j+1}^n - T_{j-1}^n}{2\Delta x} , \tag{3.24}$$

which has a truncation error of $O(\Delta x^2)$. Clearly the centred difference approximation produces a higher-order truncation error than the forward (3.19) or backward (3.20) difference approximations. Equation (3.24) evaluates $[\partial \bar{T}/\partial x]_j^n$ as the slope AC in Fig. 3.3.

Using a similar representation to (3.22) the following centred difference form for $[\partial^2 \bar{T}/\partial x^2]_j^n$ can be obtained as

$$\left[\frac{\partial^2 \bar{T}}{\partial x^2}\right]_j^n = \frac{\bar{T}_{j-1}^n - 2\bar{T}_j^n + \bar{T}_{j+1}^n}{\Delta x^2} + O(\Delta x^2) . \tag{3.25}$$

The above technique, (3.22), can be used to obtain one-sided difference formulae by expanding about an appropriate node. The same technique, (3.22), can also be used to develop multidimensional formulae or difference formulae on a non-uniform grid (Sect. 10.1.5).

3.2.3 Three-point Asymmetric Formula for $[\partial \bar{T}/\partial x]_j^n$

The general technique for obtaining algebraic formulae for derivatives (Sect. 3.2.2) is used to derive the three-point one-sided representation for $[\partial \bar{T}/\partial x]_j^n$. The starting point is the following general expression, in place of (3.22),

$$\left[\frac{\partial \bar{T}}{\partial x}\right]_j^n = a\bar{T}_j^n + b\bar{T}_{j+1}^n + c\bar{T}_{j+2}^n + O(\Delta x^m) , \tag{3.26}$$

where a, b, and c are to be determined. $\bar{T}_{j+1}^n$ and $\bar{T}_{j+2}^n$ are expanded about j as Taylor series (Sect. 3.2.1). Substituting into (3.26) and rearranging gives

$$\left[\frac{\partial \bar{T}}{\partial x}\right]_j^n = (a+b+c)\bar{T}_j^n + (b\Delta x + c2\Delta x)\left[\frac{\partial \bar{T}}{\partial x}\right]_j^n$$

$$+ \left(\frac{b\Delta x^2}{2} + \frac{c(2\Delta x)^2}{2}\right)\left[\frac{\partial^2 \bar{T}}{\partial x^2}\right]_j^n + \dots \tag{3.27}$$

Comparing the left- and right-hand sides of (3.27) indicates that the following conditions must be imposed on a, b and c to obtain the smallest error:

$$a+b+c=0 , \quad b\Delta x + c2\Delta x = 1 , \quad \frac{b\Delta x^2}{2} + \frac{c(2\Delta x)^2}{2} = 0 .$$

This gives the values

$$a = -\frac{1.5}{\Delta x} , \quad b = \frac{2}{\Delta x} , \quad \text{and} \quad c = -\frac{0.5}{\Delta x} ,$$

and

$$\left[\frac{\partial \bar{T}}{\partial x}\right]_j^n = \frac{-1.5\bar{T}_j^n + 2\bar{T}_{j+1}^n - 0.5\bar{T}_{j+2}^n}{\Delta x} - \frac{\Delta x^2}{3}\left[\frac{\partial^3 \bar{T}}{\partial x^3}\right]_j^n + \ldots ,$$

which agrees with the result given in Table 3.3. This formula has a truncation error of $O(\Delta x^2)$ like the centred difference formula (3.24).

If more terms are included in (3.26), e.g.

$$\left[\frac{d\bar{T}}{dx}\right]_j^n \approx a\bar{T}_j^n + b\bar{T}_{j+1}^n + c\bar{T}_{j+2}^n + d\bar{T}_{j+3}^n + e\bar{T}_{j+4}^n ,$$

extra conditions to determine the coefficients a to e are obtained from (3.27) extended by requiring that the coefficients multiplying higher-order derivatives are zero. However schemes based on higher-order discretisations often have more severe stability restrictions (Sect. 4.3) than those based on low-order discretisations. Consequently an alternative strategy is to choose some of the coefficients a to e to reduce the error and some to improve the stability. A similar approach is taken in constructing schemes to solve ordinary differential equations (Hamming 1973, p. 405).

3.3 Accuracy of the Discretisation Process

Discretisation is necessary to convert the governing differential equation into an equivalent system of algebraic equations that can be solved using a computer. The discretisation process invariably introduces an error unless the underlying exact solution has a very elementary analytic form. Thus the centred difference formula (3.24) is exact for polynomials up to quadratic, whereas the one-sided formulae (3.19, 20) are exact only for linear polynomials. The exactness can be inferred from the fact that all terms in the truncation error are zero for polynomials of sufficiently low order.

In general the error for a finite difference representation of a derivative can be obtained by making a Taylor series expansion about the node at which the derivative is being evaluated (Sect. 3.2.2), and the evaluation of the leading term in the remainder provides a close approximation to the error if the grid size is small. However, the complete evaluation of the terms in the Taylor series relies on the exact solution being known.

A more direct way of comparing the accuracy of various algebraic formulae for derivatives is to consider a simple analytic function, like $\bar{T} = \exp x$, and to compare the value of the derivative obtained analytically and obtained from the discretisation formula. Table 3.1 shows such a comparison for $d\bar{T}/dx$ evaluated at $x = 1$, with $\bar{T} = \exp x$ as the analytic function; the step size $\Delta x = 0.1$. Generally the three-point formulae, whether symmetric or asymmetric, are considerably more accurate than either the (two-point) forward or backward difference formulae, but considerably less accurate than the five-point symmetric formula. It is apparent (Table 3.1) that the leading term in the Taylor expansion (T.E.) gives a good estimate of the error, if Δx is sufficiently small. For this particular example all higher-order derivatives in the Taylor expansion equal $\exp x$. For a more general problem where higher-order derivatives may be larger a step size of less than 0.1 may be necessary to ensure the error is closely approximated by the leading term in the truncation error.

Table 3.1. Comparison of formulae to evaluate $d\bar{T}/dx$ at $x = 1.0$

Case	Algebraic formula	$\left[\dfrac{d\bar{T}}{dx}\right]_j$	Error	Leading term in T.E.
Exact	—	2.7183	—	—
3PT SYM	$(\bar{T}_{j+1} - \bar{T}_{j-1})/2\Delta x$	2.7228	0.4533×10^{-2}	0.4531×10^{-2}
FOR DIFF	$(\bar{T}_{j+1} - \bar{T}_j)/\Delta x$	2.8588	0.1406×10^{-0}	0.1359×10^{-0}
BACK DIFF	$(\bar{T}_j - \bar{T}_{j-1})/\Delta x$	2.5868	-0.1315×10^{-0}	-0.1359×10^{-0}
3PT ASYM	$(-1.5\bar{T}_j + 2\bar{T}_{j+1} - 0.5\bar{T}_{j+2})/\Delta x$	2.7085	-0.9773×10^{-2}	-0.9061×10^{-2}
5PT SYM	$(\bar{T}_{j-2} - 8\bar{T}_{j-1} + 8\bar{T}_{j+1} - \bar{T}_{j+2})/12\Delta x$	2.7183	-0.9072×10^{-5}	-0.9061×10^{-5}

Typical algebraic formulae for $d^2\bar{T}/dx^2$ evaluated at $x = 1.0$ for function values of $\bar{T} = \exp x$ are shown in Table 3.2. The function values are evaluated at intervals $\Delta x = 0.1$. As before, the three-point symmetric formula is accurate, but now the three-point asymmetric formula is inaccurate. As with the evaluation of the first derivative formulae, the leading term in the Taylor expansion provides an accurate estimate of the error.

The algebraic formulae for the leading term in the truncation error expressions are shown in Tables 3.3 and 3.4. These formulae are obtained by making a Taylor expansion about the jth node as in Sect. 3.2.1. In Table 3.3, $\bar{T}_{xxx} \equiv d^3\bar{T}/dx^3$, etc. For this particular example ($\bar{T} = \exp x$), $\bar{T}_{xxx} = \bar{T}_{xxxx}$, etc. Thus the magnitude of the error depends primarily on powers of Δx. Consequently, as Δx is reduced it is

Table 3.2. Comparison of formulae to evaluate $d^2\bar{T}/dx^2$ at $x = 1.0$

Case	Algebraic formula	$\left[\dfrac{d\bar{T}^2}{dx^2}\right]_j$	Error	Leading term in T.E.
Exact	—	2.7183	—	—
3PT SYM	$(\bar{T}_{j-1} - 2\bar{T}_j + \bar{T}_{j+1})/\varDelta x^2$	2.7205	0.2266×10^{-2}	0.2265×10^{-2}
3PT ASYM	$(\bar{T}_j - 2\bar{T}_{j+1} + \bar{T}_{j+2})/\varDelta x^2$	3.0067	0.2884×10^{-0}	0.2718×10^{-0}
5PT SYM	$(-\bar{T}_{j-2} + 16\bar{T}_{j-1} - 30\bar{T}_j$ $+ 16\bar{T}_{j+1} - \bar{T}_{j+2})/12\varDelta x^2$	2.7183	-0.3023×10^{-5}	-0.3020×10^{-5}

Table 3.3. Truncation error leading term (algebraic): $d\bar{T}/dx$

Case	Algebraic formula	Truncation error leading term
3PT SYM	$(\bar{T}_{j+1} - \bar{T}_{j-1})/2\varDelta x$	$\varDelta x^2 \bar{T}_{xxx}/6$
FOR DIFF	$(\bar{T}_{j+1} - \bar{T}_j)/\varDelta x$	$\varDelta x \bar{T}_{xx}/2$
BACK DIFF	$(\bar{T}_j - \bar{T}_{j-1})/\varDelta x$	$-\varDelta x \bar{T}_{xx}/2$
3PT ASYM	$(-1.5\bar{T}_j + 2\bar{T}_{j+1} - 0.5\bar{T}_{j+2})/\varDelta x$	$-\varDelta x^2 \bar{T}_{xxx}/3$
5PT SYM	$(\bar{T}_{j-2} - 8\bar{T}_{j-1} + 8\bar{T}_{j+1} - \bar{T}_{j+2})/12\varDelta x$	$-\varDelta x^4 \bar{T}_{xxxxx}/30$

Table 3.4. Truncation error leading term (algebraic): $d^2\bar{T}/dx^2$

Case	Algebraic formula	Truncation error leading term
3PT SYM	$(\bar{T}_{j-1} - 2\bar{T}_j + \bar{T}_{j+1})/\varDelta x^2$	$\varDelta x^2 \bar{T}_{xxxx}/12$
3PT ASYM	$(\bar{T}_j - 2\bar{T}_{j+1} + \bar{T}_{j+2})/\varDelta x^2$	$\varDelta x \bar{T}_{xxx}$
5PT SYM	$(-\bar{T}_{j-2} + 16\bar{T}_{j-1} - 30\bar{T}_j$ $+ 16\bar{T}_{j+1} - \bar{T}_{j+2})/12\varDelta x^2$	$-\varDelta x^4 \bar{T}_{xxxxx}/90$

expected that the truncation error, when using the five-point formula, will reduce far more quickly than the error when using the two-point forward or backward difference formulae.

The reason for the large error associated with the three-point asymmetric formula shown in Table 3.2 is apparent in Table 3.4 where the leading term in the truncation error is seen to be only first-order accurate.

Roughly the directly computed error, E, may be written

$$E = A(\varDelta x)^k$$

and the truncation error as

$$TE = B(\varDelta x)^k \dots,$$

where k is the exponent of the grid size in the leading term of the truncation error, as

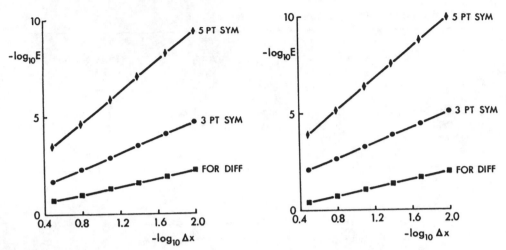

Fig. 3.4. Convergence results for the evaluation of $d\bar{T}/dx$. $E = |[d\bar{T}/dx]_{FD}/[d\bar{T}/dx]_{ex} - 1|$

Fig. 3.5. Convergence results for the valuation of $d^2\bar{T}/dx^2$. $E = |[d^2\bar{T}/dx^2]_{FD}/[d^2\bar{T}/dx^2]_{ex} - 1|$

in Tables 3.3 and 3.4 for example. Therefore it is expected that the directly computed error will reduce with Δx in the manner shown in Tables 3.3 and 3.4. This is confirmed by the results shown in Figs. 3.4 and 3.5. By plotting on a log-scale, k is given by the slope of the data, and corresponds to the exponent of the gridsize in the leading term in the truncation error. It is clear from the data plotted that the various cases are achieving the expected convergence rate implied by the truncation error leading terms in Tables 3.3 and 3.4. The convergence rate can still be estimated from the truncation error even when the exact solution is unknown. Thus one may infer from a truncation error with a fourth-order ($k = 4$) leading term (5PT SYM) that the solution error decreases at a much faster rate with grid refinement than the solution error corresponding to a truncation error with a second-order ($k = 2$) leading term (3PT SYM).

3.3.1 Higher-Order vs Low-Order Formulae

From the results presented so far it might appear that a higher-order formula on a fine grid should always be used. However this is deceptive. First, the evaluation of a higher-order formula involves more points and hence is less economical than the evaluation of a low-order formula. From a practical perspective, the accuracy that can be achieved for a given execution time, or the computational efficiency, is more important than the accuracy alone; the accuracy can always be increased by refining the grid. Computational efficiency is considered in Sect. 4.5.

Second, higher-order formulae show a relatively small accuracy advantage over low-order formulae for a coarse grid but demonstrate a much greater accuracy advantage when the grid is refined. However for a particular problem it is often the case that the general accuracy level required of the answers is appropriate to a coarse grid or that a coarse grid is necessary because of computer memory or execution time limitations. The superiority of the higher-order formulae shown in Figs. 3.4 and 3.5 is also dependent on the smoothness of the exact solution. Inviscid supersonic flows can produce discontinuous solutions, associated with the presence of shock waves (Liepmann and Roshko 1957, p. 56). If the solution is discontinuous the validity of the techniques (Sects. 3.2.1 and 3.2.2) for constructing the difference formulae is compromised since there is no guarantee that successive terms in the truncation error expansion reduce in magnitude. As a result the inclusion of more points in the finite difference expression and the cancellation of more terms in the truncation error expansions implies nothing about the corresponding solution accuracy. This can be seen for the exact solution shown in Fig. 3.6.

Using the three-point symmetric formula: $[d\bar{T}/dx]_{x=1} = -0.5/\Delta x$

Using the five-point symmetric formula: $[d\bar{T}/dx]_{x=1} = -7/(12\Delta x)$

Since the exact solution is $[d\bar{T}/dx]_{x=1} = -\infty$, the five-point formula is not appreciably more accurate than the three-point formula.

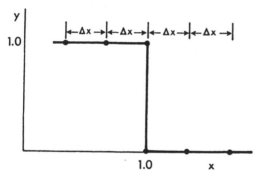

Fig. 3.6. Discontinuous (exact) solution

For viscous problems at high Reynolds number (i.e. little natural dissipation) discontinuities cannot occur but very severe gradients do occur. If the gradient is severe enough and the grid coarse enough higher-order schemes are not advantageous. This can be illustrated by the function

$$\bar{y} = \tanh[k(x-1)] \ .$$

This is plotted for three values of k in Fig. 3.7. Clearly, there is a gradient centred at $x=1$ whose severity grows with k. The first derivative dy/dx has been evaluated at $x=0.96$ using the three-point and five-point symmetric formulae (Table 3.1) for decreasing Δx with $k=5$ and 20.

The result is shown in Fig. 3.8. It is noticeable that the five-point formula only produces superior accuracy if the grid is sufficiently refined. The necessary degree of refinement increases as the gradient becomes sharper (increasing k). For some intermediate values of Δx the five-point formula produces a less accurate evaluation of the derivative. The corresponding comparison for the second derivative evaluation is shown in Fig. 3.9. The same general trend is apparent, namely that the

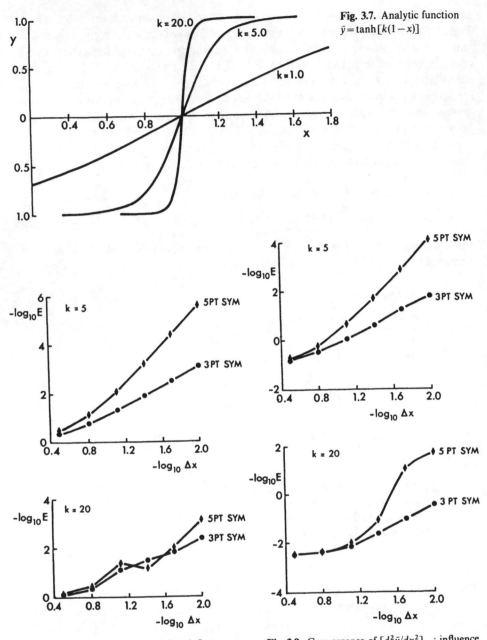

Fig. 3.7. Analytic function $\bar{y}=\tanh[k(1-x)]$

Fig. 3.8. Convergence of $[d\bar{y}/dx]_{FD}$: influence of solution smoothness.
$E=|[d\bar{y}/dx]_{FD}/[d\bar{y}/dx]_{ex}-1|$

Fig. 3.9. Convergence of $[d^2\bar{y}/dx^2]_{FD}$: influence of solution smoothness.
$E=|[d^2\bar{y}/dx^2]_{FD}/[d^2\bar{y}/dx^2]_{ex}-1|$

higher-order formula only provides a substantial improvement when the grid is refined.

When severe gradients occur, the magnitude of higher-order derivatives is much larger than that of low-order derivatives. Consequently, on a given grid higher-order terms in the truncation error expression do not diminish at such a rapid rate as when the underlying exact solution is smooth. For the same reason, unless the grid is made very fine the magnitude of the higher derivative in the leading term of the truncation error may be so large for a higher-order discretisation that the overall error is comparable to that of a low-order discretisation.

As a general comment, at least second-order discretisations should be used for reasons that are discussed in Sect. 9.4. The use of higher-order discretisations may be justified in special circumstances.

3.4 Wave Representation

Many fluid flow phenomena demonstrate a wave-like motion. Therefore it is conceptually useful to consider the exact solution as though it were broken up into its separate Fourier components. This raises the question of whether the discretisation process represents waves of short and long wavelength with the same accuracy.

3.4.1 Significance of Grid Coarseness

The finite difference method replaces a continuous function $g(x)$ with a vector of nodal values (g_j) corresponding to a vector of discrete grid points (x_j). The choice of an appropriate grid spacing Δx is dependent on the smoothness of $g(x)$. A poor choice is illustrated in Fig. 3.10a, and a reasonable choice is shown in Fig. 3.10b. To obtain an accurate representation of $g(x)$ shown in Fig. 3.10a would require a much smaller grid spacing Δx than for $g(x)$ shown in Fig. 3.10b.

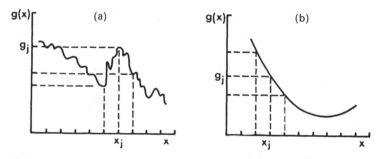

Fig. 3.10a, b. Discrete representation of $g(x)$. Grid spacing too coarse (a) and reasonable (b)

A Fourier representation of $g(x)$ (assumed periodic) in the interval $0 \le x \le 2\pi$ is

$$g(x) = \sum_{m=-\infty}^{\infty} g_m e^{imx} , \qquad (3.28)$$

where $i = (-1)^{1/2}$, m is the wave number and g_m is the amplitude of the Fourier mode of wavelength $\lambda = 2\pi/m$ given by (Hamming 1973, p. 509)

$$g_m = \frac{1}{2\pi} \int_0^{2\pi} g(x) e^{-imx} dx . \qquad (3.29)$$

The vector of nodal values, (g_j), can also be given a Fourier representation. This has the form

$$g_j = \sum_{m=1}^{J} g_m e^{imj\Delta x} , \qquad (3.30)$$

where the modal amplitudes g_m are given by

$$g_m = \Delta x \sum_{j=1}^{J} g_j e^{-imj\Delta x} . \qquad (3.31)$$

The discrete nature of the grid restricts the range of wavelengths that can be represented. In particular, wavelengths shorter than the cut-off wavelength $\lambda = 2\Delta x$ cannot be represented. Consequently (g_j) should be interpreted as a long-wave approximation of $g(x)$. Similarly, T_j^{n+1}, the approximate solution obtained from (3.5) should be considered a long-wave approximation to the exact solution of (3.1). This aspect is considered further in Sect. 9.2 and is exploited in multigrid methods (Sect. 6.3.5).

3.4.2 Accuracy of Representing Waves

The accuracy of finite difference approximations, when wave-like motion is to be expected, may be assessed by application to progressive waves such as

$$\bar{T}(x, t) = \Re \{e^{[im(x-qt)]}\} = \cos[m(x-qt)] , \qquad (3.32)$$

where $i = (-1)^{1/2}$, $\Re$ denotes the real part, m is the wave number, as in (3.28), and q is the propagation speed of the wave which is moving in the positive x direction. At a fixed point x_j the wave motion is periodic with a period $P = 2\pi/(qm)$.

At the (j, n)-th node, the exact value of the first and second derivatives of $\bar{T}$ are

$$\frac{\partial \bar{T}}{\partial x} = -m \sin[m(x_j - qt_n)] , \qquad (3.33)$$

$$\frac{\partial^2 \bar{T}}{\partial x^2} = -m^2 \cos[m(x_j - qt_n)] . \qquad (3.34)$$

Substitution of (3.32) into the three-point formula

$$\left[\frac{\partial \bar{T}}{\partial x}\right]_j^n \approx \frac{(\bar{T}_{j+1}^n - \bar{T}_{j-1}^n)}{2\Delta x}$$

gives

$$\left[\frac{\partial \bar{T}}{\partial x}\right]_j^n = \frac{1}{2\Delta x}(\cos\{[m(x_j - qt_n)] + m\Delta x\} - \cos\{[m(x_j - qt_n)] - m\Delta x\})$$

$$= \frac{-m \sin[m(x_j - qt_n)] \sin(m\Delta x)}{m\Delta x} . \qquad (3.35)$$

Thus the amplitude ratio of the first derivative representation is

$$\underset{\text{3PT}}{AR(1)} = \frac{[\partial \overline{T}/\partial x]_j^n}{[\partial \overline{T}/\partial x]} = \frac{\sin(m\Delta x)}{m\Delta x} . \tag{3.36}$$

Making use of (3.32), the central difference approximation to $\partial^2 \overline{T}/\partial x^2$ gives

$$\frac{\overline{T}_{j-1}^n - 2\overline{T}_j^n + \overline{T}_{j+1}^n}{\Delta x^2} = -m^2 \left(\frac{[\sin(m\Delta x/2)]}{(m\Delta x/2)} \right)^2 \cos[m(x_j - qt_n)] . \tag{3.37}$$

The amplitude ratio of the second derivative representation is

$$\underset{\text{3PT}}{AR(2)} = \left(\frac{\sin(0.5m\Delta x)}{0.5m\Delta x} \right)^2 . \tag{3.38}$$

An examination of (3.36) shows that the use of the finite difference approximation has introduced a change in the amplitude of the derivative. For long wavelengths, that is $\lambda > 20\Delta x$, the amplitude of the first derivative is reduced by a factor between 0.984 and 1.000 in using the centred difference approximation. However, when there are less than 4 grid spacings in one wavelength (short wavelengths) the amplitude of the derivative is less than 0.64 of its correct value. For a wavelength of $\lambda = 20\Delta x$, the centred difference representation of $\partial^2 \overline{T}/\partial x^2$ reduces the amplitude by 0.992. However, at a wavelength of $\lambda = 2\Delta x$ the amplitude of the second derivative is reduced by 0.405. As noted in Sect. 3.4.1, long wavelengths are represented more accurately than short wavelengths.

When the forward difference approximation to $[\partial \overline{T}/\partial x]_j^n$, (3.19), is compared with the exact value of the derivative, for $\overline{T}$ given by (3.32), it is found that errors are introduced in both phase and amplitude. The true amplitude is multiplied by the factor $[\sin(m\Delta x/2)/(m\Delta x/2)]$ and the phase is decreased by $m\Delta x/2$, which is equivalent to a spatial lead of $\Delta x/2$. For the above examples the amplitude and phase errors disappear as $\Delta x \to 0$, i.e. the long wavelength limit.

3.4.3 Accuracy of Higher-Order Formulae

In Sect. 3.4.2 it was indicated that the accuracy of discretisation could be assessed by looking at a progressive wave travelling with constant amplitude and speed, q. The exact solution is given by (3.32). Here this example will be used to see if higher-order difference formulae represent waves more accurately than low-order formulae. Specifically, a comparison will be made of the symmetric three-point and five-point formulae for $\partial \overline{T}/\partial x$ and $\partial^2 \overline{T}/\partial x^2$ given in Tables 3.1 and 3.2.

Following the same development as for (3.36) the amplitude ratio for the five-point symmetric representation for $\partial \overline{T}/\partial x$ (Table 3.1) is

$$\underset{\text{5PT}}{AR(1)} = \left(\frac{4}{3} - \frac{1}{3} \cos m\Delta x \right) \frac{\sin m\Delta x}{m\Delta x} . \tag{3.39}$$

The long and short wavelength behaviour of (3.39) is shown in Table 3.5. The

Table 3.5. Amplitude ratios for a progressive wave

Derivative	Scheme	Amplitude ratio	Amplitude ratio
		Long wavelength ($\lambda = 20\,\Delta x$)	Short wavelength ($\lambda = 4\Delta x$)
$\dfrac{d\bar{T}}{dx}$	3PT SYM	0.9836	0.6366
	5PT SYM	0.9996	0.8488
Derivative	Scheme	Long wavelength ($\lambda = 20\Delta x$)	Short wavelength ($\lambda = 2\Delta x$)
$\dfrac{d^2\bar{T}}{dx^2}$	3PT SYM	0.9918	0.4053
	5PT SYM	0.9999	0.5404

amplitude ratio for the five-point representation of $\partial^2 \bar{T}/\partial x^2$ (Table 3.2) is given by

$$\underset{5\mathrm{PT}}{\mathrm{AR}}(2) = \frac{4}{3}\{1 - 0.25\,[\cos(0.5m\Delta x)]^2\}\left(\frac{\sin(0.5m\Delta x)}{0.5m\Delta x}\right)^2. \tag{3.40}$$

The long and short wavelength behaviour of (3.40) is shown in Table 3.5. The results shown in Table 3.5 indicate that both schemes are more accurate for long wavelengths, with the five-point scheme being particularly accurate. However, neither scheme is very accurate for short wavelengths. This result is consistent with the previous comments (Sect. 3.3) about the relative lack of advantage in using a higher-order scheme on a coarse grid.

For a given wave being modelled, refining the grid (i.e. allowing more points to represent each wavelength) shifts the problem from a short wavelength to a long wavelength problem. Consequently the main conclusion to be drawn from Table 3.5 is the need to use a sufficiently refined grid to accurately compute wavelike motions.

3.5 Finite Difference Method

As implied in Sect. 3.1 the basis for the finite difference method is the construction of a discrete grid (Fig. 3.2), the replacement of the continuous derivatives in the governing partial differential equations with equivalent finite difference expressions and the rearrangement of the resulting algebraic equation into an algorithm, as in (3.5). In this section the above aspects are linked together and a simple finite difference program is provided.

3.5.1 Conceptual Implementation

The various steps in applying the finite difference method to a given problem are represented schematically in Fig. 3.11.

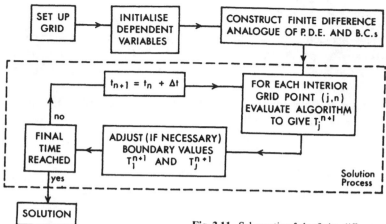

Fig. 3.11. Schematic of the finite difference solution process

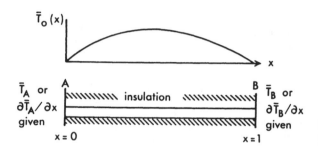

Fig. 3.12. Unsteady heat conduction in a rod

The procedure can be made specific by applying it to the following transient heat conduction (diffusion) problem. An insulated rod (Fig. 3.12) initially has a temperature of $T(x, 0) = 0°C$. At $t = 0$ hot reservoirs ($T = 100°C$) are brought into contact with the two ends, A and B. The problem is to find (numerically) the subsequent temperature $T(x, t)$ of any point in the rod.

The governing equation for this problem is (3.1)

$$\frac{\partial \bar{T}}{\partial t} - \alpha \frac{\partial^2 \bar{T}}{\partial x^2} = 0 .$$

For the solution of this equation, the initial conditions (3.3) serve to initialise the dependent variables T_j^0 (Fig. 3.2). One of the simplest finite difference analogues of the diffusion equation is obtained by replacing the time derivative with a forward difference approximation and the spatial derivative with a centred difference approximation (as in Sect. 3.1.1). This leads, via (3.5), to the forward time, centred space (FTCS) scheme

$$T_j^{n+1} = s T_{j-1}^n + (1 - 2s) T_j^n + s T_{j+1}^n , \tag{3.41}$$

where $s = \alpha \Delta t / \Delta x^2$. Equation (3.41) is applied at all internal nodes $j = 2, \ldots, J-1$. In a typical heat conduction problem the boundary values T_1^{n+1} and T_J^{n+1} are given by the boundary conditions (3.2). The solution process is repeated, advancing in time ($n = 1, 2, \ldots$), until the required final time is reached.

3.5.2 DIFF: Transient Heat Conduction (Diffusion) Problem

In Sect. 3.5.1 the implementation of the finite difference method is described qualitatively; here a corresponding computer program, DIFF, is provided. DIFF applies the FTCS scheme, (3.41), to the transient heat conduction (diffusion) problem shown in Fig. 3.12.

The listing of program DIFF is shown in Fig. 3.13 and the various parameters used by DIFF are described in Table 3.6.

```
1
2  C      SOLVES 1D TRANSIENT HEAT CONDUCTION EQUATION USING
3  C      FTCS SCHEME
4  C
5         DIMENSION TN(41),DUM(41),TD(41),X(41),TE(41)
6  C
7  C      INPUT DATA;
8  C
9  C      JMAX  = THE NUMBER OF POINTS ALONG THE ROD
10 C      MAXEX = THE NUMBER OF TERMS IN THE EXACT SOLUTION
11 C      NMAX  = THE MAXIMUM NUMBER OF TIME STEPS
12 C      ALPH  = THE THERMAL DIFFUSIVITY
13 C      S     = ALPH*DELT/DELX/DELX
14 C      TMAX  = THE MAXIMUM TIME
15 C
16        OPEN(1,FILE='DIFF.DAT')
17        OPEN(6,FILE='DIFF.OUT')
18        READ(1,1)JMAX,MAXEX,NMAX,ALPH,S,TMAX
19      1 FORMAT(3I5,E10.3,F5.2,F5.0)
20        PI = 3.1415927
21 C
22 C      TD = DIMENSIONAL TEMPERATURE
23 C      TN = NONDIMENSIONAL TEMPERATURE
24 C
25        JMAP = JMAX - 1
26        AJM = JMAP
27        DELX = 1./AJM
28        DELT = DELX*DELX*S/ALPH
29        WRITE(6,2)JMAX,MAXEX,NMAX,TMAX
30      2 FORMAT(' JMAX=',I5,' MAXEX=',I5,' NMAX=',I5,' TMAX=',F8.2)
31        WRITE(6,3)S,ALPH,DELT,DELX
32      3 FORMAT(' S=',F5.3,' ALPH =',E10.3,' DELT =',E10.3,
33     1' DELX =',E10.3,//)
34        WRITE(6,4)S
35      4 FORMAT(' FTCS(EXPLICIT) SCHEME,',5X,' S =',F5.3,//)
36 C
37 C      SET INITIAL CONDITIONS
38 C
39        DO 5 J = 1,JMAP
40      5 TN(J) = 0.
41        N = 0
42        T = 0.
43        SJ = 1.0 - 2.0*S
44 C
45 C      SET BOUNDARY CONDITIONS
46 C      EACH TIME STEP STARTS AT STATEMENT 6
47 C
48      6 TN(1) = 1.
49        TN(JMAX) = 1.
50        IF(T .LT. 0.01)TN(1) = 0.5
51        IF(T .LT. 0.01)TN(JMAX) = 0.5
52        TD(1) = 100.*TN(1)
53        TD(JMAX) = 100.*TN(JMAX)
```

Fig. 3.13. Listing of program DIFF

```
54 C
55 C       COMPUTE F.D. SOLUTION
56 C
57         DO 7 J = 2,JMAP
58         DUM(J) = SJ*TN(J) + S*(TN(J-1) + TN(J+1))
59       7 CONTINUE
60         DO 8 J = 2,JMAP
61       8 TN(J) = DUM(J)
62 C
63         DO 9 J = 2,JMAP
64       9 TD(J) = 100.*TN(J)
65         T = T + DELT
66         WRITE(6,10)T,(TD(J),J=1,JMAX)
67      10 FORMAT(' T= ',F5.0,'  TD=',11F6.2)
68 C
69 C       IF MAXIMUM TIME OR MAXIMUM NUMBER OF TIME-STEPS EXCEEDED
70 C       EXIT FROM LOOP
71 C
72         IF(N .GE. NMAX)GOTO 11
73         IF(T .LT. TMAX)GOTO 6
74 C
75 C       OBTAIN EXACT SOLUTION AND COMPARE
76 C
77      11 SUM = 0.
78         DO 13 J = 1,JMAX
79         AJ = J - 1
80         X(J) = DELX*AJ
81         TE(J) = 100.
82         DO 12 M = 1,MAXEX
83         AM = M
84         DAM = (2.*AM - 1.)
85         DXM = DAM*PI*X(J)
86         DTM = -ALPH*DAM*DAM*PI*PI*T
87 C
88 C       LIMIT THE ARGUMENT SIZE OF EXP(DTM)
89 C
90         IF(DTM .LT. - 87.)DTM = -87.0
91      12 TE(J) = TE(J) - 400./DAM/PI*SIN(DXM)*EXP(DTM)
92         SUM = SUM + (TE(J) - TD(J))**2
93      13 CONTINUE
94         WRITE(6,14)T,(TE(J),J=1,JMAX)
95      14 FORMAT(/,' T= ',F5.0,'  TE=',11F6.2,//)
96 C
97 C       RMS IS THE RMS ERROR
98 C
99         AVS = SUM/(1. + AJM)
100        RMS = SQRT(AVS)
101        WRITE(6,15)RMS
102     15 FORMAT(' RMS DIF = ',E11.4,//)
103        STOP
104        END
```

Fig. 3.13. (cont.) Listing of program DIFF

Program DIFF reads in the various control parameters (line 16), sets Δx and Δt and writes out the various parameters. The solution is computed at each time step $n+1$ (Fig. 3.11) until either NMAX or TMAX (lines 72, 73) are exceeded. Then the exact solution is computed (lines 74–91) from

$$TE(x_j, t_n) = 100 - \sum_{m=1}^{MAXEX} \left[\frac{400}{(2m-1)\pi} \right] \sin[(2m-1)\pi x_j] e^{[-\alpha(2m-1)^2\pi^2 t_n]} \ . \qquad (3.42)$$

This solution is obtained by the separation of variables approach (Sect. 2.5.2). Simultaneously the rms error is obtained from

Table 3.6. Parameters used in program DIFF

Parameter	Description
JMAX	Number of points along the rod
MAXEX	Number of terms in the exact equation
NMAX	Maximum number of time steps
ALPH	Thermal diffusivity α
S	$\alpha \Delta t / \Delta x^2$
TMAX	Maximum time
TD	Dimensional temperature array
TN	Nondimensional temperature array
TE	Exact (dimensional) temperature array
DELX	Δx
DELT	Δt
TIM	Time
X	Location, $0 \leqslant x \leqslant 1.0$
RMS	rms error in the solution

$$\text{RMS} = \left[\left(\sum_{j=1}^{\text{JMAX}} (T_j - \text{TE}_j)^2 \right) \Big/ \text{JMAX} \right]^{1/2} . \tag{3.43}$$

For the initial conditions $\bar{T}(x, 0) = 0$ and boundary conditions $\bar{T}(0, t) = \bar{T}(1, t) = 100$ the computer output generated by DIFF, corresponding to $s = 0.5$, is given in Fig. 3.14 and plotted in Fig. 3.15.

The value of s is used in DIFF to set the time step Δt; the value of $\Delta x = 0.1$. Reducing s reduces Δt and leads to a smaller rms error in the solution. This is apparent from the results shown in Table 3.7 with $\bar{T}(0, 0) = \bar{T}(1, 0) = 50$. The result corresponding to $s = 0.167$ is a special case which is discussed further in Sect. 4.1.2.

A consideration of lines 50 and 51 of DIFF (Fig. 3.13) indicates that the values of $\bar{T}(0, 0)$ and $\bar{T}(1, 0)$ are averaged between their values implied by the initial

```
JMAX=   11  MAXEX=   10  NMAX=  500  TMAX= 2999.00
S= .500  ALPH  =  .100E-04  DELT  =  .500E+03  DELX  =  .100E+00


FTCS(EXPLICIT) SCHEME,      S = .500


T=  500.  TD= 50.00 25.00   .00   .00   .00   .00   .00   .00 25.00 50.00
T= 1000.  TD=100.00 50.00 12.50   .00   .00   .00   .00   .00 12.50 50.00100.00
T= 1500.  TD=100.00 56.25 25.00  6.25   .00   .00   .00  6.25 25.00 56.25100.00
T= 2000.  TD=100.00 62.50 31.25 12.50  3.13   .00  3.13 12.50 31.25 62.50100.00
T= 2500.  TD=100.00 65.63 37.50 17.19  6.25  3.13  6.25 17.19 37.50 65.63100.00
T= 3000.  TD=100.00 68.75 41.41 21.88 10.16  6.25 10.16 21.88 41.41 68.75100.00

T= 3000.  TE=100.00 68.33 41.53 22.49 11.68  8.25 11.68 22.49 41.53 69.33100.00


RMS DIF  =   .9418E+00
```

Fig. 3.14. Typical output from DIFF; $s = 0.5$

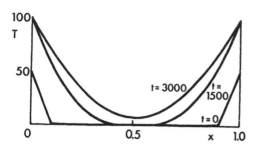

Fig. 3.15. Solution of (3.41) with $s=0.5$: stable behaviour

conditions and those implied by the boundary conditions. The effect of setting $\bar{T}(0,0)=\bar{T}(1,0)=0$, (i.e. as given by the initial conditions) is also shown in Table 3.7. Clearly this leads to a much larger solution error. If $\bar{T}(0,0)=\bar{T}(1,0)=100$ the error is of a similar magnitude to that for $\bar{T}(0,0)=\bar{T}(1,0)=0$.

Table 3.7. rms error for $t=t_{max}$

s	rms error $\bar{T}(0,0)=\bar{T}(1,0)=50$.	rms error $\bar{T}(0,0)=\bar{T}(1,0)=0$.
0.5	0.492	2.251
0.3	0.187	0.953
0.167	0.00169	0.440

3.6 Closure

Discretisation is seen to introduce an error that can be assessed by considering the truncation error, at least for the finite difference method. The truncation error is likely to be a progressively more accurate indicator of the solution error as the grid is refined (Sect. 4.1.2).

The particular formulae used to represent derivatives can be chosen by eliminating successive terms from the truncation error (Sect. 3.2.2). However such an approach often needs to be modified to choose some of the coefficients so that the resulting algorithm is stable (Sect. 4.3).

It is apparent from the specific examples considered in Sects. 3.3 and 3.4 that higher-order formulae are not likely to be significantly more accurate than low-order formulae if the exact solution contains discontinuities or severe gradients. In addition, in representing wave-like motion, short wavelengths are predicted less accurately than long wavelengths.

The provision of a simple finite difference program, DIFF, permits many of the features of the finite difference method to be illustrated. By way of example it is shown that the accuracy of the finite difference solution of the diffusion equation is quite sensitive to the way that the discontinuity in the boundary and initial condition specification is handled. Computational solutions of the diffusion equation for different schemes and boundary conditions are discussed in Chap. 7.

3.7 Problems

Approximation to Derivatives (Sect. 3.2)

3.1 Use the general technique (Sect. 3.2.2) to determine the coefficients a to d in the formula

$$\frac{d^2 T}{dx^2} \approx a T_j + b T_{j+1} + c T_{j+2} + d T_{j+3} \ .$$

What is the truncation error of this formula?

3.2 A three-level explicit discretisation of $\partial T/\partial t - \alpha \partial^2 T/\partial x^2 = 0$ can be written

$$\frac{0.5 T_j^{n-1} - 2 T_j^n + 1.5 T_j^{n+1}}{\Delta t}$$

$$- \alpha \left[\frac{(1+d)(T_{j-1} - 2T_j + T_{j+1})^n}{\Delta x^2} - \frac{d(T_{j-1} - 2T_j + T_{j+1})^{n-1}}{\Delta x^2} \right] = 0 \ .$$

(a) Expand each term as a Taylor series to determine the truncation error of the complete equation for arbitrary values of d.

(b) Use the general technique to choose d as a function of s so that the scheme is fourth-order accurate in Δx. The formulae after (4.11) are useful for this problem.

Accuracy of Discretisation (Sect. 3.3)

3.3 For $y = \sin \pi x/2$ obtain dy/dx at $x = 0.5$ with $\Delta x = 0.1$ using:

(a) $\quad \dfrac{dy}{dx} \approx \dfrac{y_{j+1} - y_{j-1}}{2\Delta x}$,

(b) $\quad \dfrac{dy}{dx} \approx \dfrac{y_{j+1} - y_j}{\Delta x}$,

(c) $\quad \dfrac{dy}{dx} \approx \dfrac{y_{j-2} - 8y_{j-1} + 8y_{j+1} - y_{j+2}}{12\Delta x}$,

and compare the accuracy of the results.

3.4 Repeat Problem 3.3 with $\Delta x = 0.05, 0.025, 0.0125$ and determine whether the convergence with Δx is consistent with the leading term of the truncation error. It is recommended that a computer program is written to solve this problem. Double precision may be required to resolve the error associated with scheme (c).

3.5 For $y = \sin \pi x/2$ obtain $d^2 y/dx^2$ at $x = 0.5$ with $\Delta x = 0.1, 0.05, 0.025, 0.0125$ using:

(a) $\quad \dfrac{d^2 y}{dx^2} \approx \dfrac{y_{j-1} - 2y_j + y_{j+1}}{\Delta x^2}$,

(b) $\dfrac{d^2y}{dx^2} \approx \dfrac{-y_{j-2} + 16y_{j-1} - 30y_j + 16y_{j+1} - y_{j+2}}{12\Delta x^2}$,

and compare the accuracy of the results and the convergence with the leading term in the truncation error. It is recommended that a computer program is written to solve this problem, using double precision if required.

Wave Representation (Sect. 3.4)

3.6 For a progressive wave

$$\bar{T}(x, t) = \cos[m(x - qt)]$$

show that the use of $\partial \bar{T}/\partial t \approx (T_j^{n+1} - T_j^n)/\Delta t$ is equivalent to

$$\frac{\partial T}{\partial t} = \left(\frac{2}{\Delta t}\right)\sin(0.5mq\Delta t)\sin\left\{m\left[x - q\left(t + \frac{\Delta t}{2}\right)\right]\right\} ,$$

and that the use of $\partial \bar{T}/\partial t \approx (1.5T_j^{n+1} - 2T_j^n + 0.5T_j^{n-1})/\Delta t$ is equivalent to

$$\frac{\partial T}{\partial t} = \left(\frac{1}{\Delta t}\right)\sin(0.5mq\Delta t)\left(3\sin\left\{m\left[x - q\left(t + \frac{\Delta t}{2}\right)\right]\right\}\right.$$
$$\left. - \sin\left\{m\left[x - q\left(t - \frac{\Delta t}{2}\right)\right]\right\}\right) .$$

Show that, in the limit $\sin(0.5mq\Delta t)\rightarrow 0.5mq\Delta t$ and $\cos(0.5mq\Delta t)\rightarrow 1.0$, the above two expressions are almost equivalent. Show further that, in the limit $0.5mq\Delta t \rightarrow 0$, the above expressions coincide with the exact evaluation of $\partial \bar{T}/\partial t$.

Finite Difference Method (Sect. 3.5)

3.7 Modify program DIFF to use the following five-point symmetric scheme for interior points:

$$\frac{\partial^2 T}{\partial x^2} \approx \frac{-T_{j-2}/12 + 4T_{j-1}/3 - 2.5T_j + 4T_{j+1}/3 - T_{j+2}/12}{\Delta x^2} .$$

For $j = 2$ use

$$\frac{\partial^2 T}{\partial x^2} \approx \frac{11T_{j-1}/12 - 5T_j/3 + 0.5T_{j+1} + T_{j+2}/3 - T_{j+3}/12}{\Delta x^2} ,$$

and the equivalent formula for $j = \text{JMAX} - 1$.
(a) Obtain solutions with $\Delta x = 0.1$ for $s = 0.3, 0.2, 0.1$.
(b) How does the solution accuracy compare with that of the FTCS scheme?

3.8 Modify program DIFF to introduce the following scheme, in place of FTCS scheme:

$$\frac{0.5T_j^{n-1}-2T_j^n+1.5T_j^{n+1}}{\Delta t}-\frac{\alpha(T_{j-1}^n-2T_j^n+T_{j+1}^n)}{\Delta x^2}=0 \ .$$

For the first time step use the forward time formula $\partial T/\partial t \approx (T^{n+1}-T^n)/\Delta t$.

(a) Obtain solutions with $\Delta x=0.1$ for $s=0.5, 0.3, 0.1$ and compare with the solutions produced by the FTCS scheme.

(b) Is the accuracy behaviour with decreasing s what you would expect? [*Hint*: Consider the truncation error expression in Problem 3.2.]

4. Theoretical Background

In practice the algebraic equations that result from the discretisation process, Sect. 3.1, are obtained on a finite grid. It is to be expected, from the truncation errors given in Sects. 3.2 and 3.3, that more accurate solutions could be obtained on a refined grid. These aspects are considered further in Sect. 4.4. However for a given required solution accuracy it may be more economical to solve a higher-order finite difference scheme on a coarse grid than a low-order scheme on a finer grid, if the exact solution is sufficiently smooth. This leads to the concept of computational efficiency which is examined in Sect. 4.5.

An important question concerning computational solutions is what guarantee can be given that the computational solution will be close to the exact solution of the partial differential equation(s) and under what circumstances the computational solution will coincide with the exact solution. The second part of this question can be answered (superficially) by requiring that the approximate (computational) solution should converge to the exact solution as the grid spacings Δt, Δx shrink to zero (Sect. 4.1). However, convergence is very difficult to establish directly so that an indirect route, as indicated in Fig. 4.1, is usually followed. The indirect route requires that the system of algebraic equations formed by the discretisation process (Sect. 3.1) should be consistent (Sect. 4.2) with the governing partial differential equation(s). Consistency implies that the discretisation process can be reversed, through a Taylor series expansion, to recover the governing equation(s). In addition, the algorithm used to solve the algebraic equations to give the approximate solution, T, must be stable (Sect. 4.3). Then the pseudo-equation

$$\text{CONSISTENCY} + \text{STABILITY} = \text{CONVERGENCE} \tag{4.1}$$

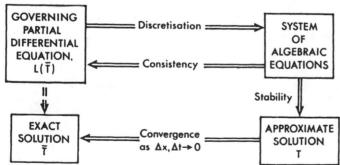

Fig. 4.1. Conceptual relationship between consistency, stability and convergence

is invoked to imply convergence. The conditions under which (4.1) can be made precise are given by the Lax equivalence theorem (Sect. 4.1.1).

It is very difficult to obtain theoretical guidance for the behaviour of the solution on a grid of finite size. Most of the useful theoretical results are strictly only applicable in the limit that the grid size shrinks to zero. However the connections that are established between convergence (Sect. 4.1), consistency (Sect. 4.2) and stability (Sect. 4.3) are also qualitatively useful in assessing computational solutions on a finite grid.

4.1 Convergence

A solution of the algebraic equations (Fig. 4.1) which approximate a given partial differential equation is said to be convergent if the approximate solution approaches the exact solution of the partial differential equation for each value of the independent variable as the grid spacing tends to zero. Thus we require $T_j^n \to \bar{T}(x_j, t_n)$, as $\Delta x, \Delta t \to 0$.

The difference between the exact solution of the partial differential equation and the exact solution of the system of algebraic equations is called the solution error, denoted by e_j^n; that is

$$e_j^n = \bar{T}(x_j, t_n) - T_j^n \; . \tag{4.2}$$

The exact solution of the system of algebraic equations is the approximate solution of the governing partial differential equation. The exact solution of the system of algebraic equations is obtained when no numerical errors of any sort, such as those due to round-off, are introduced during the computation. The magnitude of the error, e_j^n, at the (j, n)-th node typically depends on the size of the grid spacings, Δx and Δt, and on the values of the higher-order derivatives at that node, omitted from the finite difference approximations to the derivatives in the given differential equation.

Proof that a solution to the system of algebraic equations converges to the solution of the partial differential equation is generally very difficult, even for the simplest cases. For the approximate solution to the diffusion equation, using the very simple FTCS algorithm (3.41), a proof of convergence for $s \leq \frac{1}{2}$ is given by Noye (1984, pp. 117–119). Convergence is very difficult to show when the given partial differential equation is more complicated than the diffusion equation and the method of discretisation is less direct.

A few flow problems possess exact solutions so that, for these cases, convergence can be inferred by obtaining computational solutions on progressively refined grids (Sect. 4.1.2).

4.1.1 Lax Equivalence Theorem

For a restricted class of problems convergence can be established via the Lax equivalence theorem (Richtmyer and Morton 1967, p. 45): "Given a properly posed linear initial value problem and a finite difference approximation to it that satisfies the consistency condition, stability is the necessary and sufficient condition for convergence".

Although the theorem is expressed in terms of a finite difference approximation it is applicable to any discretisation procedure that leads to nodal unknowns, e.g., the finite element method. The Lax equivalence theorem is of great importance, since it is relatively easy to show stability of an algorithm and its consistency with the original partial differential equation, whereas it is usually very difficult to show convergence of its solution to that of the partial differential equation.

Most "real" flow problems are nonlinear and are boundary or mixed initial/ boundary value problems so that the Lax equivalence theorem cannot always be applied rigorously. Consequently the Lax equivalance theorem should be interpreted as providing necessary, but not always sufficient, conditions. In the form of (4.1) the Lax equivalence "equation" is useful for excluding inconsistent discretisations and unstable algorithms.

4.1.2 Numerical Convergence

For the equations that govern fluid flow, convergence is usually impossible to demonstrate theoretically. However, for problems that possess an exact solution, like the diffusion equation, it is possible to obtain numerical solutions on a successively refined grid and compute a solution error. Convergence implies that the solution error should reduce to zero as the grid spacing is shrunk to zero.

For program DIFF (Fig. 3.13), solutions have been obtained on successively refined spatial grids, $\Delta x = 0.2, 0.1, 0.05$ and 0.025. The corresponding rms errors are shown in Table 4.1 for $s = 0.50$ and 0.30. It is clear that the rms error reduces like Δx^2 approximately. Based on these results it would be a reasonable inference that refining the grid would produce a further reduction in the rms error and, in the limit of Δx (for fixed s) going to zero, the solution of the algebraic equations would converge to the exact solution.

The establishment of numerical convergence is rather an expensive process since usually very fine grids are necessary. As s is kept constant in the above example the timestep is being reduced by a factor of four for each halving of Δx. In Table 4.1 the solution error is computed at $t = 5000$ s. This implies the finest grid solution at $s = 0.30$ requires 266 time steps before the solution error is computed.

For the diffusion equation (3.1) with zero boundary values and initial value $T(x, 0) = \sin(\pi x)$, $0 \leq x \leq 1$, the rms solution error $|e|_{rms}$ is plotted against grid

Table 4.1. Solution error (rms) reduction with grid refinement

	rms error			
$s = \alpha \Delta t / \Delta x^2$	$\Delta x = 0.2$	$\Delta x = 0.1$	$\Delta x = 0.05$	$\Delta x = 0.025$
0.50	1.658	0.492	0.121	0.030
0.30	0.590	0.187	0.048	0.012

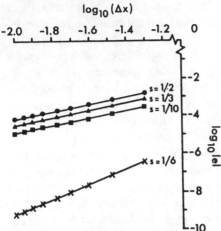

$\log_{10}(\Delta x)$

Fig. 4.2. Numerical convergence for the FTCS method

spacing Δx in Fig. 4.2. The increased rate of convergence (fourth-order convergence) for $s = \frac{1}{6}$, compared with other values of $s \leq \frac{1}{2}$ (second-order convergence), is clearly seen, i.e. the convergence rate is like Δx^4 for $s = \frac{1}{6}$, and like Δx^2 otherwise. As will be demonstrated in Sect. 4.2, the superior convergence rate for $s = \frac{1}{6}$ is to be expected from a consideration of the leading term in the truncation error. Typically, for sufficiently small grid spacings Δx, Δt, the solution error will reduce like the truncation error as Δx, $\Delta t \to 0$.

4.2 Consistency

The system of algebraic equations generated by the discretisation process is said to be consistent with the original partial differential equation if, in the limit that the grid spacing tends to zero, the system of algebraic equations is equivalent to the partial differential equation at each grid point.

Clearly, consistency is necessary if the approximate solution is to converge to the solution of the partial differential equation under consideration. However, it is not a sufficient condition (Fig. 4.1), for even though the system of algebraic equations might be equivalent to the partial differential equation as the grid spacing tends to zero, it does not follow that the solution of the system of algebraic equations approaches the solution of the partial differential equation. For instance, choosing $s > \frac{1}{2}$ in program DIFF causes the solution using the FTCS algorithm (3.41) to diverge rapidly. Thus as indicated by the Lax equivalence theorem (Sect. 4.1.1), both consistency and stability are required for convergence.

The mechanics of testing for consistency requires the substitution of the exact solution into the algebraic equations resulting from discretisation, and the expansion of all nodal values as Taylor series about a single point. For consistency the resulting expression should be made up of the original partial differential equation plus a remainder. The structure of the remainder should be such that it reduces to zero as the grid is refined.

In this section the FTCS scheme (Sect. 4.2.1) and the fully implicit scheme (Sect. 4.2.2) will be analysed to see if they are consistent discretisations of the diffusion equation (3.1).

4.2.1 FTCS Scheme

The FTCS scheme (3.41) was used in Sect. 3.5.2 to obtain the solution of the one-dimensional diffusion equation. The substitution of $\bar{T}_j^n$, the exact solution of the diffusion equation at the (j, n)-th node, for the approximation T_j^n gives

$$\bar{T}_j^{n+1} = s\bar{T}_{j-1}^n + (1 - 2s)\bar{T}_j^n + s\bar{T}_{j+1}^n \ . \tag{4.3}$$

We wish to determine how closely (4.3) corresponds to the diffusion equation (3.1) at the (j, n)-th node. Substitution of Taylor series expansions about the (j, n)-th node for each term of (4.3) produces, on simplification,

$$\left[\frac{\partial \bar{T}}{\partial t}\right]_j^n - \alpha\left[\frac{\partial^2 \bar{T}}{\partial x^2}\right]_j^n + E_j^n = 0 \ , \quad \text{where} \tag{4.4}$$

$$E_j^n = 0.5\Delta t\left[\frac{\partial^2 \bar{T}}{\partial t^2}\right]_j^n - \alpha\left(\frac{\Delta x^2}{12}\right)\left[\frac{\partial^4 \bar{T}}{\partial x^4}\right]_j^n + O(\Delta t^2, \Delta x^4) \ . \tag{4.5}$$

It can be seen that the partial differential equation (4.4) differs from the diffusion equation (3.1) by the inclusion of the extra terms contained in E_j^n, which is called the truncation error. The extra terms can be identified with the discretisation of the derivatives $\partial \bar{T}/\partial t$ and $\partial^2 \bar{T}/\partial x^2$ by (3.21) and (3.25), respectively.

Clearly, as the grid spacings become smaller and smaller, the truncation error E_j^n tends to zero at the fixed point (x_j, t_n). In the limit as $\Delta x, \Delta t \to 0$, the FTCS algorithm (3.41) is equivalent to the diffusion equation. This property is called consistency.

For the FTCS algorithm the truncation error E_j^n is generally $O(\Delta t, \Delta x^2)$. However, since $\bar{T}$ satisfies the diffusion equation (3.1) it also satisfies

$$\frac{\partial^2 \bar{T}}{\partial t^2} = \alpha\frac{\partial}{\partial t}\frac{\partial^2 \bar{T}}{\partial x^2} = \alpha\frac{\partial^2}{\partial x^2}\frac{\partial \bar{T}}{\partial t} = \alpha^2\frac{\partial^4 \bar{T}}{\partial x^4} \ . \tag{4.6}$$

Therefore the truncation error may be written

$$E_j^n = 0.5\alpha\Delta x^2\left(s - \frac{1}{6}\right)\left[\frac{\partial^4 \bar{T}}{\partial x^4}\right]_j^n + O(\Delta t^2, \Delta x^4) \ . \tag{4.7}$$

If $s = \frac{1}{6}$, the first term in the above expression vanishes and the truncation error is $O(\Delta t^2, \Delta x^4)$, or, what is the same thing for fixed s, $O(\Delta x^4)$. In this case, the truncation error goes to zero faster than for any other value of s as the grid spacing is reduced. It may be recalled from Sect. 4.1.2 that the solution error also goes to zero faster for $s = \frac{1}{6}$ than for any other value (Fig. 4.2).

From the above it is clear that the same algebraic manipulation permits both the consistency and the likely convergence rate to be established.

4.2.2 Fully Implicit Scheme

Consistency will be investigated here for the fully implicit discretisation (7.20) of the diffusion equation (3.1). The fully implicit scheme is

$$-s T_{j-1}^{n+1} + (1+2s) T_j^{n+1} - s T_{j+1}^{n+1} = T_j^n . \tag{4.8}$$

To demonstrate consistency it is easier to start from the equivalent form of (4.8), applied to the exact solution,

$$\frac{\bar{T}_j^{n+1} - \bar{T}_j^n}{\Delta t} - \alpha \frac{(\bar{T}_{j-1}^{n+1} - 2\bar{T}_j^{n+1} + \bar{T}_{j+1}^{n+1})}{\Delta x^2} = 0 . \tag{4.9}$$

First, the terms $\bar{T}_{j-1}^{n+1}$ and $\bar{T}_{j+1}^{n+1}$ are expanded about the $(j, n+1)$-th node and substituted into (4.9). The result is

$$\frac{\bar{T}_j^{n+1} - \bar{T}_j^n}{\Delta t} - \alpha \left\{ [\bar{T}_{xx}]_j^{n+1} + \left(\frac{\Delta x^2}{12}\right) [\bar{T}_{x^4}]_j^{n+1} + \left(\frac{\Delta x^4}{360}\right) [\bar{T}_{x^6}]_j^{n+1} + \ldots \right\} = 0 . \tag{4.10}$$

Second, the terms $\bar{T}_j^{n+1}$, $[\bar{T}_{xx}]_j^{n+1}$, etc. in (4.10) are expanded about the (j, n)-th node. The result is

$$[\bar{T}_t]_j^n + (0.5\Delta t)[\bar{T}_{tt}]_j^n + \left(\frac{\Delta t^2}{6}\right)[\bar{T}_{t^3}]_j^n + \ldots$$

$$- \alpha [\bar{T}_{xx}]_j^n + \Delta t [\bar{T}_{xxt}]_j^n + (0.5\Delta t^2)[\bar{T}_{xxtt}]_j^n + \ldots + \left(\frac{\Delta x^2}{12}\right)([\bar{T}_{x^4}]_j^n$$

$$+ \Delta t [\bar{T}_{x^4 t}]_j^n + \ldots) + \left(\frac{\Delta x^4}{360}\right)([\bar{T}_{x^6}]_j^n + \ldots) \ldots] = 0 . \tag{4.11}$$

From the governing equation

$$\bar{T}_t = \alpha \bar{T}_{xx} , \qquad \bar{T}_{tt} = \alpha^2 \bar{T}_{x^4} , \qquad \bar{T}_{ttt} = \alpha^3 \bar{T}_{x^6} , \tag{4.12}$$

and $s = \alpha \Delta t / \Delta x^2$ or $\Delta x^2 = \alpha \Delta t / s$. Therefore (4.11) can be simplified to

$$[\bar{T}_t - \alpha \bar{T}_{xx}]_j^n + E_j^n = 0 , \tag{4.13}$$

where the truncation error is

$$E_j^n = -0.5\Delta t \left(1 + \frac{1}{6s}\right)[\bar{T}_{tt}]_j^n - \left(\frac{\Delta t^2}{3}\right)\left(1 + \frac{1}{4s} + \frac{1}{120s^2}\right)[\bar{T}_{ttt}]_j^n + \ldots . \tag{4.14}$$

It is clear that as Δt tends to zero, E_j^n tends to zero and (4.13) coincides with the governing equation. Consequently (4.8) is consistent with the governing equation.

In (4.14) all spatial derivatives have been converted to equivalent time derivatives. Using (4.12) it would be possible to express the truncation error in terms of the spatial grid size and derivatives only, as in (4.7). A comparison of (4.14) and (4.7) indicates that there is no choice of s that will reduce the truncation error of the fully implicit scheme to $O(\Delta x^4)$.

It might appear from the above two examples that consistency can be taken for granted. However, attempts to construct algorithms that are both accurate and stable can sometimes generate potentially inconsistent discretisations, e.g. the DuFort–Frankel scheme, Sect. 7.1.2.

4.3 Stability

This is the tendency for any spontaneous perturbations (such as round-off error) in the solution of the system of algebraic equations (Figs. 3.1 and 4.1) to decay. A stable solution produced by the FTCS scheme with $s = 0.5$ is shown in Fig. 3.15. A typical unstable result ($s = 0.6$) is shown in Fig. 4.3. These results have been obtained with $\Delta x = 0.1$ and the same initial and boundary conditions as used to generate Fig. 3.15.

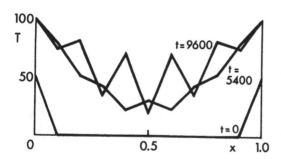

Fig. 4.3. Solution of (3.41) with $s = 0.6$; unstable behaviour

It is clear from Fig. 4.3 that an unphysical oscillation originates on the line of symmetry and propagates to the boundaries. The amplitude of the oscillation grows with increasing time.

The concept of stability is concerned with the growth, or decay, of errors introduced at any stage of the computation. In this context, the errors referred to are not those produced by incorrect logic but those which occur because the computer cannot give answers to an infinite number of decimal places. In practice, each calculation made on the computer is carried out to a finite number of significant figures which introduces a round-off error at every step of the computation. Hence the computational solution to (3.41) is not T_j^{n+1}, but $*T_j^{n+1}$, the numerical solution of the system of algebraic equations.

A particular method is stable if the cumulative effect of all the round-off errors produced in the application of the algorithm is negligible. More specifically, consider the errors

$$\xi_j^n = T_j^n - {}^*T_j^n \tag{4.15}$$

introduced at grid points (j, n), where $j = 2, 3, \ldots, J-1$ and $n = 0, 1, 2$. It is usually not possible to determine the exact value of the numerical error ξ_j^n at the (j, n)-th grid point for an arbitrary distribution of errors at other grid points. However, it can be estimated using certain standard methods, some of which will be discussed in this section. In practice, the numerical solutions are typically more accurate than these estimates indicate, because stability analyses often assume the worst possible combination of individual errors. For instance, it may be assumed that all errors have a distribution of signs so that their total effect is additive, which is not always the case.

It can be shown that, for linear algebraic equations produced by discretisation, the corresponding error terms satisfy the same homogeneous algebraic equations as the values of T. For instance, using the FTCS scheme (3.41) means that we are actually calculating $^*T_j^{n+1}$ using $^*T_{j-1}^n$, $^*T_j^n$ and $^*T_{j+1}^n$, so that

$$^*T^{n+1} = s(^*T_{j-1}^n) + (1-2s)(^*T_j^n) + s(^*T_{j+1}^n) \ . \tag{4.16}$$

Substitution of (4.15) into (4.16), followed by application of (3.41), which applies since the exact solutions of the algebraic equations, T_j^n, satisfy the FTCS algorithm, yields the homogeneous algebraic equation

$$\xi_j^{n+1} = s\xi_{j-1}^n + (1-2s)\xi_j^n + s\xi_{j+1}^n \ . \tag{4.17}$$

Assuming given boundary and initial values, the initial errors, ξ_j^0, $j = 2, 3, \ldots, J-1$, and the boundary errors ξ_1^n and ξ_J^n, $n = 0, 1, 2, \ldots$ for this equation, will all be zero. Unless some (round-off) error is introduced in calculating the value of T_j^n at some interior node, the resulting errors in the solution will remain zero.

The two most common methods of stability analysis are the matrix method and the von Neumann method. Both methods are based on predicting whether there will be a growth in the error between the true solution of the numerical algorithm and the actually computed solution, i.e. including round-off contamination.

An alternative interpretation of stability analysis is to suppose that the initial conditions are represented by a Fourier series. Each harmonic or mode in the Fourier series will grow or decay depending on the discretised equation, which typically furnishes a specific expression for the growth (or decay) factor for each mode. If a particular mode can grow without bound, the discretised equation has an unstable solution. This interpretation of stability (Richtmyer and Morton 1967, pp. 9–13) is exploited directly in the von Neumann method of stability analysis (Sects. 4.3.4 and 4.3.5). The unbounded growth of a particular mode is still possible if the discretised equations are solved exactly, i.e. with no (round-off) errors being present. If (round-off) errors are introduced the same unstable nature of the discretised equations will cause unacceptable growth of the errors. Consequently the procedures for analysing the stability of the discretised equations are the same irrespective of the manifestation of the inherent stability or instability.

4.3.1 Matrix Method: FTCS Scheme

In this technique we express the set of equations governing the error propagation in matrix form and examine the eigenvalues of the associated matrix. Initially this method will be illustrated for the FTCS scheme (3.41).

Substituting $j = 2, 3, \ldots, J-1$ into (4.17) and noting that the errors at the boundary are zero, so that $\xi_1^n = 0$ and $\xi_j^n = 0$ for all n, gives

$$\xi_2^{n+1} = (1-2s)\xi_2^n + s\xi_3^n \ ,$$

$$\xi_3^{n+1} = s\xi_2^n + (1-2s)\xi_3^n + s\xi_4^n \ ,$$

$$\xi_j^{n+1} = s\xi_{j-1}^n + (1-2s)\xi_j^n + s\xi_{j+1}^n \ , \tag{4.18}$$

$$\xi_{J-1}^{n+1} = s\xi_{J-2}^n + (1-2s)\xi_{J-1}^n \ .$$

This may be written in matrix form as

$$\xi^{n+1} = \underline{A}\,\xi^n \ , \qquad n = 0, 1, \ldots \ , \tag{4.19}$$

where $\underline{A}$ is a $(J-2)$ square matrix and ξ^n is a vector of length $(J-2)$ defined by

$$\underline{A} = \begin{bmatrix} (1-2s) & s & & & & \\ s & (1-2s) & s & & & \\ & s & & s & & \\ & & & \cdot & & \\ & & & & \cdot & \cdot \\ & & & & s & (1-2s) \end{bmatrix}, \quad \xi = \begin{bmatrix} \xi_2^n \\ \cdot \\ \xi_j^n \\ \cdot \\ \xi_{J-1}^n \end{bmatrix} .$$

It can be shown that the errors ξ^n are bounded as n increases if the eigenvalues λ_j of the matrix $\underline{A}$ are all distinct and have absolute values less than or equal to one; that is, if

$$|\lambda_j| \le 1 \ , \quad \text{for all } \lambda \ . \tag{4.20}$$

Now the eigenvalues of the tridiagonal matrix $\underline{A}$ are, from (9.48),

$$\lambda_j = 1 - 4s\sin^2\left(\frac{j\pi}{2(J-1)}\right), \quad j = 1, 2, \ldots, J-2 \ . \tag{4.21}$$

The stability condition (4.20) permits only values of s for which

$$-1 \le 1 - 4s\sin^2\left(\frac{j\pi}{2(J-1)}\right) \le 1 \ . \tag{4.22}$$

The right-hand side of this inequality is satisfied for all j and s, while the left-hand side requires

$$s \sin^2\left(\frac{j\pi}{2(J-1)}\right) \leq \frac{1}{2} ,$$

which is true for all j if $s \leq \frac{1}{2}$. The FTCS algorithm is therefore stable for $s \leq \frac{1}{2}$.

The form of (4.19) is the same if (4.18) is applied to the solution rather than the computational error. Consequently the matrix method can also be interpreted in terms of growth of the individual modes in a Fourier series representation of the initial conditions.

Alteration of the nature of the boundary conditions at $x=0$ and/or $x=1$ only slightly modifies this method, as will be indicated in Sect. 4.3.3.

4.3.2 Matrix Method: General Two-Level Scheme

The matrix method will be demonstrated in this section for the general two-level scheme applied to the diffusion equation (7.24). That is

$$\frac{T_j^{n+1} - T_j^n}{\Delta t} - \alpha\beta L_{xx}T_j^{n+1} - \alpha(1-\beta)L_{xx}T_j^n = 0 , \tag{4.23}$$

where $L_{xx}T_j = (T_{j-1} - 2T_j + T_{j+1})/\Delta x^2$.

In (4.23) α is the thermal diffusivity and β is the parameter that controls the degree of implicitness (Sect. 7.2). The numerical error (4.15) is governed by an equation equivalent to (4.23). This is

$$-s\beta\xi_{j-1}^{n+1} + (1+2s\beta)\xi_j^{n+1} - s\beta\xi_{j+1}^{n+1} = s(1-\beta)\xi_{j-1}^n + [1-2s(1-\beta)]\xi_j^n$$
$$+ s(1-\beta)\xi_{j+1}^n . \tag{4.24}$$

Equation (4.24) is appropriate to internal nodes. With Dirichlet boundary conditions no equations are required at the end-points. When (4.24) is repeated for all nodal points, $j = 2, 3, \ldots, J-1$, a matrix of equations can be written,

$$\underline{A}\,\xi^{n+1} = \underline{B}\,\xi^n , \quad \text{where} \tag{4.25}$$

$$\underline{A} = \begin{bmatrix} 1+2s\beta & -s\beta & & & \\ -s\beta & 1+2s\beta & -s\beta & & \\ & \cdot & \cdot & \cdot & \\ & & -s\beta & 1+2s\beta & -s\beta \\ & & & -s\beta & 1+2s\beta \end{bmatrix}$$

and

$$\underline{B} = \begin{bmatrix} 1-2s(1-\beta) & s(1-\beta) & & & \\ s(1-\beta) & 1-2s(1-\beta) & s(1-\beta) & & \\ & \cdot & & \cdot & \\ & & s(1-\beta) & 1-2s(1-\beta) & s(1-\beta) \\ & & & s(1-\beta) & 1-2s(1-\beta) \end{bmatrix} .$$

The algorithm (4.23) is stable if the magnitude of the eigenvalues of $\underline{A}^{-1}\underline{B}$ are bounded by unity. Because of the structure of $\underline{A}$ and $\underline{B}$ this is equivalent to

$$\left|\frac{(\lambda_B)_j}{(\lambda_A)_j}\right| \le 1.0 \ . \tag{4.26}$$

Because of the symmetric tridiagonal nature of $\underline{A}$ and $\underline{B}$, analytic expressions are available for the eigenvalues (Mitchell and Griffiths 1980, p. 10):

$$\lambda_A = 1 + 2s\beta - 2s\beta\cos\left(\frac{j\pi}{J-1}\right)$$

$$= 1 + 4s\beta\sin^2\left(\frac{j\pi}{2(J-1)}\right) \quad \text{and}$$

$$\lambda_B = 1 - 2s(1-\beta) + 2s(1-\beta)\cos\left(\frac{j\pi}{J-1}\right)$$

$$= 1 - 4s(1-\beta)\sin^2\left(\frac{j\pi}{2(J-1)}\right) \ .$$

Therefore stability requires

$$ST = \left|\frac{1 - 4s(1-\beta)\sin^2[j\pi/2(J-1)]}{1 + 4s\beta\sin^2[j\pi/2(J-1)]}\right| \le 1$$

$$\text{if } \sin\left(\frac{j\pi}{2(J-1)}\right) = 0 \ , \quad ST = 1$$

$$\text{if } \sin\left(\frac{j\pi}{2(J-1)}\right) = 1 \ , \quad ST = \left|\frac{1 - 4s(1-\beta)}{1 + 4s\beta}\right| \le 1 \ ,$$

i.e. $1 - 4s(1-\beta) < 1 + 4s\beta$, which is satisfied and $1 - 4s(1-\beta) \ge -1 - 4s\beta$ so that $2 > 4s(1-2\beta)$ or $s \le 0.5/(1-2\beta)$ if $\beta < 0.5$.

Thus if $\beta < 0.5$, stability requires that $s \le 0.5/(1-2\beta)$. If $\beta \ge 0.5$, (4.23) is unconditionally stable. The desirable feature of unconditional stability is pursued in Sect. 7.2.

4.3.3 Matrix Method: Derivative Boundary Conditions

The matrix method can also be used when Neumann (derivative) boundary conditions occur. Suppose that

$$\frac{\partial \bar{T}}{\partial x} = \delta \ , \quad \text{at} \quad x = 0 \ . \tag{4.27}$$

This can be implemented by introducing a fictitious point T_0 such that

$$\left[\frac{\partial T}{\partial x}\right]_1 = \delta = \frac{T_2 - T_0}{2\Delta x} \quad \text{or} \quad T_0 = T_2 - 2\delta\Delta x \ .$$

Applying this equation to the error distribution and combining with (4.24) gives the following equation centered at node 1:

$$(1+2s\beta)\xi_1^{n+1} - 2s\beta\xi_2^{n+1} = [1-2s(1-\beta)]\xi_1^n + 2s(1-\beta)\xi_2^n - 2s\delta\Delta x \ . \tag{4.28}$$

Consequently when all the equations are considered

$$\underline{A}\xi^{n+1} = \underline{B}\xi^n + C \ , \quad \text{where} \tag{4.29}$$

$$\underline{A} = \begin{bmatrix} 1+2s\beta & -2s\beta & & & \\ -s\beta & 1+2s\beta & -s\beta & & \\ & . & . & . & \\ & & -s\beta & 1+2s\beta & -s\beta \\ & & & -s\beta & 1+2s\beta \end{bmatrix} , \quad C = \begin{bmatrix} -2s\delta\Delta x \\ 0 \\ 0 \\ 0 \\ 0 \end{bmatrix}$$

$$\underline{B} = \begin{bmatrix} 1-2s(1-\beta) & 2s(1-\beta) & & & \\ s(1-\beta) & 1-2s(1-\beta) & s(1-\beta) & & \\ & . & . & . & \\ & & s(1-\beta) & 1-2s(1-\beta) & s(1-\beta) \\ & & & s(1-\beta) & 1-2s(1-\beta) \end{bmatrix} .$$

As before, stability requires that $|(\lambda_{\underline{A}^{-1}\underline{B}})_m| \leq 1.0$.

For the general case where explicit formulae are not available for the eigenvalues of $\underline{A}^{-1}\underline{B}$ it is convenient to form

$$\xi^{n+1} = \underline{D}\xi^n + \underline{A}^{-1}C \tag{4.30}$$

where $\underline{D} = \underline{A}^{-1}\underline{B}$.

The maximum eigenvalues of $\underline{D}$ can be found using the power method (Carnahan et al. 1969, p. 226). This requires evaluating

$$X^{n+1} = \underline{D}X^n \tag{4.31}$$

until X^{n+1} coincides with X^n. After convergence X^{n+1} is the eigenvector corresponding to the maximum eigenvalue, which is given by the maximum element of X^{n+1}. It is usual to scale the maximum element of X^n to unity at each iteration. The starting values, X^0, are arbitrary. Carnahan, Luther and Wilkes provide a program and discuss possible difficulties with the method.

For the generalised two-level scheme (4.23) the maximum eigenvalues of $\underline{D} = \underline{A}^{-1}\underline{B}$ for both Dirichlet and Neumann boundary conditions at $x=0$ are shown in Table 4.2 for a range of values of β and s. These results were generated using the power method. For the Dirichlet boundary condition at $x=0$ the results are consistent with the requirement

$$s \leq \frac{0.5}{1-2\beta} \quad \text{if} \quad \beta \leq 0.5$$

Table 4.2. Numerical stability data for generalised two-level scheme

β	s	Dirichlet b.c. at $x=0$ Max. eigenvalue	Neumann b.c. at $x=0$ Max. eigenvalue
0.00	0.50	0.9660	0.9899
	0.51	-1.0053	-1.0289
	0.52	-1.0446	-1.0686
0.30	1.20	-0.9534	0.9753
	1.30	-1.0161	-1.0273
0.60	1.20	0.9221	0.9742
	1.30	0.9159	0.9721

for stability ($\lambda_{max} \leqq 1$). It is clear that implementation of the Neumann boundary condition (4.27) slightly reduces the stability of the scheme for corresponding values of β and s.

4.3.4 Von Neumann Method: FTCS Scheme

The von Neumann analysis is the most commonly used method of determining stability criteria as it is generally the easiest to apply, the most straightforward and the most dependable. Unfortunately, it can only be used to establish necessary and sufficient conditions for the stability of linear initial value problems with constant coefficients.

Practical problems typically involve variable coefficients, nonlinearities and complicated types of boundary conditions. In which case the method can only be applied locally and with the nonlinearities temporarily frozen. For this more general situation the von Neumann method provides necessary, but not always sufficient, conditions for stability. Strictly it is only applicable for interior points. But the von Neumann method can provide heuristic information about the influence of derivative boundary conditions on the numerical stability if it is applied separately to the algorithms used at the boundaries (Trapp and Ramshaw 1976).

In the von Neumann method, the errors distributed along grid lines at one time level are expanded as a finite Fourier series. Then the stability or instability of the computational algorithm is determined by considering whether separate Fourier components of the error distribution decay or amplify in progressing to the next time level. Thus an initial error vector ξ^0 is expressed as a finite complex Fourier series, so that at x_j the error is

$$\xi_j^0 = \sum_{m=1}^{J-2} a_m e^{i\theta_m j} , \quad j = 2, 3, \dots, J-1 , \tag{4.32}$$

where $i = (-1)^{1/2}$ and $\theta_m = m\pi\Delta x$. The Fourier series (4.32) has the same form as

(3.30) except that here the errors are assumed to be periodic in the interval of interest, $0 \leq x \leq 1$. As long as the computational algorithm is linear, it is sufficient to study the propagation of the error due to just a single term, $\exp(i\theta_m j)$, of the Fourier series representation (4.32). Therefore, in the following, the subscript m has been dropped from θ_m.

In view of the form of the initial error distribution, a solution of the "error" equation (4.17) appropriate to the FTCS scheme is sought in the separation-of-variables form as

$$\xi_j^n = (G)^n e^{i\theta j} , \tag{4.33}$$

where the time dependence of this Fourier component of the error is contained in the complex coefficient $(G)^n$, and the superscript n implies G is raised to the power n. Substitution of (4.33) into the "error" equation (4.17) produces the equation

$$(G)^{n+1} e^{i\theta j} = s(G)^n e^{i\theta(j-1)} + (1-2s)(G)^n e^{i\theta j} + s(G)^n e^{i\theta(j+1)} .$$

Algebraic manipulation gives

$$G = 1 - 4s \sin^2(\theta/2) . \tag{4.34}$$

The term G can be interpreted as the amplification factor for the mth Fourier mode of the error distribution as it propagates one step forward in time, since from (4.33)

$$\frac{\xi_j^{n+1}}{\xi_j^n} = G . \tag{4.35}$$

It may be noted that G is a function of s and θ. That is, $G(s, \theta)$ depends on the size of the grid spacing and the particular Fourier mode being considered, since $s = \alpha \Delta t / \Delta x^2$ and $\theta = m\pi \Delta x$. The errors will remain bounded if the absolute value of G, also called the gain, is never greater than unity for all Fourier modes θ. Thus the general stability requirement is

$$|G| \leq 1 , \quad \text{for all } \theta . \tag{4.36}$$

From (4.34) the stability requirement for the FTCS scheme is

$$-1 \leq 1 - 4s \sin^2(\theta/2) \leq 1 , \quad \text{for all } \theta , \tag{4.37}$$

which is true if $s \leq 1/2$. This result is the same as found by the matrix method.

4.3.5 Von Neumann Method: General Two-Level Scheme

The von Neumann stability analysis will be illustrated here for the general two-level implicit scheme (7.24). The round-off errors associated with solving this scheme are propagated by the same equation,

$$\frac{\xi_j^{n+1} - \xi_j^n}{\Delta t} - \alpha \beta L_{xx} \xi_j^{n+1} - \alpha(1-\beta) L_{xx} \xi_j^n = 0 , \tag{4.38}$$

where

$$L_{xx}\xi_j = \frac{\xi_{j-1} - 2\xi_j + \xi_{j+1}}{\Delta x^2} \quad ,$$

α is the thermal diffusivity and β is a parameter that controls the degree of implicitness.

The von Neumann stability analysis proceeds by introducing a Fourier series (4.32) for the error between the true numerical solution and the numerical solution contaminated by round-off errors, ξ, defined by (4.15). Since (4.38) is linear, only a single Fourier component is introduced in the expression for ξ_j^n, that is

$$\xi_j^n = (G)^n e^{i\theta j} \quad , \tag{4.39}$$

where $i = (-1)^{1/2}$ and j is the grid number. Substitution gives

$$L_{xx}\xi_j^n = (G)^n e^{i\theta(j-1)} - 2(G)^n e^{i\theta j} + (G)^n e^{i\theta(j+1)}$$

$$= (G)^n e^{i\theta j} \frac{2(\cos\theta - 1)}{\Delta x^2} \quad .$$

Consequently (4.38) becomes

$$\frac{(G)^{n+1} e^{i\theta j} - (G)^n e^{i\theta j}}{\Delta t} - \alpha\beta(G)^{n+1} e^{i\theta j} \frac{2(\cos\theta - 1)}{\Delta x^2}$$

$$+ \alpha(1-\beta)(G)^n e^{i\theta j} \frac{2(\cos\theta - 1)}{\Delta x^2} = 0 \quad .$$

Dividing through by $(G)^n e^{i\theta j}$ gives

$$[G-1]/\Delta t - \alpha\beta G[2(\cos\theta - 1)]/\Delta x^2 + \alpha(1-\beta)[2(\cos\theta - 1)]/\Delta x^2 = 0$$

or

$$G = \frac{1 - 4s(1-\beta)\sin^2(\theta/2)}{1 + 4s\beta\sin^2(\theta/2)} \quad . \tag{4.40}$$

For stability $|G| \leq 1$ for any choice of θ. For the limiting values

$$\sin(\theta/2) = 0 \quad , \quad G = 1$$

$$\sin(\theta/2) = 1 \quad , \quad G = [1 - 4s(1-\beta)]/[1 + 4s\beta]$$

$$\text{If } G \leq 1 \quad , \quad 1 - 4s(1-\beta) \leq 1 + 4s\beta \quad \text{or} \quad s > 0 \quad .$$

$$\text{If } G > -1 \quad , \quad 1 - 4s(1-\beta) \geq -1 - 4s\beta \quad \text{or} \quad s \leq 0.5/(1-2\beta) \quad .$$

These restrictions agree with the result given in Sects. 4.3.2 and 7.2.2. For more complicated equations it may be necessary to establish stability by evaluating the equivalent of (4.40) for a range of values of θ, β and s, etc. This is often done numerically.

For algorithms that couple three time levels together, e.g. Sect. 7.2.3, a quadratic equation in G must be solved to obtain the stability restrictions. For most fluid flow problems a system of governing equations is required. The von Neumann stability analysis then leads to an amplification matrix $\underline{G}$ in place of the amplification factor G. The corresponding stability condition is

$$|\lambda_m| \leq 1.0 \quad \text{for all } m \ . \tag{4.41}$$

where λ_m are the eigenvalues of $\underline{G}$.

In this section it is assumed that the underlying physical problem is stable. However it may happen that one needs to compute flows that are concerned with physical instability; for example transition from laminar to turbulent flow. For such flows a growing solution with time must be acceptable. This can be allowed for by requiring the magnitude of the eigenvalues in the matrix method (Sect. 4.3.1) and the amplification factor in the von Neumann method (Sect. 4.3.4) to be less than $1 + O(\Delta t)$. This condition replaces the inequalities (4.20, 36). This aspect is discussed further by Richtmyer and Morton (1967, p. 70).

Stability of the discretised equations is discussed more fully by Mitchell and Griffiths (1980), Isaacson and Keller (1966) and Richtmyer and Morton (1967). It is generally accepted that the matrix method of determining stability is not so reliable as the von Neumann method. The reasons for this are discussed in relation to the one-dimensional transport equation (9.56) by Morton (1980) and Hindmarsh et al. (1984). The problem of stability of the discretised equations adjacent to boundaries lies beyond the scope of this book (see Sod 1985, Chap. 5).

4.4 Solution Accuracy

Strictly, the previous discussion of convergence, consistency and stability has been concerned with the behaviour of the approximate solution in the limit $\Delta x, \Delta t \to 0$. However in practice approximate solutions are obtained on a finite grid and the corresponding accuracy is of considerable importance.

The determination of the consistency, Sect. 4.2, produces an explicit expression for the truncation error. For an exact solution of the partial differential equation it is possible to evaluate the accuracy of the representation of the individual derivative terms as in Sect. 3.3. Combining the individual terms together gives an indication of how closely the algebraic equation agrees with the partial differential equation. As is to be expected from Sect. 3.3, evaluation of the leading term in the truncation error of the complete equation provides a close estimate of the error.

As indicated in Sect. 4.2 the order of the truncation error will usually coincide with the order of the solution error if the grid spacings are sufficiently small and if the initial and boundary conditions are sufficiently smooth. It is common for the implied improvement in accuracy (from the truncation error) of higher-order schemes over say a second-order scheme, not to be achieved due to insufficient smoothness in the initial conditions for the particular grid. Refining the grid will very often produce a superior accuracy for the higher-order scheme but at an absolute accuracy level that is far higher, and hence far more expensive, than is necessary. Thus an engineering accuracy of 1% in the rms error of the solution might be required. A fourth-order scheme may only demonstrate superior accuracy over a second-order scheme when the grid has been refined to the point where the rms error is perhaps 0.01%. An example of such an effect of non-smooth initial data on the accuracy is given by Fletcher (1983).

For problems governed by hyperbolic partial differential equations, discontinuities can form in the interior of the domain, and these effectively limit the accuracy that can be achieved, unless they are isolated from the rest of the domain and a local solution sought.

One way of determining the accuracy of a particular algorithm on a finite grid is to apply it to a related but simplified problem, which possesses an exact solution. In this regard Burgers' equation (Sect. 10.1) is useful since it models the convective and diffusive terms in the momentum equations and since exact solutions can be obtained for many combinations of initial and boundary conditions. However accuracy is also problem dependent, and an algorithm which is accurate for a model problem may not necessarily be so accurate for the (more complicated) problem of interest.

A second technique for assessing accuracy is to obtain solutions on successively refined grids (assuming the computing capacity is available) and to check that, with successive refinement, the solution is not changing to some predetermined accuracy. This assumes that the approximate solution will converge to the exact solution in the limit Δt, Δx, etc. $\rightarrow 0$ and that the approximate solution on the finest grid can be used in place of the exact solution. Since this is usually impossible to guarantee for "real" problems, it is useful to compare the computational solutions with reliable experimental results (of known accuracy) for the same problem. For example, the pressure distribution on a body surface could be used to check that the refined grid solution is reasonable. However experimental data are usually not available in sufficient detail, for a particular problem, to permit an evaluation of a global rms error.

Assuming that the accuracy of the approximate solution can be assessed it is important to consider the related question of how the accuracy may be improved. At the broadest level the answer to this question may lie in making a different choice for the dependent variables, e.g. vorticity and stream function instead of velocity and pressure. Alternatively, a different choice of independent variables may be appropriate. For example polar coordinates will produce more accurate solutions for pipe flow than Cartesian coordinates for the same number of grid points.

At a more specific level the use of higher-order schemes or grid refinement would be expected to produce more accurate solutions. However it is not very meaningful to consider such choices without taking into account the execution time and consequently the computational efficiency (Sect. 4.5). For the moment one specific technique, Richardson extrapolation, for improving the accuracy will be considered.

4.4.1 Richardson Extrapolation

It has already been shown that convergence can be demonstrated numerically (Sect. 4.1.2) by obtaining solutions on successively refined grids. For sufficiently fine grids the solution error reduces like the truncation error (Table 4.1). This last feature motivates Richardson extrapolation, which is used to improve the solution accuracy on a finite (but fine) grid. The basic idea is to add together suitably weighted solutions on successively refined grids to cancel the leading term in the truncation error. However the form of the truncation error arises from Taylor series expansions over the local grid spacing. It is not surprising, therefore, to find that Richardson extrapolation often requires a relatively fine grid to be effective.

Consider two solutions of the diffusion equation obtained using the FTCS scheme with two x-steps, Δx_a and Δx_b, but with the same value of s. A composite solution is constructed as

$$T_c = aT_a + bT_b \; , \tag{4.42}$$

where a and b are coefficients to be determined. Expansion of the FTCS scheme applied to T_c as a Taylor series about the (j, n)-th node gives

$$\left[\frac{\partial T}{\partial t}\right]_j^n - \alpha \left[\frac{\partial^2 T}{\partial x^2}\right]_j^n + a[E_a]_j^n + b[E_b]_j^n = 0 \; , \tag{4.43}$$

where, from (4.7),

$$[E_a]_j^n = 0.5\alpha\Delta x_a^2 \left(s - \frac{1}{6}\right)\left(\frac{\partial^4 \bar{T}}{\partial x^4}\right) + O(\Delta x^4) \; ,$$

$$[E_b]_j^n = 0.5\alpha\Delta x_b^2 \left(s - \frac{1}{6}\right)\left(\frac{\partial^4 \bar{T}}{\partial x^4}\right) + O(\Delta x^4) \; .$$

If $\Delta x_b = \Delta x_a/2$ the leading term in the composite truncation error can be eliminated if

$$a + 0.25b = 0 \; .$$

But from (4.42) it is clear that $a + b = 1$. Therefore

$$a = -\tfrac{1}{3} \quad \text{and} \quad b = \tfrac{4}{3} \; ,$$

and the composite solution $T_c = (4/3)T_b - (1/3)T_a$ will be expected to have a solution error of $O(\Delta x^4)$ on a sufficiently refined grid. Some typical rms errors are shown in Table 4.3 with $s = 0.5$. The error is reducing like $O(\Delta x^4)$ for the finer grids. However, although the accuracy level associated with Richardson extrapolation is superior on a fine grid, the reverse is true on a coarse grid.

The FTCS scheme is unusual in the sense that a fourth-order scheme can be obtained by the special choice $s = 1/6$ without the need for Richardson extrapolation. This is not the case for other schemes such as the fully implicit scheme (7.19) and the Crank-Nicolson scheme (7.22). But both of these schemes can be made

Table 4.3. rms errors for Richardson extrapolation (RE)

Scheme	$\Delta x = 0.1$	$\Delta x = 0.05$	$\Delta x = 0.025$
FTCS, $s = 0.5$	0.4915	0.0869	0.0153
FTCS, $s = 0.5$, RE	0.9744	0.0104	0.634×10^{-3}
FTCS, $s = 1/6$	0.0017	0.542×10^{-4}	0.339×10^{-5}

fourth-order accurate by applying Richardson extrapolation. If $\Delta x_b = \Delta x_a / 2$ the appropriate composite solution is

$$T_c = \tfrac{4}{3} T_b - \tfrac{1}{3} T_a \; , \tag{4.44}$$

as above.

In the general case the choice of the coefficients a and b is determined by the power p of the leading Δx term in the expression for the truncation error (after converting Δt terms into equivalent Δx terms) and the ratio of the grid sizes, Δx_a and Δx_b. Thus

$$a = -\frac{1}{\left[\left(\dfrac{\Delta x_a}{\Delta x_b} \right)^P - 1 \right]} \quad \text{and} \quad b = \frac{\left(\dfrac{\Delta x_a}{\Delta x_b} \right)^P}{\left(\dfrac{\Delta x_a}{\Delta x_b} \right)^P - 1} \; . \tag{4.45}$$

Implementing Richardson extrapolation for an unsteady problem like the diffusion equation, it is possible to treat the solutions T_b and T_a as independent solutions obtained on different grids at the final time at which T_c is to be computed. Since Richardson extrapolation does not affect the evolving solution, but only the final solution, Dahlquist and Bjorck (1974) refer to this as passive Richardson extrapolation. If T_c is computed at the end of every time-step and this solution forms the initial solution for computing both T_a and T_b during the next time-step one has active Richardson extrapolation (Dahlquist and Bjorck 1974, p. 340).

It is instructive to compare the various schemes shown in Table 4.3 on the basis of execution time (CPT). For an integration over a finite time, CPT can be represented as

$$\text{CPT} = \frac{k}{\Delta x \Delta t} = \frac{\alpha k}{s \Delta x^3} \; ,$$

where k is a constant of proportionality. Therefore

$$\text{CPT}_{s=0.5} = \frac{2\alpha k}{\Delta x^3} \; , \quad \text{CPT}_{s=0.5, \text{RE}} = \frac{2.25\alpha k}{\Delta x^3} \; , \quad \text{CPT}_{s=1/6} = \frac{6\alpha k}{\Delta x^3} \; .$$

Thus it can be seen that the solution corresponding to $\text{CPT}_{s=0.5, \text{RE}}$ for $\Delta x_a = 0.1$, $\Delta x_b = 0.05$ is 50% more accurate than the solution corresponding to $\text{CPT}_{s=0.5}$ at

$x = 0.025$ but has one-seventh of the execution time. However, the solution corresponding to $CPT_{s=1/6}$ for $\Delta x = 0.05$ is approximately twelve times as accurate as the solution corresponding to $CPT_{s=0.5,\,RE}$ for $\Delta x_b = 0.025$ and has one-third of the execution time. Clearly the use of Richardson extrapolation is effective but is not so efficient as the special case $s = \frac{1}{6}$. However such special cases are rare, particularly for equations and algorithms more complicated than the FTCS scheme applied to the diffusion equation.

4.5 Computational Efficiency

The question of what accuracy can be achieved with a given algorithm is closely connected with the computational efficiency of the particular algorithm. The computational efficiency may be defined as the accuracy achieved per unit execution time. Thus an algorithm that achieves a modest accuracy on a coarse grid with a small execution time may be computationally as efficient as another algorithm that achieves a higher accuracy on a refined grid requiring a large execution time.

The computational efficiency (CE) can be defined by

$$CE = \frac{k}{\varepsilon CPT} \, ,$$

where ε is the error in the approximate solution in some appropriate norm (e.g. the rms norm) and CPT is the execution (CPU) time. The computational efficiency is a useful measure with which to determine whether higher-order finite difference schemes or higher-order finite elements are worth introducing.

For most flow problems the use of high-order (say fourth-order or higher) finite difference schemes and high-order (say cubic or higher) finite elements are not justified, in terms of improved computational efficiency, particularly in three dimensions. However, at least second-order finite difference schemes or linear finite elements should be used to avoid the introduction of excessive numerical dissipation and dispersion (Sect. 9.2).

In developing different computational schemes it is preferable to compare computational efficiency rather than accuracy or economy alone. If different computational schemes are compared on the same computer it is possible to directly measure the execution time. Care must be taken to ensure that the computer is being used in a dedicated mode, if the CPU time is not available. Also the execution time to read in data and to print out the results should be excluded or allowed for.

4.5.1 Operation Count Estimates

As an alternative to the direct measurement of the CPU time, an operation count estimate can be determined from a computer listing and some knowledge of the relative execution time of basic operations. This can be done at various levels of complexity. The following approach is found to give consistent estimates.

First a timing program is run to determine the approximate execution time associated with each type of operation. A typical timing program and a discussion of the procedure is given in Appendix A.1. The main purpose is to obtain the relative execution time of the various operations, since these can vary widely between different computers, operating systems and compilers. Relative execution times are shown in Table 4.4 for typical micro, supermicro and mainframe computers running FORTRAN.

The second step is to identify the major classes of operations that account for most of the overall execution time. For the NEC-APC IV results shown in Table 4.4 the following categories of operations would be worth isolating:

(a) floating point operations (FL)
(b) fixed point operations (FX)
(c) replacements, i.e. equals signs (R)
(d) logical statements, i.e. IF, GOTO, etc. (L)
(e) mathematical library functions, i.e. SIN, EXP, etc. (M).

To reduce the operation count to a single number it is necessary to assign a relative weighting to the above operations, i.e. the operation count might be measured in equivalent floating point operations. A typical weighting for the NEC-APC IV would be (based on Table 4.4)

$$1 \, FL_{eq} = 20 \, FX = 10 \, R = 10 \, L = 0.1 \, M \ .$$

For lines 48 to 73 of DIFF (Sect. 3.5.2), and excluding WRITE statements, the

Table 4.4. Relative execution time for basic operations

Operation	Microcomputer (NEC-APC IV)[a]	Supermicro (SUN Sparc St1)[b]	Mainframe (IBM-3090)[c]
add (FL)	1.0	1.0	1.0
sub. (FL)	1.0	1.0	0.8
mult. (FL)	1.1	1.0	0.8
div. (FL)	1.4	5.8	3.1
replace (=)	0.1	1.0	0.1
IF state.	0.1	0.9	0.2
add (FX)	0.05	0.6	0.2
sub. (FX)	0.05	0.6	0.2
mult. (FX)	0.4	0.6	0.6
div. (FX)	0.5	5.4	3.2
power	2.7	20	16.0
SQRT	2.0	29	16.0
SIN	10.0	29	17.6
EXP	6.7	33	15.0

[a] Microsoft Fortran 77 running under MSDOS (8086/8087 floating point hardware in single precision)
[b] Fortran 77, unoptimised compiler running under UNIX
[c] FORTV52, unoptimised compiler running under VM/XA2 operating system

following operation count is obtained.

$$OPCT = [3FL + 6FX + 5R + 4L + (N_x - 2)(5FL + 10FX + 3R)]N_t ,$$

where N_x is the number of x-steps and N_t is the number of t-steps. If $N_x = 11$ the operation count per time step is $OPCT = 56\,FL_{eq}$. In obtaining the above operation count a subscripted array has been assigned one fixed point operation to locate the subscript.

Operation count estimates are particularly useful in identifying parts of the computer program that are expensive and in comparing the likely economy of different algorithms prior to coding and testing. Because of the variability between different computers and the effort required to obtain precise results, operation counts are most effective when used to look for order-of-magnitude differences in execution time. In addition the operation count procedure needs to be adjusted when used with vector and parallel-processing computers (Gentzsch and Neves 1988).

4.6 Closure

For the equations that govern fluid motion it is not possible to demonstrate convergence directly. However, it is usually straightforward to show that the discretised form of the equations is consistent (Sect. 4.2). It is usually possible to show that a "linearised" version of the governing equations is stable (Sect. 4.3), although this may require a computer-based computation. Consequently the Lax equivalence theorem tends to provide a necessary rather than sufficient condition for convergence. In practice verification of stability usually requires numerical evaluation.

The matrix and von Neumann methods for determining stability are strictly only applicable to linear equations, although both methods will give some guidance if the nonlinearity is frozen locally. An alternative energy method can handle some nonlinear equations directly. This method is described by Richtmyer and Morton (1967, p. 132).

It should be stressed that real flow problems are computed on finite grids and that theoretical properties, deduced from allowing the grid to converge to zero size, may not be realisable on a particular finite grid. This is more likely to be a problem with higher-order schemes.

The use of a Taylor series expansion to establish consistency (Sect. 4.2) is also important in providing an explicit expression for the truncation error. Because of the close correlation between the truncation error and the solution error, at least on a fine grid, anything that can be done to reduce the truncation error is also likely to reduce the solution error. This may entail making a specific choice for $s(= \alpha \Delta t/\Delta x^2)$ as in the FTCS scheme or it may involve combining solutions so that the leading term in the truncation error is eliminated as in Richardson extrapolation (Sect. 4.4.1).

For steady problems there is another alternative, called the deferred correction method (Smith 1965, p. 140). In this method a preliminary solution is computed and used to evaluate the leading term in the truncation error. This term is "added" to the original discretised equation as a source term and an improved solution obtained. In principle the improved solution can be used to reevaluate the leading term in the truncation error and the whole process repeated. Usually a single improved solution is computationally most efficient.

Elimination of the leading term in the truncation error can also be used to construct higher-order schemes. After the initial discretisation the leading term in the truncation error is discretised and "added" to the original scheme. An example of this approach is provided in Sect. 9.3.2.

Practical flow problems often require a computational solution in complicated three-dimensional domains, involving thousands of nodal unknowns. For such large-scale problems the assessment of the relative computational efficiency of competing numerical schemes is an essential prerequisite to the main body of the computational investigation. Finlayson (1980) provides work estimates and accuracy comparisons of the finite difference, finite element and orthogonal collocation (Sect. 5.1) methods applied to ordinary differential equations, and one- and two-dimensional partial differential equations.

4.7 Problems

Convergence (Sect. 4.1)

4.1 Modify program DIFF to use the following five-point symmetric scheme in place of the FTCS scheme. For interior points use

$$\frac{\partial^2 T}{\partial x^2} \approx \left(-\frac{1}{12} T_{j-2} + \frac{4}{3} T_{j-1} - 2.5 T_j + \frac{4}{3} T_{j+1} - \frac{1}{12} T_{j+2} \right) \Big/ \Delta x^2 \ ,$$

and for $j=2$ use

$$\frac{\partial^2 T}{\partial x^2} \approx \left(\frac{11}{12} T_{j-1} - \frac{5}{3} T_j + 0.5 T_{j+1} + \frac{1}{3} T_{j+2} - \frac{1}{12} T_{j+3} \right) \Big/ \Delta x^2 \ ,$$

and the equivalent formula for $j=\text{JMAX}-1$. These are the same modifications as required for Problem 3.7.

(a) Obtain solutions with $s=0.3$, for $\Delta x=0.2$, 0.1 and 0.05.

(b) Compare the reduction in the rms error with Δx with that of the FTCS scheme (Table 4.1).

4.2 Repeat Problem 4.1 with the following scheme in place of the FTCS scheme, in program DIFF,

$$\frac{0.5 T_j^{n-1} - 2 T_j^n + 1.5 T_j^{n+1}}{\Delta t} - \frac{\alpha (T_{j-1}^n - 2 T_j^n + T_{j+1}^n)}{\Delta x^2} = 0 \ .$$

For the first time step use the forward time difference formula

$$\frac{\partial T}{\partial t} \approx \frac{T^{n+1} - T^n}{\Delta t} .$$

These are the same modifications as required for Problem 3.8.

Consistency (Sect. 4.2)

4.3 Establish that the scheme introduced in Problem 4.1 is consistent with the governing equation $\partial \bar{T}/\partial t - \alpha \partial^2 \bar{T}/\partial x^2 = 0$. Is the order of the truncation error in agreement with the convergence rate found in Problem 4.1? Consider both interior points and points ($j=2$, JMAX-1) adjacent to the boundary.

4.4 Test the consistency of the scheme introduced in Problem 4.2. Is the convergence rate obtained in Problem 4.2 in agreement with the order of the truncation error?

Stability (Sect. 4.3)

4.5 Apply the von Neumann stability analysis to determine the stability limits of the scheme introduced in Problem 4.1 for:
(a) interior points,
(b) points adjacent to the boundary, $j=2$ and JMAX-1.
It may be convenient to evaluate numerically the expression for the amplitude ratio $|G|$ as θ and s are varied.

4.6 Use the von Neumann stability analysis to establish the stability limits for the scheme introduced in Problem 4.2. This will require the solution of a quadratic equation in $|G|$.

4.7 Confirm the stability limits obtained in Problems 4.5 and 4.6 by obtaining numerical solutions for "unstable" values of s.

4.8 Examine the stability of a discretisation of $\partial \bar{T}/\partial t - \alpha \partial^2 \bar{T}/\partial x^2 = 0$ based on a forward time representation for $\partial \bar{T}/\partial t$ and the formula obtained in Problem 3.1 for $\partial^2 \bar{T}/\partial x^2$.

4.9 Apply the matrix method to the following discrete system,

$$M_x(\Delta T_j^{n+1}/\Delta t) - \alpha[\beta L_{xx} T_j^{n+1} + (1-\beta) L_{xx} T_j^n] = 0 ,$$

where

$$M_x \Delta T_j = \delta \Delta T_{j-1} + (1-2\delta)\Delta T_j + \delta \Delta T_{j+1} .$$

This scheme is the same as Eq. (7.31) with $\gamma = 0$. Obtain the following stability restriction on s and discuss the implications:

$$s \leq (0.5 - 2\delta)/(1 - 2\beta) .$$

Solution Accuracy (Sect. 4.4)

4.10 For the scheme considered in Problem 4.1 develop a Richardson extrapolated algorithm that will eliminate the leading term in the truncation error. Determine numerically whether this scheme is more accurate and if the theoretical convergence rate can be achieved.

4.11 For the scheme considered in Problem 4.2 obtain a Richardson extrapolated algorithm and compare numerical solutions with those obtained in Problems 4.1 and 4.2.

Computational Efficiency (Sect. 4.5)

4.12 For the scheme considered in Problem 4.10 determine the operation count and compare with that of the FTCS scheme when both are achieving the same accuracy.

4.13 For the scheme considered in Problem 4.11 determine the operation count and compare with:
(a) FTCS scheme when both are achieving the same accuracy,
(b) the Richardson-extrapolated FTCS scheme.

4.14 Discuss which is the "best" scheme of those considered in this chapter, taking into account computational efficiency, stability restrictions and ease of coding.

5. Weighted Residual Methods

Weighted residual methods (WRMs) are conceptually different from the finite difference method in that a WRM assumes that the solution can be represented analytically. For example, to obtain the solution of the diffusion equation (3.1) the following approximate solution would be assumed:

$$T = \sum_{j=1}^{J} a_j(t)\phi_j(x) \; , \tag{5.1}$$

where $a_j(t)$ are unknown coefficients and $\phi_j(x)$ are known analytic functions. The terms $\phi_j(x)$ are often referred to as trial functions and (5.1) as the trial solution. By forcing the analytic behaviour to follow (5.1) some error is introduced unless J is made arbitrarily large. It may be recalled that the finite difference method defines a solution at nodal points only.

In Sect. 5.1 various weighted residual methods are indicated. In Sect. 5.1.1 some of these WRMs are applied to a simple ordinary differential equation to provide a tangible basis for comparison. The finite volume method, described in Sect. 5.2, is closely related to the subdomain method (Sect. 5.1). However in implementation, finite volume discretisation resembles a finite difference method. In Sect. 5.2.2 the particular difficulties of applying the finite volume method, when second derivatives are present, are indicated and a computer program, FIVOL is developed (Sect. 5.2.3) to solve Laplace's equation in an irregular domain.

In this chapter the finite element and spectral methods will be developed from the Galerkin method which is a member of the class of methods of weighted residuals. The description of the finite element and spectral methods will be provided primarily through the medium of worked examples, which will emphasise the mechanics of applying the methods. The examples will also show, at least for the finite element method, a connection with some of the finite difference methods to be considered in Chaps. 7–10. Both the finite element and spectral methods are discussed at greater length by Fletcher (1984), and the spectral method by Canuto et al. (1987).

In Sects. 5.3–5.5 the finite element method is considered. First the increased accuracy of linear and quadratic interpolation in one and two dimensions with grid refinement is demonstrated. Specific examples implementing the finite element method are provided in Sects. 5.4 and 5.5. This includes computer programs, STURM for the Sturm-Liouville problem (Sect. 5.4.2) and DUCT for viscous flow in a rectangular duct (Sect. 5.5.2). The ability to handle distorted computational domains is a feature of the finite element method; the use of isoparametric elements for this purpose is discussed in Sect. 5.5.3.

The spectral method applied to the diffusion equation with straightforward Dirichlet boundary conditions is developed in Sect. 5.6.1. In Sect. 5.6.2 the difficulties of applying a spectral method with Neumann boundary conditions are discussed and the more general formulation indicated. The closely-related pseudo-spectral method is described in Sect. 5.6.3.

5.1 General Formulation

Many computational methods can be interpreted as methods of weighted residuals (WRMs). A comprehensive description of WRMs and applications up to 1972 is given by Finlayson (1972). Here WRMs will be developed sufficiently to demonstrate the connection between the finite element, spectral and finite volume methods.

The starting point for a WRM is to assume an approximate solution. Equation (5.1) can be extended slightly by writing

$$T(x, y, z, t) = T_0(x, y, z, t) + \sum_{j=1}^{J} a_j(t)\phi_j(x, y, z) , \tag{5.2}$$

where $T_0(x, y, z, t)$ is chosen to satisfy the boundary and initial conditions, exactly if possible. The approximating (trial) functions $\phi_j(x, y, z)$ are known. In one spatial dimension the approximating functions might be polynomials or trigonometric functions, e.g.

$$\phi_j(x) = x^{j-1} \quad \text{or} \quad \phi_j(x) = \sin j\pi x .$$

The coefficients $a_j(t)$ are unknown and are to be determined by solving a system of equations generated from the governing equation. For time-dependent problems a system of ordinary differential equations (in time) will be solved for $a_j(t)$; for steady problems a system of algebraic equations will be solved.

It is assumed that the governing equation, for example the heat conduction equation, can be written

$$L(\bar{T}) = \frac{\partial \bar{T}}{\partial t} - \alpha \frac{\partial^2 \bar{T}}{\partial x^2} = 0 , \tag{5.3}$$

where the overbar denotes the exact solution. If the approximate solution, (5.2), is substituted into (5.3) it will not be identically zero. Thus we can write

$$L(T) = R , \tag{5.4}$$

where R is referred to as the equation residual. R is a continuous function of x, y, z and t in the general case. If J is made sufficiently large then, in principle, the coefficients $a_j(t)$ can be chosen so that R is small over the computational domain.

The coefficients $a_j(t)$ are determined by requiring that the integral of the weighted residual over the computational domain is zero, i.e.

$$\iiint W_m(x, y, z) \, R \, dx \, dy \, dz = 0 \ . \tag{5.5}$$

By letting $m = 1, \ldots, M$ a system of equations for the a_j's is generated. For the unsteady case considered here, this will be a system of ordinary differential equations. For the steady case a system of algebraic equations is generated.

It may be noted that (5.5) is closely related to the weak form of the governing equation, i.e.

$$\iiint W_m(x, y, z) \, L(\bar{T}) \, dx \, dy \, dz = 0 \ . \tag{5.6}$$

which allows discontinuities in the exact solution to appear (Lax and Wendroff 1960).

Different choices for the weight (test) function W_m in (5.5) give rise to different methods in the class of methods of weighted residuals. Some of these methods are:

(i) **Subdomain Method.** The computational domain is split up into M subdomains D_m, which may overlap, and

$$W_m = 1 \quad \text{in } D_m \tag{5.7}$$

$$= 0 \quad \text{outside } D_m \ .$$

The subdomain method coincides with the finite volume method (Sect. 5.2) in its evaluation of (5.5). Equations (5.5 and 7) provide an appropriate framework for enforcing conservation at the discretised equation level. In this way the conservation properties inherent in the governing equations are preserved. This is a particular advantage in obtaining accurate solutions for internal flows or flows with shock waves.

(ii) **Collocation Method**

$$W_m(x) = \delta(x - x_m) \ , \tag{5.8}$$

where δ is the Dirac delta function and $x = (x, y, z)$. The substitution of (5.8) into (5.5) indicates that the collocation method is obtained by setting $R(x_m) = 0$. Thus finite difference methods are typically collocation methods without the use of an approximate solution.

The orthogonal collocation method (Finlayson 1980) is a particularly effective interpretation of this method. In the orthogonal collocation method, the approximating functions in (5.2) are chosen from orthogonal polynomials and the unknowns in (5.2) are the nodal values (of T). The collocation points x_m are chosen from the roots of one of the approximating function orthogonal polynomials.

(iii) **Least-squares Method**

$$W_m = \frac{\partial R}{\partial a_m} \ . \tag{5.9}$$

Equation (5.9) is equivalent to the requirement that

$$\iiint R^2 \, dx \, dy \, dz \text{ is a minimum} .$$

(iv) Galerkin Method

$$W_m(x, y, z) = \phi_m(x, y, z) , \tag{5.10}$$

i.e. the weight functions are chosen from the same family as the approximating (trial) functions. If the approximating functions form a complete set (for polynomials a complete set would be $1, x^2, x^3, \ldots, x^M$), (5.5) indicates that the residual is orthogonal to every member of a complete set. Consequently as M tends to infinity the approximate solution, T, will converge to the exact solution $\bar{T}$.

5.1.1 Application to an Ordinary Differential Equation

A comparison of the various methods of weighted residuals is made by obtaining the solution to the ordinary differential equation

$$\frac{d\bar{y}}{dx} - \bar{y} = 0 \quad \text{for} \quad 0 \le x \le 1 , \tag{5.11}$$

with boundary condition $\bar{y} = 1$ at $x = 0$. This problem has the exact solution $\bar{y} = \exp x$.

A suitable approximate solution, equivalent to (5.2), is

$$y = 1 + \sum_{j=1}^{N} a_j x^j . \tag{5.12}$$

This approximate solution automatically satisfies the boundary condition. In contrast to the situation for (5.2), the coefficients a_j in (5.12) are constants. Substitution of (5.12) into (5.11) produces the equation residual

$$R = -1 + \sum_{j=1}^{N} a_j(jx^{j-1} - x^j) . \tag{5.13}$$

The coefficients a_j are determined by setting the integral of the weighted residual over the computational domain equal to zero. Thus, following (5.5),

$$\int_0^1 W_m(x) \, R(x) \, dx = 0 . \tag{5.14}$$

The different methods of weighted residuals are produced by different choices for the weights $W_m(x)$. For example, the Galerkin method uses $W_m(x) = x^{m-1}$ for $m = 1, N$. Evaluation of (5.14) for each value of m produces a system of equations that can be written

$$SA = D ; \tag{5.15}$$

A is the vector of unknown coefficients a_j. For the Galerkin method an element of $\underline{S}$ is given by

$$S_{mj}=\left(\frac{j}{j+m-1}-\frac{1}{j+m}\right), \tag{5.16}$$

and an element of D by $d_m=1/m$. For the choice $N=3$, (5.15) becomes

$$\begin{bmatrix} 1/2 & 2/3 & 3/4 \\ 1/6 & 5/12 & 11/20 \\ 1/12 & 3/10 & 13/30 \end{bmatrix}\begin{bmatrix} a_1 \\ a_2 \\ a_3 \end{bmatrix}=\begin{bmatrix} 1 \\ 1/2 \\ 1/3 \end{bmatrix}, \tag{5.17}$$

which has the solution

$$a_1=1.0141 , \quad a_2=0.4225 , \quad a_3=0.2817 . \tag{5.18}$$

Substitution of (5.18) into the approximate solution (5.12) produces the result

$$y=1+1.0141x+0.4225x^2+0.2817x^3 . \tag{5.19}$$

This solution is compared with the exact solution in Table 5.1.

Table 5.1. Galerkin solutions of $dy/dx-y=0$

	Approximate solution			Exact solution,
x	Linear ($N=1$)	Quadratic ($N=2$)	Cubic ($N=3$)	$\bar{y}=\exp(x)$
0.	1.0	1.0	1.0	1.0
0.2	1.4	1.2057	1.2220	1.2214
0.4	1.8	1.4800	1.4913	1.4918
0.6	2.2	1.8229	1.8214	1.8221
0.8	2.6	2.2349	2.2259	2.2251
1.0	3.0	2.7143	2.7183	2.7183
Solution err. (rms)	0.2857	0.00886	0.00046	—
R_{rms}	0.5271	0.0583	0.00486	—

The solution error for different values of N is plotted in Fig. 5.1. It is clear that the solution error reduces rapidly as N is increased. It can be seen from Table 5.1 that both the rms solution error and the rms equation residual R_{rms} decrease rapidly as N increases. For the general problem the exact solution is not available and consequently the solution error cannot be computed. However the equation residual can be evaluated and this gives a qualitative indication of the closeness to the exact solution.

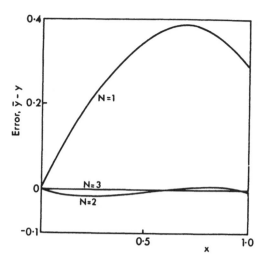

Fig. 5.1. Error distribution for Galerkin solution of $dy/dx - y = 0$

Some of the other methods of weighted residuals discussed in Sect. 5.1 are applied to (5.11). For $N = 3$ the least-squares method produces the following system of equations in place of (5.17):

$$\begin{bmatrix} 1/3 & 1/4 & 1/5 \\ 1/4 & 8/15 & 2/3 \\ 1/5 & 2/3 & 33/35 \end{bmatrix} \begin{bmatrix} a_1 \\ a_2 \\ a_3 \end{bmatrix} = \begin{bmatrix} 1/2 \\ 2/3 \\ 3/4 \end{bmatrix} . \tag{5.20}$$

The subdomain method with intervals, $0 \leq x \leq 1/3$, $1/3 \leq x \leq 2/3$ and $2/3 \leq x \leq 1.0$ produces the following system of equations:

$$\begin{bmatrix} 5/18 & 8/81 & 11/324 \\ 3/18 & 20/81 & 69/324 \\ 1/18 & 26/81 & 163/324 \end{bmatrix} \begin{bmatrix} a_1 \\ a_2 \\ a_3 \end{bmatrix} = \begin{bmatrix} 1/3 \\ 1/3 \\ 1/3 \end{bmatrix} . \tag{5.21}$$

A collocation method with the residual evaluated at $x = 0$, 0.5 and 1.0 gives the following system of equations:

$$\begin{bmatrix} 1.0 & 0. & 0. \\ 0.5 & 0.75 & 0.625 \\ 0. & 1.0 & 2.0 \end{bmatrix} \begin{bmatrix} a_1 \\ a_2 \\ a_3 \end{bmatrix} = \begin{bmatrix} 1.0 \\ 1.0 \\ 1.0 \end{bmatrix} . \tag{5.22}$$

The solutions for a_1, a_2 and a_3 corresponding to the various methods are shown in Table 5.2 and the corresponding approximate solutions (5.12) in Table 5.3.

The optimal rms solution is obtained by minimising the rms error with three unknown coefficients. It is more appropriate to compare the various weighted residual solutions with this rather than the exact solution. Clearly the Galerkin, least squares and subdomain methods are producing solutions close to optimal.

Table 5.2. Comparison of coefficients for approximate solutions of $dy/dx - y = 0$

Scheme	Coefficient a_1	a_2	a_3
Galerkin	1.0141	0.4225	0.2817
Least squares	1.0131	0.4255	0.2797
Subdomain	1.0156	0.4219	0.2813
Collocation	1.0000	0.4286	0.2857
Optimal rms	1.0138	0.4264	0.2781
Taylor series	1.0000	0.5000	0.1667

Table 5.3. Comparison of approximate solutions of $dy/dx - y = 0$

x	Galerkin	Least squares	Sub-domain	Collo-cation	Optimal rms	Taylor series	Exact
0.	1.0000	1.0000	1.0000	1.0000	1.0000	1.0000	1.0000
0.2	1.2220	1.2219	1.2213	1.2194	1.2220	1.2213	1.2214
0.4	1.4913	1.4912	1.4917	1.4869	1.4915	1.4907	1.4918
0.6	1.8214	1.8214	1.8220	1.8160	1.8219	1.8160	1.8221
0.8	2.2259	2.2260	2.2265	2.2206	2.2263	2.2053	2.2255
1.0	2.7183	2.7183	2.7187	2.7143	2.7183	2.6667	2.7183
Solution error (rms)	0.000458	0.000474	0.000576	0.004188	0.000434	0.022766	—

The accuracy of the collocation method is sensitive to the choice of the sample points. Evaluating the residual at $x = 0.1127$, 0.50 and 0.8873 produces a solution identical with the Galerkin method.

The solutions given in Table 5.3 are plotted in Fig. 5.2, and it is clear that weighted residual methods are effective (with the possible exception of the collocation method) in minimising the solution error over the whole domain. In contrast, the Taylor series expansion, which matches the exact solution at $x = 0.0$ produces a large solution error away from $x = 0.0$.

The various methods of weighted residuals are compared at greater length by Fletcher (1984, pp. 32–39) who makes the following observation: "the Galerkin method produces results of consistently high accuracy and has a breadth of application as wide as any method of weighted residuals". In the present situation the importance of the Galerkin method is that it leads directly to the Galerkin finite element method and the (Galerkin) spectral method.

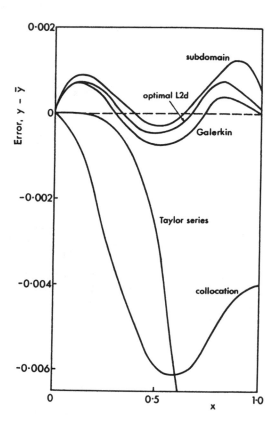

5.2 Finite Volume Method

This method is similar to the subdomain method, except that there is no explicit introduction of an approximate solution like (5.2). The method is particularly concise if only first derivatives are present (Sect. 5.2.1). If second derivatives occur (Sect. 5.2.2) then additional manipulation is required.

5.2.1 Equations with First Derivatives Only

Here the finite volume method will be illustrated for the general first-order equation

$$\frac{\partial \bar{q}}{\partial t} + \frac{\partial \bar{F}}{\partial x} + \frac{\partial \bar{G}}{\partial y} = 0 , \tag{5.23}$$

which by appropriate choice of $\bar{q}$, $\bar{F}$ and $\bar{G}$ represents the various equations of motion. For example if $\bar{q} = \bar{\varrho}$, $\bar{F} = \bar{\varrho}\bar{u}$ and $\bar{G} = \bar{\varrho}\bar{v}$ (5.23) is the two-dimensional version of the continuity equation (11.10).

Applying the subdomain method to (5.23) for the finite volume $ABCD$ shown in Fig. 5.3 gives

$$\int_{ABCD} 1 \left(\frac{\partial \bar{q}}{\partial t} + \frac{\partial \bar{F}}{\partial x} + \frac{\partial \bar{G}}{\partial y} \right) dx\, dy = 0 \ , \tag{5.24}$$

or applying Green's theorem,

$$\frac{d}{dt} \int \bar{q}\, dV + \int_{ABCD} \boldsymbol{H} \cdot \boldsymbol{n}\, ds = 0 \ , \tag{5.25}$$

where $\boldsymbol{H} = (\bar{F}, \bar{G})$. In Cartesian coordinates,

$$\boldsymbol{H} \cdot \boldsymbol{n}\, ds = \bar{F}\, dy - \bar{G}\, dx \ . \tag{5.26}$$

Equation (5.25) is just a statement of conservation. For the particular choice $\bar{q} = \bar{\varrho}$, $\bar{F} = \bar{\varrho}\bar{u}$, $\bar{G} = \bar{\varrho}\bar{v}$, (5.25) coincides with an integral statement of the conservation of mass. Consequently the finite volume method is a discretisation of the governing equation in integral form (Sect. 11.2), in contrast to the finite difference method, which is usually applied to the governing equation in differential form.

An approximate evaluation of (5.25) would be

$$\frac{d}{dt}(\mathscr{A}\, q_{j,k}) + \sum_{AB}^{DA} (F\varDelta y - G\varDelta x) = 0 \ , \tag{5.27}$$

where $\mathscr{A}$ is the area of the quadrilateral, $ABCD$ in Fig. 5.3, and the average value of q over the quadrilateral is associated with $q_{j,k}$. In (5.27),

$$\varDelta y_{AB} = y_B - y_A \ , \qquad \varDelta x_{AB} = x_B - x_A \quad \text{and}$$

$$F_{AB} = 0.5(F_{j,k-1} + F_{j,k}) \ , \qquad G_{AB} = 0.5(G_{j,k-1} + G_{j,k}) \ ,$$

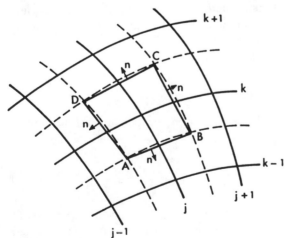

Fig. 5.3. Two-dimensional finite volume

with similar expressions for Δy_{BC}, etc. If $\mathscr{A}$ is not a function of time, (5.27) becomes

$$
\mathscr{A}\, dq_{j,k}/dt + 0.5(F_{j,k-1} + F_{j,k})\Delta y_{AB} - 0.5(G_{j,k-1} + G_{j,k})\Delta x_{AB}
$$
$$
+ 0.5(F_{j,k} + F_{j+1,k})\Delta y_{BC} - 0.5(G_{j,k} + G_{j+1,k})\Delta x_{BC}
$$
$$
+ 0.5(F_{j,k} + F_{j,k+1})\Delta y_{CD} - 0.5(G_{j,k} + G_{j,k+1})\Delta x_{CD}
$$
$$
+ 0.5(F_{j-1,k} + F_{j,k})\Delta y_{DA} - 0.5(G_{j-1,k} + G_{j,k})\Delta x_{DA} = 0 \ . \tag{5.28}
$$

If the global grid (j, k) is irregular the finite volume equation (5.28) provides a discretisation in Cartesian coordinates without the need to introduce generalised coordinates (Chap. 12). If the global grid is uniform and coincides with lines of constant x and y, (5.28) becomes

$$
\Delta x \Delta y \frac{d}{dt} q_{j,k} - 0.5(G_{j,k-1} + G_{j,k})\Delta x + 0.5(F_{j,k} + F_{j+1,k})\Delta y
$$
$$
+ 0.5(G_{j,k} + G_{j,k+1})\Delta x - 0.5(F_{j-1,k} + F_{j,k})\Delta y = 0 \ ,
$$
or
$$
\frac{d}{dt} q_{j,k} + \frac{F_{j+1,k} - F_{j-1,k}}{2\Delta x} + \frac{G_{j,k+1} - G_{j,k-1}}{2\Delta y} \ , \tag{5.29}
$$

which coincides with a centred difference representation for the spatial terms in (5.23).

The finite volume method, which has been used for both incompressible and compressible flow, has two major advantages. First it has good conservation (of mass, etc.) properties. Second it allows complicated computational domains to be discretised in a simpler way than either the isoparametric finite element formulation (Sect. 5.5.3) or generalised coordinates (Sect. 12.2).

5.2.2 Equations with Second Derivatives

In Sect. 5.2.1 the finite volume method is applied to (5.23), in which only first derivatives appear, and produces a relatively straightforward discretisation formula (5.28). If second derivatives are present in the governing equation the finite volume method requires some modification.

This situation is illustrated here by seeking solutions of Laplace's equation

$$
\frac{\partial^2 \bar{\phi}}{\partial x^2} + \frac{\partial^2 \bar{\phi}}{\partial y^2} = 0 \ , \tag{5.30}
$$

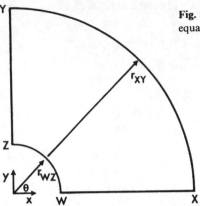

Fig. 5.4. Computational domain for the solution of Laplace's equation

in the computational domain shown in Fig. 5.4 with the following Dirichlet boundary conditions:

on WX, $\bar{\phi}=0$

on XY, $\bar{\phi}=\dfrac{\sin\theta}{r_{XY}}$ \hfill (5.31)

on YZ, $\bar{\phi}=\dfrac{1}{r_{YZ}}$

on ZW, $\bar{\phi}=\dfrac{\sin\theta}{r_{WZ}}$.

With this choice of boundary conditions, (5.30) has the exact solution

$$\bar{\phi}=\frac{\sin\theta}{r} , \tag{5.32}$$

which permits the accuracy of the computational solutions to be assessed directly.

If Laplace's equation (5.30) had been written in polar coordinates the computational domain shown in Fig. 5.4 would be regular. However, by deliberately solving this problem in Cartesian coordinates the ability of the finite volume method to handle distorted domains is illustrated, while retaining the advantage of an easily computed exact solution. Program FIVOL (Sect. 5.2.3) is sufficiently general to be applicable to other domain shapes once the interior grid points have been determined, e.g., using the techniques discussed in Sects. 13.2–4.

The finite volume method proceeds by applying the subdomain method to (5.30) for the finite volume $ABCD$ shown in Fig. 5.5, giving

$$\int\limits_{ABCD} 1\left(\frac{\partial^2\bar{\phi}}{\partial x^2}+\frac{\partial^2\bar{\phi}}{\partial y^2}\right)dx\,dy=\int\limits_{ABCD} H\cdot n\,ds=0 , \tag{5.33}$$

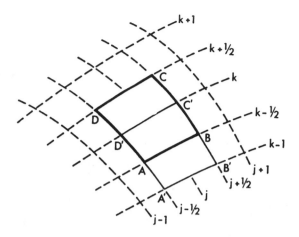

Fig. 5.5. Finite volume for a distorted grid

where

$$H \cdot n \, ds = \frac{\partial \overline{\phi}}{\partial x} dy - \frac{\partial \overline{\phi}}{\partial y} dx \ .$$

Following the same steps as in Sect. 5.2.1, (5.33) is evaluated approximately by

$$\left[\frac{\partial \phi}{\partial x} \right]_{j, k-1/2} \Delta y_{AB} - \left[\frac{\partial \phi}{dy} \right]_{j, k-1/2} \Delta x_{AB}$$
$$+ \left[\frac{\partial \phi}{\partial x} \right]_{j+1/2, k} \Delta y_{BC} - \left[\frac{\partial \phi}{\partial y} \right]_{j+1/2, k} \Delta x_{BC}$$
$$+ \left[\frac{\partial \phi}{\partial x} \right]_{j, k+1/2} \Delta y_{CD} - \left[\frac{\partial \phi}{\partial y} \right]_{j, k+1/2} \Delta x_{CD}$$
$$+ \left[\frac{\partial \phi}{\partial x} \right]_{j-1/2, k} \Delta y_{DA} - \left[\frac{\partial \phi}{\partial y} \right]_{j-1/2, k} \Delta x_{DA} = 0 \ . \tag{5.34}$$

Different techniques for evaluating $[\partial \phi / \partial x]_{j, k-1/2}$, etc. are discussed by Peyret and Taylor (1983, p. 111). Here $[\partial \phi / \partial x]_{j, k-1/2}$ is evaluated as the mean value over the area $B'BC'D'AA'B'$ in Fig. 5.5. Thus

$$\left[\frac{\partial \phi}{\partial x} \right]_{j, k-1/2} = \left(\frac{1}{S_{A'B'C'D'}} \right) \iint \left(\frac{\partial \phi}{\partial x} \right) dx \, dy = \left(\frac{1}{S_{A'B'C'D'}} \right) \int \phi \, dy \ , \tag{5.35}$$

$$\left[\frac{\partial \phi}{\partial y} \right]_{j, k-1/2} = \left(\frac{1}{S_{A'B'C'D'}} \right) \iint \left(\frac{\partial \phi}{\partial y} \right) dx \, dy = - \left(\frac{1}{S_{A'B'C'D'}} \right) \int \phi \, dx \ , \tag{5.36}$$

and

$$\int\limits_{A'B'C'D'} \phi \, dy \approx \phi_{j, k-1} \Delta y_{A'B'} + \phi_B \Delta y_{B'C'} + \phi_{j, k} \Delta y_{C'D'} + \phi_A \Delta y_{D'A'}$$

and similarly for

$$\int_{A'B'C'D'} \phi \, dx \; .$$

If the mesh is not too distorted,

$$\Delta y_{A'B'} \approx -\Delta y_{C'D'} \approx \Delta y_{AB} \quad \text{and} \quad \Delta y_{B'C'} \approx -\Delta y_{D'A'} \approx \Delta y_{k-1,k} \quad \text{and}$$

$$S_{AB} = S_{A'B'C'D'} = \Delta x_{AB} \Delta y_{k-1,k} - \Delta y_{AB} \Delta x_{k-1,k} \; . \tag{5.37}$$

Therefore (5.35) becomes

$$\left[\frac{\partial \phi}{\partial x}\right]_{j,k-1/2} = \frac{\Delta y_{AB}(\phi_{j,k-1} - \phi_{j,k}) + \Delta y_{k-1,k}(\phi_B - \phi_A)}{S_{AB}} \quad \text{and} \tag{5.38}$$

$$\left[\frac{\partial \phi}{\partial y}\right]_{j,k-1/2} = \frac{-[\Delta x_{AB}(\phi_{j,k-1} - \phi_{j,k}) + \Delta x_{k-1,k}(\phi_B - \phi_A)]}{S_{AB}} \; .$$

Developing equivalent expressions for $[\partial \phi/\partial x]_{j+1/2,k}$, etc. permits (5.34) to be written as

$$Q_{AB}(\phi_{j,k-1} - \phi_{j,k}) + P_{AB}(\phi_B - \phi_A) + Q_{BC}(\phi_{j+1,k} - \phi_{j,k}) + P_{BC}(\phi_C - \phi_B)$$

$$+ Q_{CD}(\phi_{j,k+1} - \phi_{j,k}) + P_{CD}(\phi_D - \phi_C) + Q_{DA}(\phi_{j-1,k} - \phi_{j,k}) + P_{DA}(\phi_A - \phi_D) \; ,$$

$$= 0 \; , \tag{5.39}$$

where

$$Q_{AB} = (\Delta x_{AB}^2 + \Delta y_{AB}^2)/S_{AB} \; , \qquad P_{AB} = (\Delta x_{AB}\Delta x_{k-1,k} + \Delta y_{AB}\Delta y_{k-1,k})/S_{AB}$$

$$Q_{BC} = (\Delta x_{BC}^2 + \Delta y_{BC}^2)/S_{BC} \; , \qquad P_{BC} = (\Delta x_{BC}\Delta x_{j+1,j} + \Delta y_{BC}\Delta y_{j+1,j})/S_{BC}$$

$$Q_{CD} = (\Delta x_{CD}^2 + \Delta y_{CD}^2)/S_{CD} \; , \qquad P_{CD} = (\Delta x_{CD}\Delta x_{k+1,k} + \Delta y_{CD}\Delta y_{k+1,k})/S_{CD} \quad \text{and}$$

$$Q_{DA} = (\Delta x_{DA}^2 + \Delta y_{DA}^2)/S_{DA} \; , \qquad P_{DA} = (\Delta x_{DA}\Delta x_{j-1,j} + \Delta y_{DA}\Delta y_{j-1,j})/S_{DA} \; .$$

In (5.39) ϕ_A, x_A, y_A are evaluated as the average of the four surrounding nodal values. Thus

$$\phi_A = 0.25(\phi_{j,k} + \phi_{j-1,k} + \phi_{j-1,k-1} + \phi_{j,k-1}) \; .$$

Substitution into (5.39) generates the following nine-point discretisation of (5.30):

$$0.25(P_{CD} - P_{DA})\phi_{j-1,k+1} + [Q_{CD} + 0.25(P_{BC} - P_{DA})]\phi_{j,k+1}$$

$$+ 0.25(P_{BC} - P_{CD})\phi_{j+1,k+1} + [Q_{DA} + 0.25(P_{CD} - P_{AB})]\phi_{j-1,k}$$

$$- (Q_{AB} + Q_{BC} + Q_{CD} + Q_{DA})\phi_{j,k} + [Q_{BC} + 0.25(P_{AB} - P_{CD})]\phi_{j+1,k}$$

$$+ 0.25(P_{DA} - P_{AB})\phi_{j-1,k-1} + [Q_{AB} + 0.25(P_{DA} - P_{BC})]\phi_{j,k-1}$$

$$+ 0.25(P_{AB} - P_{BC})\phi_{j+1,k-1} = 0 \; . \tag{5.40}$$

In (5.40) Q_{AB} and P_{AB}, etc. can be evaluated once and for all when the grid points are determined.

Equation (5.40) is solved conveniently using successive over-relaxation (SOR), Sect. 6.3. Equation (5.40) is manipulated to give an estimate of $\phi_{j,k}^{n+1}$, thus

$$
\begin{aligned}
\phi_{j,k}^* = \{ & 0.25(P_{CD}-P_{DA})\phi_{j-1,k+1} + [Q_{CD}+0.25(P_{BC}-P_{DA})]\phi_{j,k+1} \\
& + 0.25(P_{BC}-P_{CD})\phi_{j+1,k+1} + [Q_{DA}+0.25(P_{CD}-P_{AB})]\phi_{j-1,k} \\
& + [Q_{BC}+0.25(P_{AB}-P_{CD})]\phi_{j+1,k} \\
& + 0.25(P_{DA}-P_{AB})\phi_{j-1,k-1} + [Q_{AB}+0.25(P_{DA}-P_{BC})]\phi_{j,k-1} \\
& + 0.25(P_{AB}-P_{BC})\phi_{j+1,k-1}\}^n/(Q_{AB}+Q_{BC}+Q_{CD}+Q_{DA})
\end{aligned}
\tag{5.41}
$$

and the improved solution is

$$
\phi_{j,k}^{n+1} = \phi_{j,k}^n + \lambda(\phi_{j,k}^*-\phi_{j,k}^n) ,
\tag{5.42}
$$

where λ is the relaxation parameter (Sect. 6.3).

An interesting feature of the finite volume method is that Neumann (derivative) boundary conditions can be handled as readily as Dirichlet boundary conditions by direct substitution into (5.34).

5.2.3 FIVOL: Finite Volume Method Applied to Laplace's Equation

The above method has been incorporated into a computer program, FIVOL, a listing of which is given in Fig. 5.6. The main parameters used in FIVOL are described in Table 5.4.

FIVOL reads in and writes out the control parameters (lines 1–25). The grid is set up and the initial solution (for SOR) and the exact solution computed (lines 38–53). Since only the converged solution is of interest, the exact solution is also used as the initial solution for SOR. This leads to rapid convergence of the SOR scheme. The boundary conditions are set (lines 57–64); strictly this part of the code is not necessary since the boundary conditions have been evaluated, implicitly, by line 51.

The grid related parameters (Q_{AB}, P_{AB}, etc.) for each side are set (lines 68–123). The SOR iterative procedure evaluates (5.41) (lines 135–143), (5.42) at line 146 computes $\|\phi-\phi_n\|_{rms}$ (line 149) and exits the iteration loop (lines 127–151) if the tolerance (EPS) is satisfied. The final solution and exact solution are written out (lines 166–167) and $\|\phi-\bar{\phi}\|_{rms}$ computed and written out (lines 171–172).

Typical output for a 6×6 grid is shown in Fig. 5.7. By starting from the exact solution it is apparent that convergence of the SOR scheme is rapid. The large value of $\|\phi-\bar{\phi}\|_{rms}$ is mainly due to the large change in the solution close to r_{WZ}.

The accuracy increases as the grid is refined (Table 5.5) but more SOR iterations are required to reach convergence. The discretised equation (5.40) reduces to the centred finite difference scheme on a uniform rectangular grid

$$
\frac{\phi_{j-1,k}-2\phi_{j,k}+\phi_{j+1,k}}{\Delta x^2} + \frac{\phi_{j,k-1}-2\phi_{j,k}+\phi_{j,k+1}}{\Delta y^2} = 0 ,
\tag{5.43}
$$

```
1
2   C
3   C      FIVOL APPLIES THE FINITE VOLUME METHOD TO THE SOLUTION OF
4   C      LAPLACES EQUATION IN CARTESIAN COORDINATES ON A POLAR GRID.
5   C      THE DISCRETISED EQUATION IS SOLVED BY SOR
6   C
7          DIMENSION X(21,21),Y(21,21),QAB(21,21),PAB(21,21),QBC(21,21),
8         1PBC(21,21),QCD(21,21),PCD(21,21),QDA(21,21),PDA(21,21),
9         2PHI(21,21),PHIX(21,21)
10  C
11         OPEN(1,FILE='FIVOL.DAT')
12         OPEN(6,FILE='FIVOL.OUT')
13         READ(1,1)JMAX,KMAX,NMAX
14         READ(1,2)RW,RX,RY,RZ,THEB,THEN,EPS,OM
15       1 FORMAT(8I5)
16       2 FORMAT(8E10.3)
17  C
18         WRITE(6,3)
19         WRITE(6,4)JMAX,KMAX,NMAX,EPS,OM
20         WRITE(6,5)RW,RX,RY,RZ,THEB,THEN
21       3 FORMAT(' LAPLACE EQUATION BY FINITE VOLUME METHOD',//)
22       4 FORMAT(' JMAX=',I2,'  KMAX=',I2,'   NMAX=',I5,
23         15X,' EPS=',E10.3,'  OM=',F5.3)
24       5 FORMAT('   RW=',F5.3,'   RX=',F5.3,'   RY=',F5.3,'   RZ=',F5.3,
25         15X,'  THEB=',F5.1,'  THEN=',F5.1,//)
26  C
27         JMAP = JMAX - 1
28         KMAP = KMAX - 1
29         AJM = JMAP
30         AKM = KMAP
31         DRWX = (RX - RW)/AJM
32         DRZY = (RY - RZ)/AKM
33         DTH = (THEN-THEB)/AKM
34         PI = 3.1415927
35  C
36  C      SET X, Y, EXACT AND INITIAL PHI
37  C
38         DO 7 K = 1,KMAX
39         AK = K - 1
40         THK = (THEB + AK*DTH)*PI/180.
41         CK = COS(THK)
42         SK = SIN(THK)
43         DR = DRWX + (DRZY - DRWX)*AK/AKM
44         RWZ = RW + (RZ - RW)*AK/AKM
45         DO 6 J = 1,JMAX
46         AJ = J - 1
47         R = RWZ + AJ*DR
48         X(J,K) = R*CK
49         Y(J,K) = R*SK
50         PHIX(J,K) = SK/R
51         PHI(J,K) = PHIX(J,K)
52       6 CONTINUE
53       7 CONTINUE
54  C
55  C      SET BOUNDARY VALUES OF PHI
56  C
57         DO 8 J = 1,JMAX
58         PHI(J,1) = 0.
59         PHI(J,KMAX) = PHIX(J,KMAX)
60       8 CONTINUE
```

Fig. 5.6. Listing of program FIVOL

```
61          DO 9 K = 1,KMAX
62          PHI(1,K) = PHIX(1,K)
63          PHI(JMAX,K) = PHIX(JMAX,K)
64        9 CONTINUE
65 C
66 C        SET GRID RELATED PARAMETERS
67 C
68          DO 11 K = 2,KMAP
69          KM = K - 1
70          KP = K + 1
71          DO 10 J = 2,JMAP
72          JM = J - 1
73          JP = J + 1
74          XA = 0.25*(X(J,K) + X(JM,K) + X(JM,KM) + X(J,KM))
75          YA = 0.25*(Y(J,K) + Y(JM,K) + Y(JM,KM) + Y(J,KM))
76          XB = 0.25*(X(J,K) + X(J,KM) + X(JP,KM) + X(JP,K))
77          YB = 0.25*(Y(J,K) + Y(J,KM) + Y(JP,KM) + Y(JP,K))
78          XC = 0.25*(X(J,K) + X(JP,K) + X(JP,KP) + X(J,KP))
79          YC = 0.25*(Y(J,K) + Y(JP,K) + Y(JP,KP) + Y(J,KP))
80          XD = 0.25*(X(J,K) + X(J,KP) + X(JM,KP) + X(JM,K))
81          YD = 0.25*(Y(J,K) + Y(J,KP) + Y(JM,KP) + Y(JM,K))
82 C
83 C        SIDE AB
84 C
85          DXA = XB - XA
86          DYA = YB - YA
87          DXK = X(J,K) - X(J,K-1)
88          DYK = Y(J,K) - Y(J,K-1)
89          SAB = ABS(DXA*DYK - DXK*DYA)
90          QAB(J,K) = (DXA*DXA + DYA*DYA)/SAB
91          PAB(J,K) = (DXA*DXK + DYA*DYK)/SAB
92 C
93 C        SIDE BC
94 C
95          DXB = XC - XB
96          DYB = YC - YB
97          DXJ = X(J,K) - X(J+1,K)
98          DYJ = Y(J,K) - Y(J+1,K)
99          SBC = ABS(DYJ*DXB - DXJ*DYB)
100         QBC(J,K) = (DXB*DXB + DYB*DYB)/SBC
101         PBC(J,K) = (DXB*DXJ + DYB*DYJ)/SBC
102 C
103 C       SIDE CD
104 C
105         DXC = XD - XC
106         DYC = YD - YC
107         DXK = X(J,K) - X(J,K+1)
108         DYK = Y(J,K) - Y(J,K+1)
109         SCD = ABS(DXC*DYK - DYC*DXK)
110         QCD(J,K) = (DXC*DXC + DYC*DYC)/SCD
111         PCD(J,K) = (DXC*DXK + DYC*DYK)/SCD
112 C
113 C       SIDE DA
114 C
115         DXD = XA - XD
116         DYD = YA - YD
117         DXJ = X(J,K) - X(J-1,K)
118         DYJ = Y(J,K) - Y(J-1,K)
```

Fig. 5.6. (cont.) Listing of program FIVOL

```
119          SDA = ABS(DXJ*DYD - DYJ*DXD)
120          QDA(J,K) = (DXD*DXD + DYD*DYD)/SDA
121          PDA(J,K) = (DXD*DXJ + DYD*DYJ)/SDA
122       10 CONTINUE
123       11 CONTINUE
124    C
125    C     ITERATE USING SOR
126    C
127          DO 14 N = 1,NMAX
128          SUM = 0.
129          DO 13 K = 2,KMAP
130          KM = K - 1
131          KP = K + 1
132          DO 12 J = 2,JMAP
133          JM = J - 1
134          JP = J + 1
135          PHD = 0.25*(PCD(J,K)-PDA(J,K))*PHI(JM,KP)
136          PHD = PHD + (QCD(J,K) + 0.25*(PBC(J,K)-PDA(J,K)))*PHI(J,KP)
137          PHD = PHD + 0.25*(PBC(J,K)-PCD(J,K))*PHI(JP,KP)
138          PHD = PHD + (QDA(J,K) + 0.25*(PCD(J,K)-PAB(J,K)))*PHI(JM,K)
139          PHD = PHD + (QBC(J,K) + 0.25*(PAB(J,K)-PCD(J,K)))*PHI(JP,K)
140          PHD = PHD + 0.25*(PDA(J,K) - PAB(J,K))*PHI(JM,KM)
141          PHD = PHD + (QAB(J,K) + 0.25*(PDA(J,K)-PBC(J,K)))*PHI(J,KM)
142          PHD = PHD + 0.25*(PAB(J,K) - PBC(J,K))*PHI(JP,KM)
143          PHD = PHD/(QAB(J,K)+QBC(J,K)+QCD(J,K)+QDA(J,K))
144          DIF = PHD - PHI(J,K)
145          SUM = SUM + DIF*DIF
146          PHI(J,K) = PHI(J,K) + OM*DIF
147       12 CONTINUE
148       13 CONTINUE
149          RMS = SQRT(SUM/(AJM-1.)/(AKM-1.))
150          IF(RMS .LT. EPS)GOTO 16
151       14 CONTINUE
152          WRITE(6,15)NMAX,RMS
153       15 FORMAT(' CONVERGENCE NOT ACHIEVED IN',I5,' STEPS',5X,'  RMS=',
154          1E12.5)
155    C
156    C     COMPARE SOLUTION WITH EXACT
157    C
158       16 SUM = 0.
159          DO 21 K = 1,KMAX
160          WRITE(6,17)K
161       17 FORMAT(/,'  K=',I2)
162          DO 18 J = 1,JMAX
163          DIF = PHI(J,K) - PHIX(J,K)
164          SUM = SUM + DIF*DIF
165       18 CONTINUE
166          WRITE(6,19)(PHI(J,K),J=1,JMAX)
167          WRITE(6,20)(PHIX(J,K),J=1,JMAX)
168       19 FORMAT('  PHI=',10F7.4)
169       20 FORMAT('  PHX=',10F7.4)
170       21 CONTINUE
171          RMS = SQRT(SUM/(AJM-1.)/(AKM-1.))
172          WRITE(6,22)N,RMS
173       22 FORMAT(/,' CONVERGED AFTER ',I3,'  STEPS',4X,'  RMS=',E12.5)
174          STOP
175          END
```

Fig. 5.6. (cont.) Listing of program FIVOL

Table 5.4. Parameters used in program FIVOL

Parameter	Description
JMAX	Number of points in the radial direction
KMAX	Number of points in the circumferential direction
NMAX	Maximum number of iterations
RW	r_W (Fig. 5.4)
RX	r_X (Fig. 5.4)
RY	r_Y (Fig. 5.4)
RZ	r_Z (Fig. 5.4)
THEB	θ_{WX} (Fig. 5.4)
THEN	θ_{ZY} (Fig. 5.4)
EPS	Tolerance for convergence of SOR
OM	Relaxation parameter λ in (5.42)
PHI	ϕ
PHIX	$\bar{\phi}$
PHD	ϕ^*
XA, YA	x_A, y_A
DXA	Δx_{AB}
DXK	$\Delta x_{k-1,k}$
RMS	$\|\phi^* - \phi^n\|_{\text{rms}}$, $\|\phi^{n+1} - \bar{\phi}\|_{\text{rms}}$

```
LAPLACE EQUATION BY FINITE VOLUME METHOD


JMAX= 6  KMAX= 6   NMAX=   50          EPS=  .100E-04  OM=1.500
RW= .100   RX=1.000   RY=1.000  RZ= .100       THEB=    .0  THEN= 90.0



K= 1
PHI=  .0000   .0000   .0000   .0000   .0000   .0000
PHX=  .0000   .0000   .0000   .0000   .0000   .0000

K= 2
PHI= 3.0902 1.2439   .7445   .5205   .3929   .3090
PHX= 3.0902 1.1036   .6718   .4828   .3768   .3090

K= 3
PHI= 5.8779 2.3473 1.4036   .9828   .7441   .5878
PHX= 5.8779 2.0992 1.2778   .9184   .7168   .5878

K= 4
PHI= 8.0902 3.1786 1.8990 1.3345 1.0163   .8090
PHX= 8.0902 2.8893 1.7587 1.2641   .9866   .8090

K= 5
PHI= 9.5106 3.6209 2.1680 1.5349 1.1802   .9511
PHX= 9.5106 3.3966 2.0675 1.4860 1.1598   .9511

K= 6
PHI=10.0000 3.5714 2.1739 1.5625 1.2195 1.0000
PHX=10.0000 3.5714 2.1739 1.5625 1.2195 1.0000

CONVERGED AFTER  15  STEPS   RMS=  .13259E+00
```

Fig. 5.7. Typical output from FIVOL

Table 5.5. Finite volume solution errors with grid refinement
$(r_W = r_Z = 0.1, r_X = r_Y = 1.0, \theta_{WX} = 0, \theta_{ZY} = 90, \lambda = 1.5)$

Grid	$\| \phi - \bar{\phi} \|_{rms}$	No of iterations to convergence
6 × 6	0.1326	15
11 × 11	0.0471	19
21 × 21	0.0138	51

which has second-order convergence, i.e. halving the grid (if fine enough) reduces the error by a factor of 4 (2^2). For the non-uniform grid results shown in Table 5.5 the convergence rate is less than second order. Increasing grid distortion is expected to reduce the convergence rate further (see Problem 5.4).

The finite volume method is widely used for transonic inviscid flow (Sect. 14.3) and for viscous flow (Sect. 17.2.3).

5.3 Finite Element Method and Interpolation

The finite element method was developed initially as an adhoc engineering procedure for constructing matrix solutions to stress and displacement calculations in structural analysis. The method was placed on a sound mathematical foundation by considering the potential energy of the system and giving the finite element method a variational interpretation.

However, very few fluid dynamic (or heat transfer) problems can be expressed in a variational form. But for many situations, the Galerkin method is equivalent to the Ritz method for solving variational problems. Consequently most of the finite element applications in fluid dynamics have used the Galerkin finite element formulation. Here we will focus exclusively on the Galerkin formulation of the finite element method.

A traditional engineering interpretation of the finite element method is given by Zienkiewicz (1977). A more mathematical perspective of the method is provided by Strang and Fix (1973), Oden and Reddy (1976) and Mitchell and Wait (1977). Computational techniques for mainly structural applications of the finite element method can be found in Bathe and Wilson (1976). The application of the traditional finite element method to fluid mechanics is treated by Thomasset (1981) and Baker (1983).

In comparison with the (traditional) Galerkin method defined in Sect. 5.1, the Galerkin finite element method has two particularly important features. Firstly the approximate solution (5.2) is written directly in terms of the nodal unknowns, i.e.

$$T = \sum_{j=1}^{J} T_j \phi_j(x, y, z) . \tag{5.44}$$

Equation (5.44) can be interpreted as an interpolation of the local nodal point solution T_j. By working directly with the nodal unknowns one level of computation is saved at the equation-solving stage (Chap. 6) and the nodal solution has direct physical significance. Also the similarity with the finite difference method is easier to see and the interpretation of the finite element method as a means of discretising (Sect. 3.1) the (continuous) governing partial differential equation becomes more obvious.

The approximating functions, $\phi_j(x, y, z)$ in (5.44), are often called trial or interpolation functions in the mathematical literature and shape functions (with the symbol N_j replacing ϕ_j) in the engineering literature. The second important feature is that the approximating functions $\phi_j(x, y, z)$ are chosen, almost exclusively, from low-order piecewise polynomials restricted to contiguous elements. This produces relatively few non-zero terms which, with appropriate ordering (Jennings 1977, Chap. 5), can be located close to the main diagonal of the matrix of equations. This is important at the equation-solving stage (Chap. 6) since it implies that the solution of the matrix of equations will be more economical.

The finite element method achieves discretisation in two stages, both of which introduce errors (but not always additive errors). First a piecewise interpolation is introduced over discrete or finite elements to connect the local solution to the nodal values. This aspect will be examined in this section. The second stage uses a weighted residual construction (Sect. 5.1) to obtain algebraic equations connecting the solution nodal values. The mechanics of implementing the second stage and the solution behaviour are illustrated in developing programs STURM (Sect. 5.4.2) and DUCT (Sect. 5.5.2).

5.3.1 Linear Interpolation

Linear one-dimensional approximating functions, $\phi_j(x)$, are shown in Fig. 5.8. The function $\phi_j(x)$ takes the values

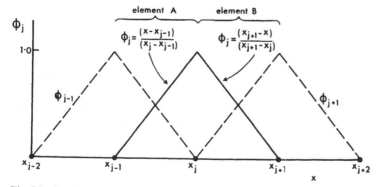

Fig. 5.8. One-dimensional linear approximating functions

$$\phi_j = 0 \qquad \text{for } x < x_{j-1} \ ,$$

$$\phi_j = \frac{x - x_{j-1}}{x_j - x_{j-1}} \qquad \text{for } x_{j-1} \leqq x \leqq x_j \text{ (element } A) \ , \tag{5.45}$$

$$\phi_j = 1 \qquad \text{for } x = x_j \ ,$$

$$\phi_j = \frac{x_{j+1} - x}{x_{j+1} - x_j} \qquad \text{for } x_j \leqq x \leqq x_{j+1} \text{ (element } B) \ , \tag{5.46}$$

$$\phi_j = 0 \qquad \text{for } x > x_{j+1} \ .$$

Thus ϕ_j is non-zero only for the range $x_{j-1} < x < x_{j+1}$. In a similar manner the approximating function ϕ_{j+1} is non-zero only between nodes j and $j+2$.

For linear approximating functions, (5.44) has the form

$$T = T_{j-1} \, \phi_{j-1} + T_j \phi_j \quad \text{in element } A \ , \tag{5.47}$$

$$T = T_j \phi_j + T_{j+1} \, \phi_{j+1} \quad \text{in element } B \ ,$$

i.e. only two terms contribute in each element due to the local nature of the approximating functions (Fig. 5.8). In element A, ϕ_j is given by (5.45) and ϕ_{j-1} by

$$\phi_{j-1} = \frac{x - x_j}{x_{j-1} - x_j} \ . \tag{5.48}$$

In element B, ϕ_j is given by (5.46) and ϕ_{j+1} by

$$\phi_{j+1} = \frac{x - x_j}{x_{j+1} - x_j} \ . \tag{5.49}$$

The above construction will be used to obtain a linear interpolation of the function

$$\bar{y} = 1 + \cos(0.5\pi x) + \sin(\pi x) \tag{5.50}$$

in the range $0 \leqq x \leqq 1$. Equation (5.50) is shown in Fig. 5.9. The nodal values y_j are obtained from the exact solution $y_j = \bar{y}(x_j)$. In a particular element the local interpolated solution y_{in} is given by

$$y_{\text{in}} = y_{j-1} \, \phi_{j-1}(x) + y_j \phi_j(x) \quad \text{in element } A \quad \text{and} \tag{5.51}$$

$$y_{\text{in}} = y_j \phi_j(x) + y_{j+1} \, \phi_{j+1}(x) \quad \text{in element } B \ .$$

In element A, ϕ_{j-1} and ϕ_j are given by (5.48) and (5.45), respectively. In element B, ϕ_j and ϕ_{j+1} are given by (5.46) and (5.49).

The linearly interpolated solution (5.51) is plotted in Fig. 5.9. For a two-element division of the interval $0 \leqq x \leqq 1$ the interpolated solution is clearly rather inaccurate at the element midpoints, e.g. at

$$x = 0.25 \ , \quad \bar{y} = 2.63099 \ , \quad y_{\text{in}} = 2.35356 \ ,$$

$$x = 0.75 \ , \quad \bar{y} = 2.08979 \ , \quad y_{\text{in}} = 1.85355 \ .$$

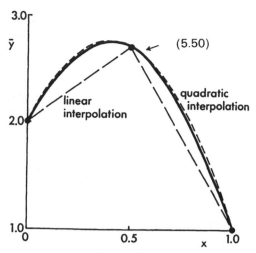

Fig. 5.9 One-dimensional finite element interpolation

However, the accuracy improves rapidly as the number of elements subdividing the interval $0 \leq x \leq 1$ is increased (Table 5.6). The rms error for linear interpolation, Table 5.6, is reducing like Δx^2, where Δx is the span of each element. This is typical of linear interpolation.

Table 5.6. Finite element interpolation errors for (5.50)

Linear interpolation		Quadratic interpolation	
Number of elements	rms error	Number of elements	rms error
2	0.18662	1	0.04028
4	0.04786	2	0.01599
6	0.02138	3	0.00484
8	0.01204	4	0.00206

At this point the reader seeing this material for the first time may prefer to jump over quadratic interpolation and two-dimensional interpolation, and to pursue the discretisation process and equation solution for the Sturm-Liouville equation (Sect. 5.4).

5.3.2 Quadratic Interpolation

The use of linear interpolation, as in Figs. 5.8, 9, imposes a constraint on the approximate solution that it must vary linearly between the nodal points. We see (Table 5.6) that this introduces an error on a finite grid. Estimates for such (interpolation) errors are given by Mitchell and Wait (1977, p. 119), which are useful when the exact solution is not available. For the same grid spacing we would expect a smaller interpolation error if quadratic interpolation is used.

As a one-dimensional quadratic approximating function, ϕ_j (Fig. 5.10) takes the values

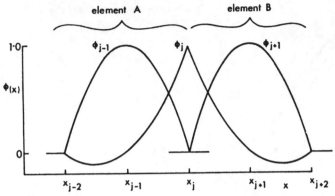

Fig. 5.10. One-dimensional quadratic approximating functions

$$\phi_j = 0 \qquad\qquad\qquad \text{for } x < x_{j-2} \ ,$$

$$\phi_j = \left(\frac{x - x_{j-2}}{x_j - x_{j-2}}\right)\left(\frac{x - x_{j-1}}{x_j - x_{j-1}}\right) \quad \text{for } x_{j-2} \leqq x \leqq x_j \text{ (element } A) \ , \qquad (5.52)$$

$$\phi_j = \left(\frac{x - x_{j+1}}{x_j - x_{j+1}}\right)\left(\frac{x - x_{j+2}}{x_j - x_{j+2}}\right) \quad \text{for } x_j \leqq x \leqq x_{j+2} \text{ (element } B) \ , \qquad (5.53)$$

$$\phi_j = 0 \qquad\qquad\qquad \text{for } x > x_{j+2} \ .$$

The approximating functions are recognisable as Lagrange interpolation functions. Thus $\phi_j = 1$ at node j and $\phi_j = 0$ at nodes $j-2, j-1, j+1$ and $j+2$. For quadratic interpolation (5.44) takes the following form in element A:

$$T = T_{j-2}\phi_{j-2} + T_{j-1}\phi_{j-1} + T_j\phi_j \ , \qquad (5.54)$$

and in element B

$$T = T_j\phi_j + T_{j+1}\phi_{j+1} + T_{j+2}\phi_{j+2} \ , \qquad (5.55)$$

i.e. only three terms contribute in each element.

For the sample problem (5.50) a suitable quadratic interpolation is

$$\begin{aligned} y_{\text{in}} &= y_{j-2}\,\phi_{j-2}(x) + y_{j-1}\,\phi_{j-1}(x) + y_j\,\phi_j(x) \quad \text{in element } A \ , \\ y_{\text{in}} &= y_j\,\phi_j(x) + y_{j+1}\,\phi_{j+1}(x) + y_{j+2}\,\phi_{j+2}(x) \quad \text{in element } B \ , \end{aligned} \qquad (5.56)$$

where. in element A $(x_{j-2} \leqq x \leqq x_j)$,

$$\phi_{j-2} = \left(\frac{x - x_{j-1}}{x_{j-2} - x_{j-1}}\right)\left(\frac{x - x_j}{x_{j-2} - x_j}\right) \ ,$$

$$\phi_{j-1} = \left(\frac{x - x_{j-2}}{x_{j-1} - x_{j-2}}\right)\left(\frac{x - x_j}{x_{j-1} - x_j}\right) \ , \qquad (5.57)$$

$$\phi_j = \left(\frac{x - x_{j-2}}{x_j - x_{j-2}}\right)\left(\frac{x - x_{j-1}}{x_j - x_{j-1}}\right) \ ,$$

and similarly for element B $(x_j \leqq x \leqq x_{j+2})$.

The case of a single quadratic element spanning the interval $0 \leq x \leq 1$ is shown in Fig. 5.9. The quadratically interpolated solution is clearly more accurate than the linearly interpolated solution with the same number of nodes. The effect of grid refinement (increasing the number of quadratic elements) is to rapidly reduce the rms error (Table 5.6). The rate of reduction shown in Table 5.6 is like Δx^3 for quadratic elements. Consequently quadratic interpolation becomes progressively more accurate, compared with linear interpolation, as the grid is refined.

In principle, cubic and higher-order interpolation are also available, in practice it is unusual to use interpolation that is of higher order than quadratic. Although higher-order interpolation is more accurate it generates systems of equations that have many more non-zero terms than when low-order interpolation is used; consequently higher-order interpolation is computationally more expensive, particularly in two and three dimensions. Achieving the proper balance between accuracy and economy, i.e. in choosing methods that are computationally efficient (Sect. 4.5), is one of the more interesting strategic considerations in computational fluid dynamics.

We can expect the finite element method solution error to reduce at the same rate as the interpolation error, with grid refinement. Generally the solution error will be larger on a given grid, since there is an additional error due to the nodal point solution not coinciding with the exact solution. As a rough guide, the use of linear approximating functions generates solutions of about the same accuracy as produced by second-order finite difference methods and the use of quadratic approximating functions provides about the same accuracy as third-order finite difference methods.

5.3.3 Two-Dimensional Interpolation

The concept of finite elements, from which the name of the method is taken, becomes more useful in more than one dimension. Thus in two dimensions, the local solution is interpolated, separately, in each of the four elements A, B, C and D, which surround the (j, k)-th node (Fig. 5.11). The approximate solution, equivalent to (5.44), is expressed conveniently in terms of an element-based coordinate system (ξ, η). For bilinear interpolation the approximate solution can be written

$$T = \sum_{l=1}^{4} T_l \phi_l(\xi, \eta) \,, \tag{5.58}$$

where in each element $-1 \leq \xi \leq 1$, $-1 \leq \eta \leq 1$ (Fig. 5.11). From (5.58) four terms contribute in each element. The approximating function $\phi_l(\xi, \eta)$ takes the form

$$\phi_l(\xi, \eta) = 0.25(1 + \xi_l \xi)(1 + \eta_l \eta) \,. \tag{5.59}$$

For example, in element C of Fig. 5.15

Fig. 5.11. Global/local nodal numbering

$$\phi_1 = 0.25(1 - \xi)(1 - \eta) \ ,$$
$$\phi_2 = 0.25(1 + \xi)(1 - \eta) \ ,$$
$$\phi_3 = 0.25(1 + \xi)(1 + \eta) \ ,$$
$$\phi_4 = 0.25(1 - \xi)(1 + \eta) \ ,$$

(5.60)

so that $\phi_l = 1$ when $\xi = \xi_l$, $\eta = \eta_l$ and $\phi_l = 0$ at all other nodes. For lines of constant ξ or η the approximating functions, ϕ_l, demonstrate the same type of variation as the one-dimensional approximating functions (Fig. 5.8).

A three-dimensional view of a bilinear interpolating function centered at the (j, k)-th node is shown in Fig. 5.12. An examination of (5.58) and Fig. 5.12 indicates

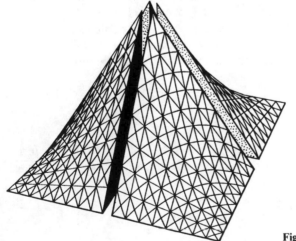

Fig. 5.12. Bilinear interpolation function

that T is continuous as element boundaries are crossed; however derivatives of T are not.

Just as in one dimension, two-dimensional (bi)quadratic interpolating functions may be defined within a particular element (Fig. 5.13). For Lagrange elements the approximate solution, equivalent to (5.44), takes the form

$$T = \sum_{l=1}^{9} T_l \phi(\xi, \eta) ,$$
(5.61)

where the biquadratic Lagrange approximating functions are given by

corner nodes: $\qquad\qquad \phi_l(\xi, \eta) = 0.25\, \xi_l \xi (1 + \xi_l \xi) \eta_l \eta (1 + \eta_l \eta)$,

midside nodes $(\xi_l = 0)$: $\quad \phi_l(\xi, \eta) = 0.5(1 - \xi^2)\eta_l \eta(1 + \eta_l \eta)$,

midside nodes $(\eta_l = 0)$: $\quad \phi_l(\xi, \eta) = 0.5(1 - \eta^2)\xi_l \xi(1 + \xi_l \xi)$,   (5.62)

internal node: $\qquad\qquad \phi_l(\xi, \eta) = (1 - \xi^2)(1 - \eta^2)$.

The quadratic Lagrange approximating functions are interpolatory as before, i.e.

$$\phi_l = 1 \quad \text{when} \quad \xi = \xi_l , \quad \eta = \eta_l ,$$

$$\phi_l = 0 \quad \text{at all other nodes} .$$

The form of the approximating function (5.62) can be visualised by considering constant values of ξ or η; then the approximating function resembles a one-dimensional approximating function, Fig. 5.10. This follows from the fact that the two-dimensional Lagrange approximating functions are just products of the cor-

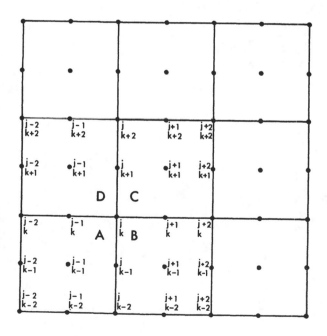

Fig. 5.13. Two-dimensional quadratic Lagrange elements

responding one-dimensional Lagrange approximating functions. This has important ramifications for interpolation in more than one dimension, particularly in
relation to constructing dimensionally split schemes (Sect. 8.3).

Two-dimensional interpolation on finite elements will be illustrated here for the
function

$$\bar{f} = [1 - 0.8\cos(0.5\pi x)][\cos(0.5\pi y)] \ , \tag{5.63}$$

which is shown in Fig. 5.14 for the domain $-1 \leq x \leq 1, \ -1 \leq y \leq 1$.

This domain can be conveniently split into four elements (Fig. 5.15) over which
the solution is to be interpolated (bi)linearly in terms of the nine nodal values $f_l = \bar{f}_l$.

In a particular element, say C, the interpolated solution f_{in} is given by

$$f_{in} = \sum_{l=1}^{4} f_l \phi_l(\xi, \eta) \ , \tag{5.64}$$

where (ξ, η) is a local element-based coordinate system. For the element C,

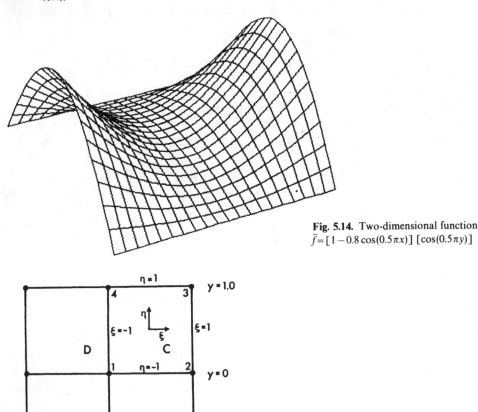

Fig. 5.14. Two-dimensional function
$\bar{f} = [1 - 0.8\cos(0.5\pi x)][\cos(0.5\pi y)]$

Fig. 5.15. Linear elements in two dimensions

$\xi = 2x - 1$ and $\eta = 2y - 1$.

The (bi)linear interpolating functions ϕ_l are given by (5.60) and more generally by (5.59).

The interpolated solution with only four elements is not very accurate. For example at

$$x = 0.5 , \quad y = 0.5 , \quad \bar{f} = 0.45711 , \quad f_{in} = 0.375 .$$

The complete interpolated solution of (5.63) is shown in Fig. 5.16. However the accuracy of the solution increases rapidly as the number of elements spanning the domain is increased (Table 5.7). The bilinear interpolation results shown in Table 5.7 indicate that the rms error is reducing like $\Delta x^2 (= \Delta y^2)$.

A more accurate interpolation may be obtained using (bi)quadratic interpolation. Each element then has nine nodal values (Fig. 5.13) which are interpolated by

$$f_{in} = \sum_{l=1}^{9} f_l \phi_l(\xi, \eta) , \tag{5.65}$$

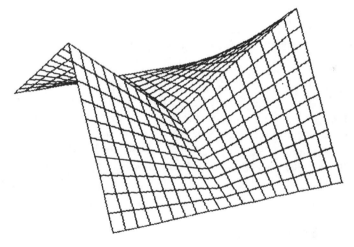

Fig. 5.16. (Bi)linear interpolation of (5.63)

Table 5.7. Errors for two-dimensional finite element interpolation

Bilinear interpolation		Biquadratic interpolation	
Number of elements	rms error	Number of elements	rms error
4	0.08527	—	—
16	0.02199	4	0.01199
64	0.00557	16	0.00151
		64	0.00019

where the interpolating functions $\phi_l(\xi, \eta)$ are given by (5.62). As indicated in Table 5.7 elements with biquadratic interpolation are more accurate than elements with bilinear interpolation, for the same number of nodal points. With reference to Table 5.7, 16 contiguous linear elements contain the same number of nodal points as 4 contiguous quadratic elements. As more quadratically interpolated elements are used to span the domain $-1 \leq x \leq 1$, $-1 \leq y \leq 1$, the rms error reduces like $\Delta x^3 (= \Delta y^3)$.

Lagrange approximating functions can be introduced in three dimensions. For trilinear interpolation eight-node brick elements replace four-node rectangular elements in two dimensions. For triquadratic interpolation, 27-node brick elements replace 9-node rectangular elements in two dimensions (Zienkiewicz 1977, p. 171). In two and three dimensions, cubic and higher-order interpolation is possible but not of much interest due to the large number of nodes that become coupled (Fletcher 1984, p. 96).

It is possible to eliminate the central node from the quadratic rectangular element shown in Fig. 5.13; this leads to Serendipity elements and interpolation in which no internal nodes are present. Such elements are possible for quadratic or higher-order interpolation in any number of dimensions. Serendipity elements are described by Zienkiewicz (1977, p. 155).

5.4 Finite Element Method and the Sturm-Liouville Equation

In Sect. 5.1.1 the Galerkin method is defined as a member of the class of methods of weighted residuals. In introducing the form of the approximate (interpolating) solution for the finite element method, i.e. (5.44), it is stressed that the nodal values are the unknowns and the approximating functions are low-order piecewise polynomials. Here the Galerkin finite element method will be applied to the Sturm-Liouville equation, to illustrate the mechanics of the method.

5.4.1 Detailed Formulation

A simplified Sturm-Liouville equation can be written as

$$\frac{d^2 \bar{Y}}{dx^2} + \bar{Y} = F = - \sum_{l=1}^{L} a_l \sin(l - 0.5)\pi x \tag{5.66}$$

with boundary conditions

$$\bar{Y}(0) = 0 \quad \text{and} \quad \frac{d\bar{Y}}{dx}(1) = 0 . \tag{5.67}$$

For the particular form of F shown in (5.66), the exact solution is

$$\bar{Y} = - \sum_{l=1}^{L} \frac{a_l}{1 - [(l - 0.5)\pi]^2} \sin[(l - 0.5)\pi x] . \tag{5.68}$$

The Galerkin finite element method starts by introducing a trial solution, equivalent to (5.44), for the approximate solution Y as

$$Y = \sum_{j=1}^{J} Y_j \phi_j(x) ,$$

(5.69)

and $\phi_j(x)$ are linear approximating functions that can be expressed, conveniently, in an element-based coordinate system $x(\xi)$ as (Fig. 5.17):

in element A, $\phi_j(\xi) = 0.5(1 + \xi)$, $\xi = \dfrac{2[x - (x_{j-1} + x_j)/2]}{\Delta x_j}$;

in element B, $\phi_j(\xi) = 0.5(1 - \xi)$, $\xi = \dfrac{2[x - (x_j + x_{j+1})/2]}{\Delta x_{j+1}}$.

(5.70)

Substitution of (5.69) into (5.5) produces an equation residual

$$R = \frac{d^2 Y}{dx^2} + Y - F .$$

(5.71)

Application of the weighted residual equation (5.5) with $W_m(x) = \phi_m(x)$ gives

$$\int_0^1 \phi_m(x) \left(\frac{d^2 Y}{dx^2} + Y - F \right) dx = 0 .$$

(5.72)

It may be noted that the integration is made over the domain $0 \le x \le 1$. Since the approximate solution is expressed, in (5.69) in terms of nodal values Y_j, the highest derivative permitted to appear in (5.72) is the first (Mitchell and Wait 1977, p. 51). Therefore an integration by parts is applied as follows:

$$\int_0^1 \left(\phi_m \frac{d^2 Y}{dx^2} \right) dx = \left[\phi_m \frac{dY}{dx} \right]_0^1 - \int_0^1 \left(\frac{d\phi_m}{dx} \right) \left(\frac{dY}{dx} \right) dx .$$

(5.73)

Since $Y(0) = 0$, no equation need be formed with $m = 0$ (Fig. 5.17). For $m \ge 1$, $\phi_m(0) = 0$. At $x = 1$, $dY/dx = 0$ therefore (5.73) becomes

$$\int_0^1 \left(\phi_m \frac{d^2 Y}{dx^2} \right) dx = - \int_0^1 \left(\frac{d\phi_m}{dx} \right) \left(\frac{dY}{dx} \right) dx .$$

(5.74)

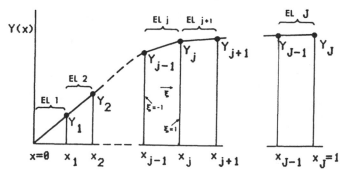

Fig. 5.17. Sturm-Liouville problem

Substituting (5.74 and 69) into (5.72), and rearranging, gives

$$\sum_{j=1}^{J} \left[\int_0^1 \left(\frac{-d\phi_m}{dx} \frac{d\phi_j}{dx} + \phi_m \phi_j \right) dx \right] Y_j = \int_0^1 \phi_m F \, dx \ , \tag{5.75}$$

$$m = 1, \ldots, J \ .$$

Since all the terms in the integrals are known, (5.75) is really a system of algebraic equations that can be written

$$\underline{B} Y = G \ , \tag{5.76}$$

where $b_{m,j}$ is an element of $\underline{B}$ given by

$$b_{m,j} = \int_0^1 \left(\frac{-d\phi_m}{dx} \frac{d\phi_j}{dx} + \phi_m \phi_j \right) dx \ , \tag{5.77}$$

Y_j is an element of Y and g_m is an element of G, i.e.

$$g_m = \int_0^1 \phi_m F \, dx \ . \tag{5.78}$$

Equation (5.77) is most easily evaluated by converting the integral to one over ξ for each element in turn (Fig. 5.17). Because ϕ_m is also given by (5.70) only the elements adjacent to the $j = m$th node produce a non-zero contribution. For node $m = J$ only one element contributes to the evaluation of (5.77 and 78). Thus $\underline{B}$ is tridiagonal, i.e. only the following terms are non-zero ($1 \leq j \leq J - 1$):

$$b_{j,j-1} = \frac{1}{\Delta x_j} + \frac{\Delta x_j}{6} \ ,$$

$$b_{j,j} = -\left(\frac{1}{\Delta x_j} + \frac{1}{\Delta x_{j+1}} \right) + \frac{\Delta x_j + \Delta x_{j+1}}{3} \ , \tag{5.79}$$

$$b_{j,j+1} = \frac{1}{\Delta x_{j+1}} + \frac{\Delta x_{j+1}}{6} \ ,$$

and for node $m = J$

$$b_{J,J-1} = \frac{1}{\Delta x_J} + \frac{\Delta x_J}{6} \ ,$$

$$b_{J,J} = \frac{-1}{\Delta x_J} + \frac{\Delta x_J}{3} \ ,$$

$$(b_{J,J+1} = 0) \ .$$

The step sizes, Δx_j and Δx_{j+1}, are shown in Fig. 5.18.

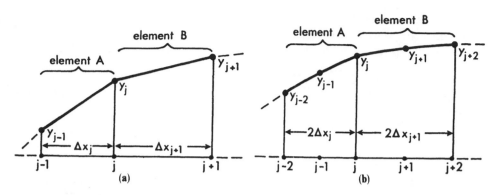

Fig. 5.18 a, b. Element and nodal contributions to the discretised equations. (a) Linear interpolation. (b) Quadratic interpolation, corner node

At first sight, since F is known from (5.66), (5.78) can be evaluated directly. However, in the more general case where F is known but a complicated function of x, it is convenient to interpolate F in the same manner as Y, i.e.

$$F = \sum_{j=1}^{J} F_j \phi_j(x) \; , \tag{5.80}$$

where the nodal values F_j are given directly by (5.66). Substituting (5.80) into (5.78) gives

$$g_m = \sum_{j=1}^{J} F_j \int_0^1 \phi_m \phi_j \, dx \; . \tag{5.81}$$

For linear approximating functions, (5.81) can be evaluated as $(1 \leq j \leq J - 1)$

$$g_j = \left(\frac{\Delta x_j}{6}\right) F_{j-1} + \left(\frac{\Delta x_j + \Delta x_{j+1}}{3}\right) F_j + \left(\frac{\Delta x_{j+1}}{6}\right) F_{j+1} \tag{5.82}$$

and for $m = J$, $g_J = (\Delta x_J/6) F_{J-1} + (\Delta x_J/3) F_J$.

For the special case $\Delta x_j = \Delta x_{j+1} = \Delta x$ it is instructive to write out (5.76) in full, and divide through by Δx. The result is $(1 \leq j \leq J - 1)$

$$\frac{Y_{j-1} - 2Y_j + Y_{j+1}}{\Delta x^2} + \left[\left(\frac{1}{6}\right) Y_{j-1} + \left(\frac{2}{3}\right) Y_j + \left(\frac{1}{6}\right) Y_{j+1}\right]$$

$$= \left[\left(\frac{1}{6}\right) F_{j-1} + \left(\frac{2}{3}\right) F_j + \left(\frac{1}{6}\right) F_{j+1}\right] . \tag{5.83}$$

Clearly the first group of terms coincides with a finite difference representation of $d^2 \bar{Y}/dx^2$. The second and third terms indicate an evaluation of $\bar{Y}$ and F which is a weighted average of three adjacent nodal values. The weighting $(\frac{1}{6}, \frac{2}{3}, \frac{1}{6})$ is characteristic of linear elements applied on a uniform grid.

Since the system of equations (5.76) is tridiagonal, an economical solution is available using the Thomas algorithm (Sect. 6.2.2).

Solutions have been obtained with the following values of a_l in (5.66):

$$a_1 = 1.0 , \quad a_2 = -0.5 , \quad a_3 = 0.3 , \quad a_4 = -0.2 , \quad a_5 = 0.1 .$$

For equally spaced points the rms error for various values of J is shown in Table 5.8. The rms error is defined by

$$\| Y - \bar{Y} \|_{rms} = \left[\sum_{j=1}^{J} \frac{(Y_j - \bar{Y}_j)^2}{J} \right]^{1/2}$$

The reduction in error with reducing Δx shown in Table 5.8 is proportional to Δx^2. This agrees with the theoretically predicted result for linear interpolation and is also the same order as for three-point centered finite difference formulae (e.g. Table 7.1).

Table 5.8. rms error for the simplified Sturm-Liouville equation

J	Δx	$\| Y - Y \|_{rms}$
6	0.2	7.830×10^{-3}
11	0.1	1.860×10^{-3}
21	0.05	0.447×10^{-3}
41	0.025	0.109×10^{-3}
81	0.0125	0.027×10^{-3}

5.4.2 STURM: Computation of the Sturm-Liouville Equation

In Sect. 5.4.1 the finite element method, with linear interpolation, is applied to the simplified Sturm-Liouville equation (5.66) with boundary conditions given by (5.67). In this subsection a computer program, STURM, will be described to obtain finite element solutions to (5.66 and 67) with linear and quadratic interpolation.

For quadratic interpolating functions (5.57), the equations, (5.77 and 81) for $b_{m,j}$ and g_m have the following form for corner nodes in the interior:

$$b_{j,j-2} = \frac{-1}{6\Delta x_j} - \frac{\Delta x_j}{15} ,$$

$$b_{j,j-1} = \frac{4}{3\Delta x_j} + \frac{2\Delta x_j}{15} ,$$

$$b_{j,j} = -\frac{7}{6}\left(\frac{1}{\Delta x_j} + \frac{1}{\Delta x_{j+1}} \right) + \frac{4}{15}(\Delta x_j + \Delta_{j+1}) , \qquad (5.84)$$

$$b_{j,j+1} = \frac{4}{3\Delta x_{j+1}} + \frac{2\Delta x_{j+1}}{15} ,$$

$$b_{j,j+2} = \frac{-1}{6\Delta x_{j+1}} - \frac{\Delta x_{j+1}}{15} ,$$

and

$$
g_j = -\left(\frac{\Delta x_j}{15}\right)F_{j-2} + \left(\frac{2\Delta x_j}{15}\right)F_{j-1} + \frac{4}{15}(\Delta x_j + \Delta x_{j+1})F_j
$$
$$
+ \left(\frac{2\Delta x_{j+1}}{15}\right)F_{j+1} - \left(\frac{\Delta x_{j+1}}{15}\right)F_{j+2} \; .
\tag{5.85}
$$

The specific form of (5.84 and 85) results from the restrictions $x_{j-1} = 0.5(x_{j-2} + x_j)$ and $x_{j+1} = 0.5(x_j + x_{j+2})$. These restrictions can be relaxed but this leads to more complicated algebraic expressions.

At the boundary node $m = J$, (5.77 and 81) take the form

$$
b_{J, J-2} = \frac{-1}{6\Delta x_J} - \frac{\Delta x_J}{15} \; ,
$$

$$
b_{J, J-1} = \frac{4}{3\Delta x_J} + \frac{2\Delta x_J}{15} \; ,
\tag{5.86}
$$

$$
b_{J, J} = \frac{-7}{6\Delta x_J} + \frac{4\Delta x_J}{15} \; , \quad \text{and}
$$

$$
g_J = -\left(\frac{\Delta x_J}{15}\right)F_{J-2} + \left(\frac{2\Delta x_J}{15}\right)F_{J-1} + \left(\frac{4\Delta x_J}{15}\right)F_J \; .
\tag{5.87}
$$

For equations obtained using quadratic interpolation but centered at midside nodes, (5.77, 81) have the form

$$
b_{j, j-1} = \frac{4}{3\Delta x_j} + \frac{2\Delta x_j}{15} \; ,
$$

$$
b_{j, j} = -\frac{8}{3\Delta x_j} + \frac{16\Delta x_j}{15} \; ,
\tag{5.88}
$$

$$
b_{j, j+1} = \frac{4}{3\Delta x_j} + \frac{2\Delta x_j}{15} \; , \quad \text{and}
$$

$$
g_j = \left(\frac{2\Delta x_j}{15}\right)F_{j-1} + \left(\frac{16\Delta x_j}{15}\right)F_j + \left(\frac{2\Delta x_j}{15}\right)F_{j+1} \; .
\tag{5.89}
$$

The system of equations (5.76) is tridiagonal if linear interpolation is used and alternating tridiagonal and pentadiagonal if quadratic interpolation is used. Efficient modifications of Gauss elimination are available for solving (5.76) for either situation and are described in Sects. 6.2.2 and 6.2.3.

Program STURM applies the Galerkin finite element method to the Sturm-Liouville equation (5.66). A listing of STURM is given in Fig. 5.19. The main parameters used in STURM are described in Table 5.9.

STURM reads in (and writes out) the control parameters (lines 11–21). The x locations, exact solution and nodal values of the right-hand side functions, F in

```
1
2 C
3 C      STURM COMPUTES THE SOLUTION OF THE STURM-LIOUVILLE PROBLEM,
4 C      P = Q = 1, F = -SIG(AL*SIN(L-0.5)*PI*X), USING THE FINITE ELEMENT
5 C      METHOD ON A UNIFORM GRID WITH EITHER LINEAR OR QUADRATIC
6 C      INTERPOLATION.
7 C      THE BANDED SYSTEM IS SOLVED WITH BANFAC, BANSOL (SECTION 6.2).
8 C
9        DIMENSION Y(65),YEX(65),X(65),B(5,65),G(65),F(65),A(5),YD(65)
10 C
11       OPEN(1,FILE='STURM.DAT')
12       OPEN(6,FILE='STURM.OUT')
13       READ(1,1)INT,IPR,NX,A
14     1 FORMAT(3I5,5E10.3)
15 C
16       IF(INT .EQ. 1)WRITE(6,2)
17       IF(INT .EQ. 2)WRITE(6,3)
18       WRITE(6,4)NX,A
19     2 FORMAT(' STURM-LIOUVILLE PROBLEM,  FEM:LINEAR INTERPOLATION')
20     3 FORMAT(' STURM-LIOUVILLE PROBLEM,  FEM:QUADRATIC INTERPOLATION')
21     4 FORMAT(' NX=',I3,'  A=',5E10.3,//)
22 C
23       PI = 3.1415927
24       NXP = NX - 1
25       ANX = NXP
26       DX = 1./ANX
27       DXS = DX*DX
28 C
29 C      SET GRID, EXACT SOLUTION AND RHS(F)
30 C
31       DO 5 I = 1,NX
32       AI = I - 1
33       X(I) = AI*DX
34       F(I) = 0.
35       YEX(I) = 0.
36       DO 5 L = 1,5
37       AL = L
38       AL = (AL - 0.5)*PI
39       DUM = A(L)*SIN(AL*X(I))
40       F(I) = F(I) - DUM
41       YEX(I) = YEX(I) - DUM/(1. - AL*AL)
42     5 CONTINUE
43 C
44 C      SET UP ARRAYS, B AND G
45 C
46       DO 7 I = 2,NX,INT
47       IM = I - 1
48       IF(INT .EQ. 2)GOTO 6
49       B(1,IM) = 0.
50       B(2,IM) = 1./DXS + 1./6.
51       B(3,IM) = -2./DXS + 2./3.
52       B(4,IM) = B(2,IM)
53       B(5,IM) = 0.
54       G(IM) = (F(I-1) + 4.*F(I) + F(I+1))/6.
55       GOTO 7
```

Fig. 5.19. Listing of program STURM

(5.66), are set (lines 28–42). The contributions to $\underline{B}$ and G in (5.76) are evaluated (lines 43–69) and an adjustment made to account for the boundary conditions. The banded system of equations (5.76) is solved by a call to BANFAC (line 83) to factorise $\underline{B}$ into $\underline{L} \cdot \underline{U}$ form and a call to BANSOL (line 86) to solve the factorised system. BANFAC and BANSOL are listed and described in Sect. 6.2.3. The rms error between Y, the solution returned by BANSOL, and the exact solution is computed and the output generated (lines 87–103).

```
56      6 B(1,IM) = 0.
57        B(2,IM) = 4.*(1./DXS + 0.1)/3.
58        B(3,IM) = 4.*(-2./DXS + 0.8)/3.
59        B(4,IM) = B(2,IM)
60        B(5,IM) = 0.
61        G(IM) = 0.4*(F(I-1) + 8.*F(I) + F(I+1))/3.
62        B(1,I) = 2.*(-0.25/DXS - 0.1)/3.
63        B(2,I) = 2.*(2./DXS + 0.2)/3.
64        B(3,I) = 2.*(-3.5/DXS + 0.8)/3.
65        B(4,I) = B(2,I)
66        B(5,I) = B(1,I)
67        G(I) = -0.1*(F(I-1)+F(I+3)) + 0.2*(F(I)+F(I+2)) + 0.8*F(I+1)
68        G(I) = 2.*G(I)/3.
69      7 CONTINUE
70 C
71 C      ADJUST FOR BOUNDARY CONDITIONS
72 C
73        B(2,1) = 0.
74        B(1,2) = 0.
75        B(3,NXP) = 0.5*B(3,NXP)
76        B(4,NXP) = 0.
77        B(5,NXP) = 0.
78        IF(INT .EQ. 1)G(NXP) = (F(NXP) + 2.*F(NX))/6.
79        IF(INT .EQ. 2)G(NXP) = 2.*(-0.1*F(NXP-1) + 0.2*F(NXP) + 0.4*F(NX))
80 C
81 C      SOLVE BANDED SYSTEM OF EQUATIONS
82 C
83        CALL BANFAC(B,NXP,INT)
84 C
85 C
86        CALL BANSOL(G,YD,B,NXP,INT)
87 C
88 C
89 C      COMPUTE RMS ERROR
90 C
91        SUM = 0.
92        DO 9 I = 2,NX
93        Y(I) = YD(I-1)
94        DY = Y(I) - YEX(I)
95        SUM = SUM + DY*DY
96        IF(IPR .EQ. 0)GOTO 9
97        WRITE(6,8)I,X(I),Y(I),YEX(I),DY
98      8 FORMAT(' I=',I3,'  X=',F7.5,'  Y=',F7.5,'  YEX=',F7.5,'  DY=',
99       1F7.5)
100     9 CONTINUE
101       RMS = SQRT(SUM/ANX)
102       WRITE(6,10)RMS,NX
103    10 FORMAT(//,' RMS=',E10.3,'   NX=',I3)
104       STOP
105       END
```

Fig. 5.19. (cont.) Listing of program STURM

Typical output for linear interpolation with $\Delta x = 0.125$ is shown in Fig. 5.20. This solution corresponds to the following choice of coefficients a_i in (5.66):

$$a_1 = 1.000 , \quad a_2 = -1.000 , \quad a_3 = 1.000 , \quad a_4 = -1.000 , \quad a_5 = 1.000$$

The results shown in Fig. 5.20 indicate that the solution error is generally smaller close to a Dirichlet boundary condition than close to a Neumann boundary condition.

A comparison of the rms solution errors for linear and quadratic interpolation with grid refinement is shown in Table 5.10.

Table 5.9. Parameters used in program STURM

Parameter	Description
INT	= 1, linear interpolation; = 2, quadratic interpolation
IPR	= 0, only print rms error; = 1, also print solution
NX	number of points in the x direction, including $x = 0$ and $x = 1.0$
A	coefficients, a_l, in (5.66)
F	nodal values of right-hand side of (5.66)
YEX	exact solution, $\bar{Y}$, (5.68)
B	$\underline{B}$ in (5.76)
G	$\underline{G}$ in (5.76)
BANFAC	see Sect. 6.2.3
BANSOL	see Sect. 6.2.3
RMS	rms error

```
STURM-LIOUVILLE PROBLEM,  FEM:LINEAR INTERPOLATION
NX=  9  A=  .100E+01 -.100E+01  .100E+01 -.100E+01  .100E+01


I=  2  X= .12500  Y= .11631  YEX= .11721  DY=-.00089
I=  3  X= .25000  Y= .22691  YEX= .22733  DY=-.00042
I=  4  X= .37500  Y= .33732  YEX= .33833  DY=-.00100
I=  5  X= .50000  Y= .44403  YEX= .44633  DY=-.00230
I=  6  X= .62500  Y= .53796  YEX= .53943  DY=-.00147
I=  7  X= .75000  Y= .62765  YEX= .62899  DY=-.00134
I=  8  X= .87500  Y= .71190  YEX= .71739  DY=-.00549
I=  9  X=1.00000  Y= .74987  YEX= .75848  DY=-.00861


RMS=  .380E-02   NX=  9
```

Fig. 5.20. Typical output from STURM

Noteworthy features of Table 5.10 are that:

i) Higher-order interpolation (i.e. quadratic) is only slightly more accurate on a very coarse grid than low-order (linear) interpolation.

ii) The accuracy of higher-order interpolation increases at a faster rate than low-order interpolation as the grid is refined.

iii) Theoretical rates of convergence are achieved approximately (Δx^2 for linear interpolation; Δx^3 for quadratic interpolation).

Table 5.10. Solution errors for the Sturm-Liouville problem

Grid size, Δx	rms error	
	Linear interpolation	Quadratic interpolation
1/8	0.00380	0.00148
1/12	0.00163	0.000178
1/16	0.000904	0.000056
1/20	0.000562	0.000023

It may be recalled that for the finite element method the approximating functions and weight functions are non-zero only in the small region surrounding the particular node. Consequently the finite element method is a local method. As with the finite difference method the (algebraic) equations generated by the finite element method connect together nodal values in a small region only. However, the number of connected nodes for the finite element method in multidimensions is considerably greater than for the finite difference method (Sect. 8.3 and Fletcher, 1984).

5.5 Further Applications of the Finite Element Method

In this section the finite element method is applied to the diffusion equation (Sect. 5.5.1), the Poisson equation (Sect. 5.5.2) and Laplace's equation (Sect. 5.5.3). The first problem illustrates the usual finite element practice of discretising only the spatial terms. The time derivative term is discretised as a separate step. The second problem introduces two spatial coordinates and, as a result, a more complicated structure in the discretised equations. The third problem outlines the isoparametric formulation which is useful if the computational domain is irregular.

5.5.1 Diffusion Equation

In this subsection the Galerkin finite element method is applied to the one-dimensional diffusion (heat conduction) equation

$$\frac{\partial \bar{T}}{\partial t} - \alpha \frac{\partial^2 \bar{T}}{\partial x^2} = 0 \; , \tag{5.90}$$

over the interval $0 \leq x \leq 1$ and $t \geq 0$. Initial conditions of the form $\bar{T}(x, 0) = T_0(x)$ and boundary conditions $\bar{T}(0, t) = a$ and $\bar{T}(1, t) = b$ are assumed.

Introducing the same linear approximating functions (5.70) as used for the Sturm-Liouville problem and applying the Galerkin finite element method in the same way produces the following result for internal nodes, on a uniform grid:

$$\frac{1}{6}\left[\frac{dT}{dt}\right]_{j-1} + \frac{2}{3}\left[\frac{dT}{dt}\right]_{j} + \frac{1}{6}\left[\frac{dT}{dt}\right]_{j+1} - \frac{\alpha(T_{j-1} - 2T_j + T_{j+1})}{\Delta x^2} = 0 \; . \tag{5.91}$$

For the Dirichlet boundary conditions indicated above no discretised equations are required at the boundary nodes.

Since the finite element formulation is applied in space only, the treatment of the term $\partial \bar{T}/\partial t$ in (5.90) is the same as the treatment of the undifferentiated (in space) term Y in (5.66). Not surprisingly the same $(1/6, 2/3, 1/6)$ weighting occurs in both (5.91) and in (5.83). If dT/dt is replaced by $\Delta T^{n+1}/\Delta t$ and the right-hand side of (5.91) is evaluated as a weighted average between the nth and the $(n+1)$-th time levels, as in (7.24), the result is

$$\frac{1}{6}\left(\frac{\Delta T_{j-1}^{n+1}}{\Delta t}\right)+\frac{2}{3}\left(\frac{\Delta T_j^{n+1}}{\Delta t}\right)+\frac{1}{6}\left(\frac{\Delta T_{j+1}^{n+1}}{\Delta t}\right)$$

$$-\alpha\left(\frac{(1-\beta)(T_{j-1}^n-2T_j^n+T_{j+1}^n)}{\Delta x^2}+\frac{\beta(T_{j-1}^{n+1}-2T_j^{n+1}+T_{j+1}^{n+1})}{\Delta x^2}\right)\,, \qquad (5.92)$$

where β is a parameter controlling the degree of implicitness. Equation (5.92) can be written compactly as

$$\frac{M_x\Delta T_j^{n+1}}{\Delta t}=\alpha[(1-\beta)L_{xx}T_j^n+\beta L_{xx}T_j^{n+1}]=0\,, \qquad (5.93)$$

where $\Delta T_j^{n+1}=T_j^{n+1}-T_j^n$ and the mass and difference operators, M_x and L_{xx} respectively, are given by

$$M_x=(\tfrac{1}{6},\tfrac{2}{3},\tfrac{1}{6}) \quad \text{and} \quad L_{xx}=(1/\Delta x^2,\,-2/\Delta x^2,\,1/\Delta x^2)\,.$$

A comparison with the implicit finite difference scheme (7.24) indicates that the only difference is the spreading of the time derivative term over adjacent nodes in the finite element formulation. Equation (7.24) can be put in the same form as (5.93) if the following mass operator is defined:

$$M_x^{fd}=(0,\,1,\,0)\,.$$

The similarity in structure of (5.93) and the equation it is created from, (5.90), suggests that the finite element method can be interpreted as a term-by-term discretisation process like the finite difference method. This interpretation is examined in Appendix A.2 and used in Sect. 8.3.

Equation (5.93) can be rearranged to give the implicit algorithm

$$(M_x-\Delta t\alpha\beta L_{xx})T_j^{n+1}=[M_x+\Delta t\alpha(1-\beta)L_{xx}]T_j^n\,. \qquad (5.94)$$

The significance of the structure of (5.94) may be better appreciated after reading Sect. 7.2. Repeated application of (5.94) centred at each node produces a tridiagonal system of equations which can be solved using the Thomas algorithm (Sect. 6.2.2).

It will be seen that for the finite difference scheme, (7.24), setting $\beta=0$ generates an explicit algorithm. This is not the case for the present scheme due to the 'implicit' nature of M_x. However, due to the symmetric nature of operators M_x and L_{xx} it is possible to obtain an explicit scheme from (5.94) for the particular choice $\Delta t=\Delta x^2/(6\alpha\beta)$.

Equation (5.94) is consistent with (5.90) with a truncation error of $O(\Delta t^2,\Delta x^2)$ if $\beta=0.5$ and unconditionally stable if $\beta\geq0.5$. For the particular choice $\beta=0.5$, (5.94) has been applied to the example considered in Sect. 7.2.5 and the results are presented in Table 7.6.

5.5.2 DUCT: Viscous Flow in a Rectangular Duct

Here the finite element method will be used to discretise a representative two-dimensional problem. Fully developed viscous (laminar) flow in a rectangular duct (Fig. 5.21) is governed by the z momentum equation

$$\frac{\partial p}{\partial z} = \mu \left(\frac{\partial^2 \bar{w}}{\partial x^2} + \frac{\partial^2 \bar{w}}{\partial y^2} \right) , \tag{5.95}$$

where the axial velocity $\bar{w}(x, y)$ is to be determined for a given pressure gradient $\partial p/\partial z$ and viscosity μ.

It is computationally advantageous to nondimensionalise (5.95) by introducing

$$x_{nd} = \frac{x_d}{a} , \qquad y_{nd} = \frac{y_d}{b} , \qquad \bar{w}_{nd} = \bar{w}_d \left(-\frac{\mu}{b^2 \dfrac{\partial p}{\partial z}} \right) . \tag{5.96}$$

Equation (5.95) becomes (dropping the subscript nd)

$$\left(\frac{b}{a} \right)^2 \frac{\partial^2 \bar{w}}{\partial x^2} + \frac{\partial^2 \bar{w}}{\partial y^2} + 1 = 0 \tag{5.97}$$

with boundary conditions

$$\bar{w} = 0 \quad \text{at} \quad x = \pm 1 , \quad y = \pm 1 . \tag{5.98}$$

Following a similar procedure as for the Sturm-Liouville equation, Sect. 5.4.1, a trial solution is introduced for $\bar{w}$ as

$$w = \sum_{i=1}^{I} w_i \phi_i(x, y) , \tag{5.99}$$

where $\phi_i(x, y)$ are two-dimensional interpolating functions, typically given by the equivalent of (5.60) or (5.62). For the present problem (bi)linear interpolation will be introduced over individual elements, as in Fig. 5.15.

Substitution of (5.99) into (5.97), application of the Galerkin formulation, (5.5, 10), and some manipulation produces a system of algebraic equations

$$\underline{B} W = G , \tag{5.100}$$

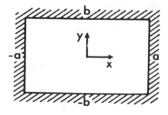

Fig. 5.21. Viscous flow in a rectangular duct

where W contains the unknown nodal values w_i in (5.99), an element of $\underline{B}$ is given by

$$b_{m,i} = -\int_{-1}^{1}\int_{-1}^{1}\left[\left(\frac{b}{a}\right)^2\frac{\partial\phi_i}{\partial x}\frac{\partial\phi_m}{\partial x} + \frac{\partial\phi_i}{\partial y}\frac{\partial\phi_m}{\partial y}\right]dx\,dy\;, \tag{5.101}$$

and an element of G is given by

$$g_m = \int_{-1}^{1}\int_{-1}^{1}\phi_m 1\,dx\,dy\;. \tag{5.102}$$

The parameter m ranges over all the internal nodes, i.e. $2\le j\le NX-1,\,2\le k\le NY-1$. The integrals in (5.101) are best evaluated over each element, individually, using element-based coordinates $(\xi,\,\eta)$ as in (5.64). Only the four contiguous rectangular elements surrounding node m make nonzero contributions to the global integrals in (5.101 and 102).

The structure of (5.101 and 102) is similar to that of (5.77 and 78), except that now the integration is over two dimensions instead of one. The form of the interpolating functions $\phi_i(x,\,y)$ is such that contributions to the integration can be determined as products of one-dimensional integrations (Appendix A.2).

For a uniform grid (5.100) takes a particularly concise form

$$\left[\left(\frac{b}{a}\right)^2 M_y\otimes L_{xx} + M_x\otimes L_{yy}\right]w_{j,k} = -1\;, \tag{5.103}$$

where the subscripts $j,\,k$ are part of the global grid, Fig. 5.11. In (5.103) L_{xx} and L_{yy} are directional difference operators

$$L_{xx} = (1/\Delta x^2,\,-2/\Delta x^2,\,1/\Delta x^2)\;,$$
$$L_{yy} = (1/\Delta y^2,\,-2/\Delta y^2,\,1/\Delta y^2)^T\;, \tag{5.104}$$

M_x and M_y are directional mass operators

$$M_x = M_y^T = (\tfrac{1}{6},\,\tfrac{2}{3},\,\tfrac{1}{6}) \tag{5.105}$$

and $\otimes$ is the tensor product. The operators have the following meaning:

$$L_{xx}w_{j,k} = \frac{w_{j-1,k} - 2w_{j,k} + w_{j+1,k}}{\Delta x^2}\;,$$

$$M_y w_{j,k} = \tfrac{1}{6}w_{j,k-1} + \tfrac{2}{3}w_{j,k} + \tfrac{1}{6}w_{j,k+1}\;, \quad\text{and}$$

$$M_y\otimes L_{xx}w_{j,k} = \frac{1}{6}\left\{\frac{w_{j-1,k+1} - 2w_{j,k+1} + w_{j+1,k+1}}{\Delta x^2}\right\} + \frac{2}{3}\left\{\frac{w_{j-1,k} - 2w_{j,k} + w_{j+1,k}}{\Delta x^2}\right\}$$
$$+ \frac{1}{6}\left\{\frac{w_{j-1,k-1} - 2w_{j,k-1} + w_{j+1,k-1}}{\Delta x^2}\right\}\;. \tag{5.106}$$

The construction of $M_x\otimes L_{yy}w_{j,k}$ follows a similar path.

An equivalent three-point finite difference discretisation of (5.97) is

$$\left(\frac{b}{a}\right)^2 \frac{w_{j-1,k} - 2w_{j,k} + w_{j+1,k}}{\Delta x^2} + \frac{w_{j,k-1} - 2w_{j,k} + w_{j,k+1}}{\Delta y^2} = -1 \ , \tag{5.107}$$

which can be written in the form of (5.103) by defining

$$M_{x,\mathrm{fd}} = M_{y,\mathrm{fd}}^T = (0,\ 1,\ 0) \ .$$

The individual mass and difference operators appearing in (5.103) are identical with the operators appearing in (5.93). What is different is that the integration over two dimensions has introduced the (tensor) product of one-dimensional operators. Use is made of the above difference and mass operator notation in Sect. 8.3. This notation is discussed more fully in Appendix A.2. The significance of the notation is that it permits a term-by-term finite element discretisation in a manner equivalent to that of the finite difference method.

The system of equations formed by applying (5.103) at each internal node ($j=2$, $J-1$; $k=2$, $K-1$) is solved very conveniently using SOR (Sect. 6.3), in a manner like that used in Sect. 5.2.3. To fit in with the listing of program DUCT (Fig. 5.22), the algorithm is written as

$$w^*_{j,k} = 0.75[1. + (PAR1/6)(w^n_{j-1,k+1} + w^n_{j+1,k+1} + w^{n+1}_{j-1,k-1} + w^{n+1}_{j+1,k-1})$$
$$+ PAR2(w^n_{j,k+1} + w^{n+1}_{j,k-1}) + PAR3(w^{n+1}_{j-1,k} + w^n_{j+1,k})]/PAR1 \ , \tag{5.108}$$

where $PAR1 = (b/a)^2/\Delta x^2 + 1/\Delta y^2$,
$\qquad PAR2 = [-(b/a)^2/\Delta x^2 + 2/\Delta y^2]/3$,
$\qquad PAR3 = [2(b/a)^2/\Delta x^2 - 1/\Delta y^2]/3$.

Equation (5.108) provides an estimate and (5.109) an improvement,

$$w^{n+1}_{j,k} = w^n_{j,k} + \lambda(w^*_{j,k} - w^n_{j,k}) \ , \tag{5.109}$$

where λ is the SOR relaxation parameter.

The above formulation is implemented by program DUCT (Fig. 5.22) to provide the velocity distribution in a rectangular duct (Fig. 5.21). DUCT reads in (and writes out) the various control parameters (lines 10–21). The exact and initial solutions are set up (lines 37–51). Since the number of iterations (SOR) to reach convergence is of no interest in this problem the exact solution is used as the initial solution for SOR. Subroutine VEL (line 47 and Fig. 5.23) computes the exact solution, obtained using the separation of variables method (Sect. 2.5.2), as

$$\bar{w} = \left(\frac{8}{\pi^2}\right)^2 \sum_{i=1,3,5,\dots}^{NEM} \sum_{j=1,3,5,\dots}^{NEM} \left(\frac{(-1)^{(i+j)/2-1}}{ij(ib/a)^2 + j^2} \cos(0.5i\pi x)\cos(0.5j\pi y)\right) \ , \tag{5.110}$$

$$q_l = 2\left(\frac{8}{\pi^2}\right)^3 \sum_{i=1,3,5,\dots}^{NEM} \sum_{j=1,3,5,\dots}^{NEM} (1/\{i^2 j^2 [(ib/a)^2 + j^2]\}) \ , \tag{5.111}$$

where the total flow rate $q_l = \iint \bar{w}\, dx\, dy$.

```
1
2  C
3  C        DUCT.FOR COMPUTES THE FULLY DEVELOPED VISCOUS LAMINAR FLOW IN A
4  C        RECTANGULAR CHANNEL OF ASPECT RATIO, B/A (= BAR), BY FEM (LINEAR
5  C        ELEMENTS) AND FDM (3PT CENTERED DIFFERENCES).
6  C        SOR IS USED TO OBTAIN  ITERATIVE SOLUTION
7  C
8          DIMENSION WEX(41,41),WA(41,41)
9  C
10         OPEN(1,FILE='DUCT.DAT')
11         OPEN(6,FILE='DUCT.OUT')
12         READ(1,1)NX,ME,NEM,IPR,ITMX,BAR,EPS,OM
13       1 FORMAT(5I5,3E10.3)
14 C
15         IF(ME .EQ. 1)WRITE(6,2)
16         IF(ME .EQ. 2)WRITE(6,3)
17       2 FORMAT(' DUCT FLOW BY FEM: LINEAR ELEMENTS')
18       3 FORMAT(' DUCT FLOW BY FDM: 3PT CEN. DIFF.')
19         WRITE(6,4)NX,NEM,BAR,ITMX,EPS,OM
20       4 FORMAT(' NX=',I3,'  NEM=',I2,'  B/A=',F5.2,'  MAX ITER =',I4,
21         1'  CONV TOL = ',E10.3,'  OM = ',F6.3,//)
22 C
23         PI = 3.1415927
24         NH = NX/2 + 1
25         NY = NX
26         NXP = NX - 1
27         NYP = NXP
28         ANX = NXP
29         DX = 2./ANX
30         DXS = DX*DX
31         DY = DX
32         DYS = DXS
33         BAS = BAR*BAR/DXS
34         PAR1 = BAS + 1./DYS
35         PAR2 = (2./DYS - BAS)/3.
36         PAR3 = (2.*BAS - 1./DYS)/3.
37 C
38 C        SET UP EXACT AND INITIAL SOLUTIONS
39 C
40         DO 7 J = 1,NX
41         AJ = J - 1
42         X = -1. + AJ*DX
43         DO 5 K = 1,NY
44         AK = K - 1
45         Y = -1. + AK*DY
46 C
47         CALL VEL(X,Y,NEM,PI,BAR,WD,QLX)
48 C
49         WEX(J,K) = WD
50         WA(J,K) = WD
51       5 CONTINUE
52         IF(IPR .EQ. 0)GOTO 7
53         WRITE(6,6)(WEX(J,K),K=1,NY)
54       6 FORMAT(' WE=',10F7.4)
55       7 CONTINUE
56         WRITE(6,8)
57       8 FORMAT(/)
```

Fig. 5.22. Listing of program DUCT

```
58  C
59  C       ITERATE USING SOR
60  C
61          DO 13 I = 1,ITMX
62          SUM = 0.
63          DO 12 J = 2,NXP
64          JM = J - 1
65          JP = J + 1
66          DO 11 K = 2,NYP
67          KM = K - 1
68          KP = K + 1
69          IF(ME .EQ. 2)GOTO 9
70          DUM = 1. + PAR1/6.*(WA(JM,KP) + WA(JP,KP) + WA(JM,KM) + WA(JP,KM))
71          DUM = DUM + PAR2*(WA(J,KP) + WA(J,KM)) + PAR3*(WA(JM,K) +WA(JP,K))
72          WDD = 0.75*DUM/PAR1
73          GOTO 10
74        9 DUM = 1. + BAS*(WA(JM,K) + WA(JP,K)) + (WA(J,KM) + WA(J,KP))/DYS
75          WDD = 0.5*DUM/PAR1
76       10 DIF = WDD - WA(J,K)
77          SUM = SUM + DIF*DIF
78          WA(J,K) = WA(J,K) + OM*DIF
79       11 CONTINUE
80       12 CONTINUE
81          RMS = SQRT(SUM/(ANX-1.)/(ANX-1.))
82          IF(RMS .LT. EPS)GOTO 15
83       13 CONTINUE
84          WRITE(6,14)ITMX,RMS
85       14 FORMAT(' CONVERGENCE NOT ACHIEVED IN',I5,' STEPS',5X,
86         1' ITER RMS = ',E12.5)
87  C
88  C       COMPARE SOLUTION WITH EXACT
89  C
90       15 SUM = 0.
91          QL = 0.
92          DO 18 J = 2,NXP
93          DO 16 K = 2,NYP
94          DIF = WA(J,K) - WEX(J,K)
95          SUM = SUM + DIF*DIF
96          QL = QL + WA(J,K)
97       16 CONTINUE
98          IF(IPR .EQ. 0)GOTO 18
99          WRITE(6,17)(WA(J,K),K=2,NYP)
100      17 FORMAT(' WA=',11F7.4)
101      18 CONTINUE
102         QL = QL*DX*DY
103         RMS = SQRT(SUM/(ANX-1.)/(ANX-1.))
104         WRITE(6,19)I,RMS
105         WRITE(6,21)WA(NH,NH),WEX(NH,NH)
106         WRITE(6,20)QL,QLX
107      19 FORMAT(/,' CONVERGED AFTER ',I3,' STEPS',5X,'SOL RMS =',E12.5,//)
108      20 FORMAT(' COMP. FLOW RATE = ',E12.5,'    EXACT FLOW RATE = ',E12.5)
109      21 FORMAT(' W(C/L)=',E12.5,'    WEX(C/L)=',E12.5,//)
110         STOP
111         END
```

Fig. 5.22. (cont.) Listing of program DUCT

DUCT (lines 58–83) evaluates (5.108) and (5.109) iteratively until $\|w_{j,k}^* - w_{j,k}\|_{rms}$ is less than the tolerance EPS. Subsequently (lines 87–106) the solution is compared with the exact solution and written out. The main parameters used by the program DUCT are given in Table 5.11. Typical output produced by program DUCT for a linear finite element solution obtained on a 6×6 grid is shown in Fig. 5.24.

Fig. 5.23. Listing of subroutine VEL

```
1
2          SUBROUTINE VEL(X,Y,N,PI,BAR,W,QL)
3   C
4   C      FOR GIVEN X,Y COMPUTES AXIAL VELOCITY IN CHANNEL
5   C      AND COMPUTES THE FLOW RATE
6   C
7          CON = (8./PI/PI)**2
8          XAN = 0.5*PI*X
9          YAN = 0.5*PI*Y
10         W = 0.
11         QL = 0.
12         DO 2 I = 1,N,2
13         AI = I
14         CX = COS(AI*XAN)
15         DO 1 J = 1,N,2
16         AJ = J
17         CY = COS(AJ*YAN)
18         IJ = (I+J)/2 - 1
19         IF(IJ .EQ. 0)SIG = 1.0
20         IF(IJ .GT. 0)SIG = (-1.)**IJ
21         DEN = AI*AJ*((AI*BAR)**2 + AJ*AJ)
22         DUM = SIG*CX*CY/DEN
23         W = W + DUM
24         QL = QL + 1./DEN/AJ/AI
25       1 CONTINUE
26       2 CONTINUE
27         W = CON*W
28         QL = (16./PI/PI)*CON*QL
29         RETURN
30         END
```

Table 5.11. Parameters used in program DUCT

Parameter	Description
NX	($=$NY), number of points in the x and y directions
ME	$=1$, linear FEM; $=2$, 3PT FDM
NEM	number of terms in the equation for the exact solution
IPR	print control parameter
ITMX	maximum number of iterations (SOR)
EPS	tolerance (SOR)
OM	relaxation parameter (SOR), λ
X, Y	nondimensional coordinates x, y
VEL	subroutine to compute the exact velocity at (x, y) and the total flow rate
WEX	exact axial velocity, $\bar{w}$
WA	computed axial velocity, w
WDD	$w_{j,k}^{*}$
RMS	$\|w_{j,k}^{*} - w_{j,k}\|_{rms}$, $\|w_{j,k} - \bar{w}_{j,k}\|_{rms}$
QL	computed total flow rate
QLX	exact total flow rate, q_l

The effect of grid refinement for both the finite element and finite difference methods is shown in Table 5.12. It is apparent that both methods are converging like Δx^2, Δy^2 with the finite difference method producing a slightly smaller rms error but a slightly larger error in the flow rate, q_l.

```
DUCT FLOW BY FEM: LINEAR ELEMENTS
NX=  6   NEM=21   B/A= 1.00   MAX ITER =   30   CONV TOL =    .100E-04   OM =   1.500


WE=   .0000   .0000   .0000   .0000   .0000   .0000
WE=   .0000   .1386   .1926   .1926   .1386   .0000
WE=   .0000   .1926   .2750   .2750   .1926   .0000
WE=   .0000   .1926   .2750   .2750   .1926   .0000
WE=   .0000   .1386   .1926   .1926   .1386   .0000
WE=   .0000   .0000   .0000   .0000   .0000   .0000


WA=   .1453   .1990   .1989   .1453
WA=   .1990   .2842   .2842   .1989
WA=   .1989   .2842   .2842   .1989
WA=   .1453   .1989   .1989   .1453

CONVERGED AFTER   10 STEPS      SOL RMS =   .72390E-02


W(C/L)=   .28421E+00    WEX(C/L)=   .27502E+00


COMP. FLOW RATE =   .52952E+00    EXACT FLOW RATE =   .56227E+00
```

Fig. 5.24. Typical output from DUCT

Table 5.12. Accuracy of DUCT solutions with grid refinement ($b/a = 1$)

Grid	6×6		11×11		21×21		Exact
Method	rms	q_l	rms	q_l	rms	q_l	q_l
FEM	0.007239	0.5295	0.001571	0.5540	0.000256	0.5598	0.5623
FDM	0.006480	0.4949	0.001480	0.5446	0.000205	0.5583	0.5623

5.5.3 Distorted Computational Domains: Isoparametric Formulation

When the finite element method first became popular, arbitrarily 1965–1970, triangular elements were used extensively because they permitted complicated geometric shapes to be represented. Triangular elements and the appropriate interpolating formulae are described by Zienkiewicz (1977, p. 164).

The ability to handle non-uniform and distorted computational domains has always been an important feature of the finite element method. However, distorted rectangular elements can also be introduced to represent complicated computational domains, by exploiting the isoparametric transformation. Rectangular elements are usually preferred to triangular elements since they are easier to interface with grid generation programs (Chap. 13), particularly in three dimensions. However, Jameson (1989) describes an effective procedure based on tetrahedral elements for representing the domain external to a complete aircraft.

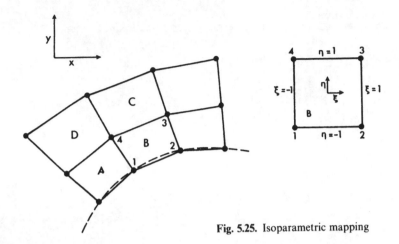

Fig. 5.25. Isoparametric mapping

A typical situation, requiring an isoparametric transformation of rectangular elements, is illustrated in Fig. 5.25. The elements shown in the physical space (x, y) are distorted. However, these can be readily transformed into uniform elements in (ξ, η) space by defining (for bilinear interpolation)

$$x = \sum_{l=1}^{4} \phi_l(\xi, \eta)x_l \quad \text{and} \quad y = \sum_{l=1}^{4} \phi_l(\xi, \eta)y_l , \tag{5.112}$$

where (x_l, y_l) are the coordinates of the lth corner node of B in physical space and $\phi_l(\xi, \eta)$ are the same bilinear approximating functions as defined by (5.59). The use of the isoparametric formulation affects the evaluation of the weighted residual integral (5.5).

For example, the application of the Galerkin finite element method to the problem of solving Laplace's equation in a distorted domain (Sect. 5.2.3) produces a linear system of equations, equivalent to (5.100), i.e.

$$\underline{B}W = G . \tag{5.113}$$

An element of $\underline{B}$ is given by

$$b_{m,j} = \iint_{WXYZ} \left(\frac{\partial \phi_j}{\partial x} \frac{\partial \phi_m}{\partial x} + \frac{\partial \phi_j}{\partial y} \frac{\partial \phi_m}{\partial y} \right) dx\, dy , \tag{5.114}$$

where m is the node at which the weight function is centred and j is the contributing node. In (5.114) the integration is carried out over the area WXYZ (Fig. 5.4). However, direct evaluation of (5.114) in physical space would be impossible to generalise to arbitrary domains.

The philosophy behind the isoparametric formulation is that it is straightforward to carry out the integration (numerically) in the transformed (ξ, η) space where the elements are uniform. To illustrate this, attention will be focussed on the first term in (5.114), i.e.

$$I = \iint \left(\frac{\partial \phi_j}{\partial x}\right)\left(\frac{\partial \phi_m}{\partial x}\right) dx\, dy \ . \tag{5.115}$$

The derivatives $\partial \phi_j / \partial x$ and $\partial \phi_j / \partial y$ are related to $\partial \phi_j / \partial \xi$ and $\partial \phi_j / \partial \eta$ by

$$\begin{pmatrix} \partial \phi_j / \partial \xi \\ \partial \phi_j / \partial \eta \end{pmatrix} = [J] \begin{pmatrix} \partial \phi_j / \partial x \\ \partial \phi_j / \partial y \end{pmatrix} \ , \tag{5.116}$$

where the Jacobian $[J]$ is given by

$$[J] = \begin{pmatrix} \partial x / \partial \xi & \partial y / \partial \xi \\ \partial x / \partial \eta & \partial y / \partial \eta \end{pmatrix} \ . \tag{5.117}$$

The terms in the Jacobian are evaluated from (5.112). For example

$$\frac{\partial y}{\partial \xi} = \sum_{l=1}^{4} \left(\frac{\partial \phi_l}{\partial \xi}(\xi, \eta)\right) y_l \ . \tag{5.118}$$

From (5.116) the following explicit expressions for $\partial \phi_j / \partial x$ (and $\partial \phi_m / \partial x$) in (5.115) are available:

$$\frac{\partial \phi_j}{\partial x} = \left[\left(\frac{\partial y}{\partial \xi}\right)\left(\frac{\partial \phi_j}{\partial \eta}\right) - \left(\frac{\partial y}{\partial \eta}\right)\left(\frac{\partial \phi_j}{\partial \xi}\right) \right] \bigg/ \det J \ . \tag{5.119}$$

In (5.115), $dx\, dy = \det J\, d\xi\, d\eta$. Substitution of (5.119) into (5.115) produces the following contribution to $b_{m,\,j}$ from each element:

$$I_e = \int_{-1}^{1} \int_{-1}^{1} \left(\frac{1}{\det J}\right)\left[\left(\frac{\partial y}{\partial \xi}\right)\left(\frac{\partial \phi_j}{\partial \eta}\right) \right.$$
$$\left. - \left(\frac{\partial y}{\partial \eta}\right)\left(\frac{\partial \phi_j}{\partial \xi}\right) \right]\left[\left(\frac{\partial y}{\partial \xi}\right)\left(\frac{\partial \phi_m}{\partial \eta}\right) - \left(\frac{\partial y}{\partial \eta}\right)\left(\frac{\partial \phi_m}{\partial \xi}\right) \right] d\xi\, d\eta \ . \tag{5.120}$$

The advantage of the isoparametric formulation is that all terms in (5.120) are functions of (ξ, η) and the elements are uniform in (ξ, η) space, as indicated in Fig. 5.25. Thus the integration can be carried out, conveniently, using Gauss quadrature (Isaacson and Keller 1966, p. 327). The isoparametric formulation is described, in greater detail, by Zienkiewicz (1977, p. 188).

5.6 Spectral Method

The spectral method uses the same form of approximate solution as the (traditional) Galerkin method, (5.1). Like the (traditional) Galerkin method the approximating functions and weight functions are non-zero throughout the computational domain. In this sense the spectral and (traditional) Galerkin methods are global methods (Fletcher 1984). In contrast to the situation for the finite element method, the unknown coefficients $a_j(t)$ in the approximate solution (5.1) cannot be identified with nodal unknowns.

The most important difference between the spectral method and the (traditional) Galerkin method is that the spectral method uses orthogonal functions for the approximating and weight functions. The approximating and weight functions are orthogonal if

$$\iiint \phi_j(x, y, z)\phi_m(x, y, z)\,dx\,dy\,dz \neq 0 \quad \text{when } m=j$$
$$=0 \quad \text{when } m \neq j \ .$$

Fourier series, Legendre polynomials and Chebyshev polynomials are examples of orthogonal functions.

It will be seen in the example to be presented shortly that the use of orthogonal approximating and weight functions simplifies the structure of the (ordinary differential) equations created by the spectral method. In addition, if the boundary conditions are appropriate the spectral method can generate solutions of very high accuracy with relatively few terms in the approximate solution, if the exact solution is sufficiently smooth. These aspects are discussed by Fletcher (1984, Chap. 5) and by Canuto et al. (1987).

5.6.1 Diffusion Equation

The mechanics of implementing the spectral method can be most easily understood by considering an example. The diffusion equation

$$\frac{\partial \bar{T}}{\partial t} - \alpha \frac{\partial^2 \bar{T}}{\partial x^2} = 0 \tag{5.121}$$

in the computational domain $0 \leq x \leq 1$ will be examined so that direct comparison with the finite difference and finite element methods will be possible. The following boundary and initial conditions are assumed:

$$\bar{T}(0, t)=0 \ , \quad \bar{T}(1, t)=1 \quad \text{and} \quad \bar{T}(x, 0)=\sin(\pi x)+x \ . \tag{5.122}$$

The starting point for the spectral method is the introduction of the approximate solution

$$T=\sin(\pi x)+x+ \sum_{j=1}^{J} a_j(t)\sin(j\pi x) \ , \tag{5.123}$$

where $a_j(t)$ are the unknown coefficients. The form of (5.123) automatically satisfies the boundary and initial conditions. Substitution of (5.123) into (5.121) produces a residual

$$R= \sum_{j=1}^{J} \left[\frac{da_j}{dt} + \alpha(j\pi)^2 a_j\right]\sin(j\pi x)+\alpha\pi^2\sin(\pi x) \ .$$

Evaluation of the weighted integral, i.e. (5.5) with (5.10),

$$\int\limits_0^1 [\sin(m\pi x)R]\,dx = 0 \; , \quad \text{gives}$$

$$\sum_{j=1}^J \left[\frac{da_j}{dt} + \alpha(j\pi)^2 a_j \right] \int\limits_0^1 [\sin(m\pi x)\sin(j\pi x)]\,dx$$

$$+ \alpha\pi^2 \int\limits_0^1 [\sin(m\pi x)\sin(\pi x)]\,dx = 0 \; . \tag{5.124}$$

The advantage of using orthogonal functions in (5.124) is that the first integral is non-zero only if $j = m$ and the second integral is non-zero only if $m = 1$. Consequently (5.124) becomes

$$\frac{da_m}{dt} + \alpha(m\pi)^2 a_m + r_m = 0 \; , \quad m = 1, \ldots, J \; , \tag{5.125}$$

where

$$\begin{aligned} r_m &= \alpha\pi^2 &&\text{if } m = 1 \; , \\ &= 0 &&\text{if } m \neq 1 \; . \end{aligned} \tag{5.126}$$

The solution of (5.125) can be written down directly as

$$a_1 = e^{-\alpha\pi^2 t} - 1 \; ,$$

$$a_m = 0 \; , \qquad\qquad m = 2, \ldots, J \; .$$

Equation (5.123) then becomes

$$T = \sin(\pi x)e^{-\alpha\pi^2 t} + x \; , \tag{5.127}$$

which is, in fact, the exact solution of (5.121). This has occurred, fortuitously, because of the particular choice of boundary and initial conditions (5.122). A more realistic situation is provided if the initial condition is given by

$$T(x, 0) = 5x - 4x^2 \tag{5.128}$$

and if $5x - 4x^2$ replaces $\sin(\pi x) + x$ in the approximate solution (5.123). The application of the spectral method with (5.128) is similar to that given above except that (5.126) becomes

$$\begin{aligned} r_m &= \frac{16\alpha}{m} \; , &&m = 1, 3, 5, \ldots, \\ &= 0 \; , &&m = 2, 4, 6, \ldots \; . \end{aligned} \tag{5.129}$$

If da_m/dt in (5.125) is replaced with a forward time difference formula $(a_m^{n+1} - a_m^n)/\Delta t$, the following explicit algorithm can be obtained:

$$a_m^{n+1} = a_m^n - \Delta t[(\alpha m\pi)^2 a_m^n + r_m] \ . \tag{5.130}$$

For this case, (5.130) has been integrated with a step-size $\Delta t = 0.001$. The results for different J are shown in Table 5.13. Also included in Table 5.13 are the exact solution T and an approximate solution T_b that only contains errors associated with the numerical integration scheme. The exact solution is obtained by the separation of variables technique (Sect. 2.5.2) as

$$\bar{T} = x + 4 \sum_{j=1}^{\infty} \{2/[(2j-1)\pi]\}^3 \sin[(2j-1)\pi x] \exp\{-\alpha[(2j-1)\pi]^2 t\} \ . \tag{5.131}$$

The approximate solution T_b is obtained by assuming a solution of the form

$$T_b = x + 4 \sum_{j=1}^{\infty} \{2/[(2j-1)\pi]\}^3 \sin[(2j-1)\pi x] X_j(t) \ . \tag{5.132}$$

Substitution into (5.121) then requires the solution of

$$\frac{dX_j}{dt} + [(2j-1)\pi]^2 X_j = 0 \ . \tag{5.133}$$

This is obtained using the same time differencing as for a_m in (5.130) and the same time-step, $\Delta t = 0.001$. Substitution into (5.132) gives the solution for T_b shown in Table 5.13.

The results shown in Table 5.13 indicate that T becomes more accurate as J is increased. However, it should be remembered that the error between T and the exact solution $\bar{T}$ comes partly from the approximation inherent in the spectral

Table 5.13 Spectral solutions, at $x = 0.5$, of (5.121)

	Approximate solution T			Approximate	Exact solution
t	$J=1$	$J=3$	$J=5$	solution T_b	$\bar{T}$
0.00	1.500	1.500	1.500	1.500	1.500
0.02	1.314	1.347	1.338	1.340	1.341
0.04	1.162	1.199	1.191	1.193	1.194
0.06	1.037	1.075	1.067	1.069	1.071
0.08	0.935	0.973	0.965	0.967	0.969
0.10	0.851	0.889	0.881	0.883	0.885
0.12	0.782	0.820	0.812	0.814	0.816
0.14	0.725	0.764	0.755	0.757	0.759
0.16	0.679	0.717	0.709	0.711	0.713
0.18	0.641	0.679	0.671	0.673	0.675
0.20	0.610	0.648	0.640	0.642	0.643

method and partly from the error in the time differencing introduced to obtain numerical solutions of (5.125). The difference between the two approximate solutions, T and T_b, is due to the error in the spectral method, only. Clearly the use of relatively few unknowns in the approximate solution provides a solution of high accuracy. This is typical of spectral methods (Fletcher 1984, Chap. 6).

Applications of the spectral method are given in Sects. 15.3.3 and 17.1.6. The traditional areas of application of the spectral method are in global meteorological modelling (Bourke et al. 1977) and in the direct simulation of turbulence (Hussaini and Zang 1987). Both of these areas require governing equations that include nonlinear terms particularly for the convection.

In order for the spectral method to generate an efficient computational algorithm it is necessary to introduce transform techniques to handle the nonlinear terms. Conceptually this can be thought of as introducing local nodal unknowns T_l in place of the $a_j(t)$ coefficients in (5.123) performing the nonlinear products in terms of T_l and re-evaluating the equivalent a_j coefficients. Transform techniques are discussed by Fletcher (1984, Chap. 5) and by Canuto et al. (1987).

5.6.2 Neumann Boundary Conditions

In Sect. 5.6.1 the spectral method was applied to the diffusion equation (5.121) with Dirichlet boundary conditions and this led to explicit ordinary differential equations (5.125) for the unknown coefficients a_m in the trial solution.

Here a problem with more difficult boundary conditions will be considered to demonstrate that the spectral method often entails additional manipulation, during the discretisation stage. Specifically, the diffusion equation (5.121) is to be solved in the domain $0 \le x \le 1.0$, $t \ge 0$, for initial and boundary conditions

$$\bar{T}(x,0) = 3 - 2x - 2x^2 + 2x^3 , \tag{5.134}$$

$$\frac{\partial \bar{T}}{\partial x}(0, t) = -2 \quad \text{and} \quad \bar{T}(1, t) = 1.0 . \tag{5.135}$$

Rather than attempting to incorporate the initial and boundary conditions into the approximate solution, as in (5.123), a general approximate solution is introduced as

$$T(x,t) = b_0(t) + \sum_{j=1}^{J} [a_j(t) \sin(j2\pi x) + b_j(t) \cos(j2\pi x)] . \tag{5.136}$$

For (5.136) to satisfy the boundary conditions (5.135) the coefficients must be related as

$$\sum_{j=1}^{J} a_j(2\pi j) = -2 , \quad \sum_{j=0}^{J} b_j = 1 . \tag{5.137}$$

Using (5.137), a_j and b_j can be eliminated from (5.136) to give

$$T = \cos(J2\pi x) + \left(-\frac{2}{2\pi J}\right)\sin(J2\pi x) + \sum_{j=1}^{J-1} a_j \left\{ \sin[j(2\pi x)] - \left(\frac{j}{J}\right)\sin[J(2\pi x)] \right\}$$

$$+ \sum_{j=0}^{J-1} b_j \{ \cos[j(2\pi x)] - \cos[J(2\pi x)] \} \ . \tag{5.138}$$

Implementation of the Galerkin spectral method (Fletcher 1984, p. 30), requires that the weight functions in (5.5) are identical with the functions in (5.138), i.e.

$$W_m(x) = \sin[m(2\pi x)] - \left(\frac{m}{J}\right)\sin[J(2\pi x)] \ , \qquad 1 \le m \le J-1 \ , \tag{5.139}$$
$$W_m(x) = \cos[m(2\pi x)] - \cos[J(2\pi x)] \ , \qquad 0 \le m \le J-1 \ .$$

Substitution of (5.138) into the diffusion equation (5.121) and evaluation of (5.5) based on (5.139) produces the following systems of ordinary differential equations:

$$\frac{da_m}{dt} + [\alpha(2\pi m)^2]a_m + \frac{m}{J}\sum_{j=1}^{J-1}\frac{j}{J}\left\{\frac{da_j}{dt} + [\alpha(2\pi J)^2]a_j\right\} = -2\alpha(2\pi m) \ , \quad 1 \le m \le J-1 \ , \tag{5.140}$$

$$\frac{db_m}{dt} + [\alpha(2\pi m)^2]b_m + \sum_{j=0}^{J-1}\left\{\frac{db_j}{dt} + [\alpha(2\pi J)^2]b_j\right\} = \alpha(2\pi J)^2 \ , \quad 1 \le m \le J-1 \ , \tag{5.141}$$

and

$$\frac{2db_0}{dt} = \sum_{j=0}^{J-1}\left\{\frac{db_j}{dt} + [\alpha(2\pi J)^2]b_j\right\} = \alpha(2\pi J)^2 \ . \tag{5.142}$$

Equations (5.140) form a system of coupled ordinary differential equations for the a_j coefficients that are independent of the ordinary differential equations (5.141, 142) for the b_j coefficients. These two systems are marched in time to provide values of $a_j(t)$ and $b_j(t)$ that are substituted into (5.138) to give the computational solution $T(x,t)$.

However, whatever marching algorithm is used to advance the a_j, b_j coefficients in time, it will be necessary to undertake an initial factorisation and a matrix multiplication at every time-step. Since the current problem is linear and accurate solutions can be obtained with relatively few terms in the approximate solution (5.136), the additional execution time associated with the coupling is not excessive.

The starting values, a_j^0, b_j^0 are obtained by fitting the approximate solution (5.136) to the initial solution (5.134), i.e.

$$b_0 = \frac{11}{6} \ , \quad a_j = 2\int_0^1 \bar{T}(x,0)\sin(j2\pi x)dx \ , \quad b_j = 2\int_0^1 \bar{T}(x,0)\cos(j2\pi x)dx \ . \tag{5.143}$$

The numerical solution of this problem is to be sought in Problem 5.17.

An alternative procedure called the tau method, which is closely related to the Galerkin method, is also available. In the tau method the weight functions are chosen to match the approximating functions in (5.136), i.e.

$$W_{m(x)} = \sin[m(2\pi x)] , \qquad 1 \leq m \leq J-1 ,$$
$$W_{m(x)} = \cos[m(2\pi x)] , \qquad 1 \leq m \leq J . \tag{5.144}$$

Substitution of (5.136) into (5.121) and evaluation of (5.5) based on (5.144) produces the following system of ordinary differential equations:

$$\frac{da_m}{dt} + \alpha(2\pi m)^2 a_m = 0 , \qquad 1 \leq m \leq J-1 ,$$
$$\frac{db_m}{dt} + \alpha(2\pi m)^2 b_m = 0 , \qquad 1 \leq m \leq J . \tag{5.145}$$

The system of equations (5.145) are explicit and have solutions

$$a_m(t) = a_m(0)e^{-\alpha(2\pi m)^2 t} , \qquad 1 \leq m \leq J-1 ,$$
$$b_m(t) = b_m(0)e^{-\alpha(2\pi m)^2 t} , \qquad 1 \leq m \leq J , \tag{5.146}$$

where $a_m(0)$ and $b_m(0)$ are given by (5.143) and $a_J(t)$ and $b_0(t)$ are given by (5.137).

For the current problem, use of (5.136, 137 and 144) provides a more efficient algorithm than the spectral Galerkin formulation, based on (5.138 and 139). In general the tau method handles difficult boundary conditions with more flexibility than does the Galerkin spectral method (Sect. 5.6.1). The tau method is described by Fletcher (1984, p. 207) and Canuto et al. (1987).

5.6.3 Pseudospectral Method

The Galerkin spectral method (Sect. 5.6.1) produces very accurate solutions with relatively few unknown coefficients a_j in the approximate solution (5.123). However, when nonlinear terms are involved the evaluation of products of approximate solutions becomes very time-consuming. This lack of economy motivates the use of collocation (5.8) instead of a Galerkin formulation. Collocation facilitates seeking the solution in terms of nodal unknowns, like the finite difference and finite element methods, rather than in terms of the unknown coefficients a_j in the approximate solution. The explicit use of nodal unknowns also permits boundary conditions to be incorporated more efficiently than does the Galerkin spectral method. In the literature the collocation spectral method is usually referred to as the pseudospectral method (Orszag 1971).

Here the pseudospectral method will be illustrated for the diffusion equation

$$\frac{\partial \bar{u}}{\partial t} - \alpha \frac{\partial^2 \bar{u}}{\partial x^2} = 0 , \tag{5.147}$$

with boundary and initial conditions

$$\bar{u}(-1, t) = -1 , \quad \bar{u}(1, t) = 1 , \quad \bar{u}(x, 0) = \sin(\pi x) + x . \quad (5.148)$$

The time derivative in (5.147) is discretised using a finite difference approximation. Using Euler time differencing, as in Sect. 5.6.1, produces the algorithm,

$$u^{n+1} = u^n + \left(\alpha \frac{\partial^2 u}{\partial x^2} \right)^n \Delta t . \quad (5.149)$$

The pseudospectral method requires that (5.149) be satisfied at each collocation point (x_j) and introduces an approximate solution

$$u(x, t) = \sum_{k=1}^{N+1} a_k(t) T_{k-1}(x) \quad (5.150)$$

that will allow spatial derivatives to be evaluated. The approximate functions in (5.150) are Chebyshev polynomials, defined in the interval $-1 \leq x \leq 1$. Chebyshev polynomials permit accurate determination of spatial derivatives and are generally preferred to trigonometric functions if u is nonperiodic. Chebyshev polynomials and their general use are described by Fox and Parker (1968). Specific recurrence relations that arise in conjunction with the spectral and pseudospectral methods are given by Gottlieb and Orszag (1977, pp. 159–161).

The direct use of (5.150) leads to the following expression for $\partial^2 u/\partial x^2|_j$ in (5.147):

$$\frac{\partial^2 u}{\partial x^2}\bigg|_j = \sum_{k=1}^{N+1} a_k \frac{\partial^2 T_{k-1}(x_j)}{\partial x^2} . \quad (5.151)$$

Thus the steps in the pseudospectral method to advance the solution from time level n to $n+1$ can be expressed as:

i) given u_j^n evaluate a_k^n from (5.150);
ii) given a_k^n evaluate $[\partial^2 u/\partial x^2]_j^n$ from (5.151);
iii) evaluate u_j^{n+1} from (5.149).

It can be seen that although step (ii) takes place in spectral space, i.e. in terms of a_k, step (iii) takes place in physical space, i.e. in terms of u_j. Step (i) is the transformation from physical space to spectral space.

The efficiency of the pseudospectral method depends on the economical execution of the three steps. The evaluation of a_k^n from (5.150) appears to require the solution of a linear system $\underline{T} a^n = u^n$, with $\underline{T}$ being dense. The direct use of Gauss elimination (Sect. 6.2.1) to obtain a^n implies $O((N+1)^3)$ operations. Fortunately, the use of Chebyshev polynomials evaluated at collocation points x_j given by

$$x_j = \cos\left[\frac{\pi(j-1)}{N} \right] , \quad (5.152)$$

allows a fast Fourier transform (FFT) to be used to evaluate a^n, in $O((N+1)$ $\log(N+1))$ operations. Fast Fourier transforms (Cooley and Tukey 1965) are described in more detail by Brigham (1974).

The direct evaluation of $\partial^2 u/\partial x^2$ from (5.151) at all grid points x_j requires $O((N+1)^2)$ operations. However, the use of Chebyshev polynomials makes available various recurrence relationships leading to a lower operation count to complete step (ii). In particular, spatial derivatives can be defined directly in terms of undifferentiated Chebyshev polynomials, i.e.

$$\frac{\partial u}{\partial x} = \sum_{k=1}^{N} a_k^{(1)} T_{k-1}(x) \tag{5.153a}$$

and

$$\frac{\partial^2 u}{\partial x^2} = \sum_{k=1}^{N-1} a_k^{(2)} T_{k-1}(x) . \tag{5.153b}$$

Explicit relationships between $a_k^{(1)}$, $a_k^{(2)}$ and a_k in (5.150) are given by Gottlieb and Orszag (1977, p. 160). Specifically, the following recurrence relations permit all the $a_k^{(1)}$, $a_k^{(2)}$ coefficients to be obtained in $O(N)$ operations

$$a_k^{(1)} = a_{k+2}^{(1)} + 2ka_{k+1} , \qquad 2 \le k < N ,$$
$$a_k^{(2)} = a_{k+2}^{(2)} + 2ka_{k+1}^{(1)} , \qquad 2 \le k \le N-1 , \tag{5.154}$$

and

$$a_1^{(1)} = 0.5a_3^{(1)} + a_2 , \qquad a_1^{(2)} = 0.5a_3^{(2)} + a_2^{(1)}$$

and

$$a_{N+1}^{(1)} = a_{N+2}^{(1)} = 0 , \qquad a_N^{(2)} = a_{N+1}^{(2)} = 0 .$$

Consequently $\partial^2 u/\partial x^2|_j$ can be evaluated using an FFT in $O((N-1)\log(N-1))$ operations. Since each step of the pseudospectral algorithm is economical the overall algorithm is very efficient.

Nonlinear terms in the governing equation are handled by the pseudospectral method in a particularly economical manner. Consider the inviscid Burgers equation (10.2)

$$\frac{\partial \bar{u}}{\partial t} + \bar{u}\frac{\partial \bar{u}}{\partial x} = 0 . \tag{5.155}$$

Following the same procedure as above, the pseudospectral method consists of the following steps to advance the solution from time level n to $n+1$:

i) given u_j^n evaluate a_k^n from (5.150);
ii) given a_k^n evaluate $[\partial u/\partial x]_j^n$ from (5.153a, 154);

iii) evaluate $w_j^n = u_j^n [\partial u / \partial x]_j^n$;

iv) evaluate $u_j^{n+1} = u_j^n - w_j^n \Delta t$.

The evaluation of the nonlinear term, step (iii), takes place in physical space, which avoids the expensive multiplication of individual coefficients in the expansions such as (5.150) and (5.153).

The concept of transforming between spectral space and physical space using the FFT to efficiently evaluate nonlinear terms is also used with the Galerkin spectral method (Orszag 1971). A detailed example in a meteorological context is provided by Haltiner and Williams (1980, pp. 193–196).

The pseudospectral method has one disadvantage compared with the Galerkin spectral method. In treating nonlinear terms the pseudospectral method allows aliasing errors to occur (Fletcher 1984). That is, the product of two series solutions, as is implicit in step (iii) above, causes new higher frequencies, $k > N + 1$, to arise. However, since the collocation process only considers $N - 1$ discrete internal points the high frequencies are not properly accounted for. In practice a slight corruption of the lower frequency amplitudes occurs.

For physical problems with little natural dissipation a time integration can become unstable due to the accumulation of aliasing errors producing a nonlinear instability. This is often a problem in weather prediction. For "engineering" flows there is usually sufficient dissipation, for example due to turbulence, that aliasing errors can be ignored. In this case the pseudospectral method is generally preferred to the Galerkin spectral method because of its greater economy and more effective treatment of boundary conditions.

The Galerkin spectral method (Sect. 5.6.1) obtains the solution entirely in spectral space, i.e. in terms of the coefficients a_j in (5.123). The pseudospectral method, as described above, works partly in spectral space and partly in physical space, with the FFT used to shift between spectral and physical space. It is possible to reinterpret the pseudospectral method so that it resembles a discretisation in physical space like the finite difference and finite element methods. The essential steps in this interpretation (due to Ku and Hatziavramidis 1984) are described below.

The nodal values u_j^n are related to the spectral coefficients a_k through (5.150), which can be written collectively as

$$u = \underline{T} a . \tag{5.156}$$

The coefficients a^n are given explicitly in terms of u^n by

$$a = \underline{\hat{T}} u , \tag{5.157}$$

where an element of $\underline{\hat{T}}$ is

$$\hat{T}_{kj} = \frac{2}{N} \frac{1}{\bar{c}_k} \frac{1}{\bar{c}_j} T_{k-1}(x_j) , \tag{5.158}$$

and

$$\bar{c}_1 = \bar{c}_{N+1} = 2 , \quad \bar{c}_i = 1 , \quad 1 < i < N+1 .$$

Equations (5.153a) evaluated at all grid points x_j can be written

$$\frac{\partial u}{\partial x} = \underline{T} \underline{a}^{(1)} . \tag{5.159}$$

But $a^{(1)}$ can be linked to a via (5.154),

$$\underline{a}^{(1)} = \underline{G}^{(1)} \underline{a} , \tag{5.160}$$

where $\underline{G}^{(1)}$ is an $(N+1) \times (N+1)$ matrix with elements

$$G_{kl}^{(1)} = 0 , \qquad \text{if } k \geq 1 \quad \text{or} \quad k+1 \text{ even} , \tag{5.161}$$
$$= 2(l-1)/c_k , \qquad \text{otherwise} .$$

The coefficients $c_1 = 2$, $c_k = 1$ for $k > 1$.

Clearly, from (5.159, 160 and 157) one obtains

$$\frac{\partial u}{\partial x} = \underline{T} \underline{G}^{(1)} a = \underline{T} \underline{G}^{(1)} \hat{\underline{T}} u = \hat{\underline{G}}^{(1)} u . \tag{5.162}$$

Consideration of a single element of the vector $\partial u / \partial x$ indicates that (5.162) provides, directly, a discretisation formula for $\partial u / \partial x |_j$ in terms of all the u_j nodal values. This may be contrasted with the tridiagonal structure of $\hat{G}^{(1)}$ if a centred finite difference expression were used to represent $\partial u / \partial x |_j$.

In an equivalent way, (5.153b) can be written

$$\frac{\partial^2 u}{\partial x^2} = \underline{T} \underline{a}^{(2)} \tag{5.163}$$

and

$$\underline{a}^{(2)} = \underline{G}^{(1)} \underline{a}^{(1)} = \underline{G}^{(1)} \underline{G}^{(1)} \underline{a} = \underline{G}^{(2)} \underline{a} . \tag{5.164}$$

Elements of $\underline{G}^{(2)}$ can be evaluated from (5.161). From (5.163, 164 and 157)

$$\frac{\partial^2 u}{\partial x^2} = \underline{T} \underline{G}^{(2)} a = \underline{T} \underline{G}^{(2)} \hat{\underline{T}} u = \hat{\underline{G}}^{(2)} u . \tag{5.165}$$

The matrices $\hat{G}^{(1)}$ and $\hat{G}^{(2)}$ can be evaluated once and for all and stored. Subsequently all operations, e.g. at each time level, are executed in physical space.

The application of the present construction to the diffusion equation (5.147) would replace steps (i) and (ii) with the evaluation of $[\partial^2 u / \partial x^2]_j^n$ from (5.165). It may be noted that this is an $O(N^2)$ process whereas the evaluation of steps (i) and (ii) involve two fast Fourier transforms and an $O(N)$ recurrence. Because of the overheads involved in applying the FFT the use of (5.165) is competitive for N typically less than 100 (Ku et al. 1987).

For the rather extreme case of $N=2$ in (5.150) the above pseudospectral procedure produces the same discretisation of the diffusion equation (5.147) as the FTCS scheme (3.4).

The explicit matrix interpretation of the pseudospectral method will be considered further in Sect. 17.1.6 in relation to computing solutions for incompressible viscous flow. More advanced treatments of various aspects of spectral methods may be found in Gottlieb and Orszag (1977), Fletcher (1984), Voigt et al. (1984), Peyret (1986) and Canuto et al. (1987).

5.7 Closure

The method of weighted residuals provides a general framework for comparing many approximate methods (Fletcher 1984, pp. 32–39). Here the Galerkin finite element, Galerkin spectral and finite volume methods have been considered, in more detail.

Like finite difference methods, both the finite volume and finite element methods are local methods. Consequently both methods produce discretised equations that are sparse, although not as sparse, typically, as a finite difference method. Both the finite element method, through the isoparametric formulation, and the finite volume method are suitable for irregular computational domains. Both methods can also be applied with generalised coordinates (Chap. 12).

The finite volume method has the additional advantage of discretising, directly, the conservation form (Sect. 11.2) of the governing equations. This implies that the discretised equations preserve the conservation laws, allowing for the discretisation error.

By contrast the spectral method is a global method. In practice this makes it a more difficult method to implement but does permit approximate solutions of high accuracy with relatively few terms in the approximate solution, (5.2). The relative merits of the finite difference, finite element and spectral methods are compared by Fletcher (1984, Chap. 6). Another powerful global method, orthogonal collocation, is compared with the finite difference and finite element methods by Finlayson (1980, Chaps. 4–6).

5.8 Problems

Weighted Residual Methods: General Formulation (Sect. 5.1)

5.1   The equation $d^2\bar{y}/dx^2 + \bar{y} = [1 - (5\pi/6)^2]\sin(5\pi x/6)$, subject to the boundary conditions $\bar{y}(0)=0$, $\bar{y}(1)=0.5$, is to be solved in the domain $0 \leq x \leq 1.0$ using the following methods of weighted residuals:

(a) Galerkin,

(b) subdomain $(0 \leq x \leq 0.5; 0.5 \leq x \leq 1.0)$,

(c) least-squares.

Assume an approximate solution of the form

$$y = a_1(x - x^3) + a_2(x^2 - x^3) + 0.5x^3 .$$

Compare the computational solutions with the exact solution $\bar{y} = \sin(5\pi x/6)$.

5.2 With a suitable nondimensionalisation the diffusion or heat conduction equation can be written

$$\frac{\partial \bar{\theta}}{\partial t} - \frac{\partial^2 \bar{\theta}}{\partial x^2} = 0 .$$

For initial conditions $\bar{\theta}(x, 0) = \sin(\pi x) + x$ and boundary conditions $\bar{\theta}(0, t) = 0$, $\bar{\theta}(1, t) = 1.0$, solutions to the above equation are to be obtained in the computational domain $0 \leq x \leq 1.0$, $0 \leq t \leq 0.20$, using the following methods:

(a) Galerkin,

(b) subdomain,

(c) collocation.

The following approximate solution should be used

$$\theta = \sin(\pi x) + x + \sum_{j=1}^{N} a_j(t)(x^j - x^{j+1}) .$$

Application of the above methods of weighted residuals to the nondimensional heat conduction equation produces a system of ordinary differential equations for the coefficients $a_j(t)$. These equations are marched in time via numerical integration.

Obtain solutions (after writing a computer program) for $N = 3, 5, 7$ for the three methods and compare the accuracy and rate of convergence. The exact solution for this problem is

$$\bar{\theta} = \sin(\pi x) e^{-\pi^2 t} + x .$$

5.3 Viscous flow in a rectangular duct is governed by (5.97)

$$\left(\frac{b}{a}\right)^2 \frac{\partial^2 \bar{w}}{\partial x^2} + \frac{\partial^2 \bar{w}}{\partial y^2} + 1 = 0 ,$$

subject to boundary conditions $\bar{w} = 0$ at $x = \pm 1$, $y = \pm 1$. The 'exact' solution for this problem is given by (5.110), if NEM is large enough. Approximate solutions are to be obtained using the

(a) Galerkin,

(b) subdomain,

(c) collocation

methods. As an approximate solution, use

$$w = \sum_{j=1}^{N} a_j (1-x^2)^j (1-y^2)^j ,$$

which is suggested by the appropriate form of the limiting one-dimensional velocity profiles ($b/a=0$ and ∞). Obtain solutions for $N=1$, 2 and 3 and compare with the finite element solutions obtained in Sect. 5.5.2. Comment on the accuracy per nodal unknown in relation to the appropriateness of the assumed approximate solution. In this regard include (5.110) with small values of NEM, since it is also a Galerkin solution (Fletcher 1984, p. 13).

Finite Volume Method (Sect. 5.2)

5.4 Obtain solutions to Laplace's equation in the region shown in Fig. 5.4, using FIVOL, for the following parameter values:

(a) $r_W=0.10$, $r_X=4.00$, $r_Y=1.00$, $r_Z=0.10$, $\theta_{WX}=0$, $\theta_{ZY}=90$;
(b) $r_W=0.10$, $r_X=8.00$, $r_Y=1.00$, $r_Z=0.10$, $\theta_{WX}=0$, $\theta_{ZY}=90$.

For each case obtain results for the three grids, JMAX=KMAX=6, 11, 21. Compare the accuracy and rate of convergence with the results shown in Table 5.5.

5.5 Extend program FIVOL to solve the equation

$$\frac{\partial \bar{\phi}}{\partial x} + \frac{\partial \bar{\phi}}{\partial y} - \alpha \left(\frac{\partial^2 \bar{\phi}}{\partial x^2} + \frac{\partial^2 \bar{\phi}}{\partial y^2} \right) = S .$$

The source term S may be treated in the same way as $\partial q/\partial t$ in (5.23). Test the program for the case $S=[\cos(2\theta)-\sin(2\theta)]/r^2$ by obtaining solutions for the same parameter values as used to obtain Table 5.5 and $\alpha=1$. Compare the solutions with the exact solution $\bar{\phi}=(\sin\theta)/r$. The boundary conditions for the above choice of S are given by (5.31).

5.6 The finite volume method, as described in Sect. 5.2, is to be applied to the situation where a Neumann boundary condition is given on WX, e.g.

$$\left. \frac{\partial \bar{\phi}}{\partial y} \right|_{WX} = \frac{1}{r_{WX}^2} .$$

The rest of the formulation is the same and the problem has the same exact solution as in Sect. 5.2.

To obtain equations on WX follow the same procedure as shown in Fig. 5.5 but use a "half-sized" finite volume centred on WX and only extending into the computational domain. The equivalent of (5.34) can be obtained directly, and approximated as in FIVOL except that $\partial\phi/\partial x|_{WX}$ may be obtained from the boundary condition. The term $\partial\phi/\partial x|_{WX}$ can be evaluated conveniently from a three-point finite difference approximation since WX coincides with the x direction. Make the necessary modifications to FIVOL and test for the same conditions as shown in Table 5.5.

Finite Element Method and Interpolation (Sect. 5.3)

5.7 Apply linear and quadratic finite element interpolation to the function

$$\bar{y} = (1 + \sin \pi x)e^{-2x}$$

over the range $0 \le x \le 1.0$. Obtain the interpolated function values for 3, 5, 9 and 17 equally spaced nodal points spanning the computational range. Calculate the rms error based on 20 equally spaced points in the range $0 \le x \le 1.0$, and demonstrate that the theoretically predicted convergence rates are achieved.

5.8 For the function considered in Problem 5.7 carry out linear and quadratic finite element interpolation on a geometrically growing grid with a growth factor of 1.2 for the following cases $(0 \le x \le x_{max})$:
(a) $\Delta x = 0.2, 0.24, 0.288, 0.3456, \ldots$;
(b) $\Delta x = 0.1, 0.12, 0.144, \ldots$;
(c) $\Delta x = 0.05, 0.06, 0.072, \ldots$.
Terminate the grid as soon as $x > 1.0$. For quadratic interpolation adjust each midpoint so that $x_{j-1} = 0.5(x_{j-2} + x_j)$, etc. Compute the rms errors based on 201 equally spaced points and compare with the equally spaced interpolated solution for approximately the same number of points spanning the domain. Consider the solution behaviour and contributions to the rms error in different parts of the computational domain and determine whether an interpolated solution of a prescribed accuracy could be obtained more economically by selective placement of the nodal points.

5.9 Obtain linear and quadratic interpolated solutions, on finite elements, of the function

$$\bar{y} = (1 + \sin \pi x)e^{-2x}\sin(\pi y)$$

in the computational domain $0 \le x \le 1, 0 \le y \le 1$. Compute the rms error on a uniform grid five times finer than the finest nodal-point grid. Obtain the rms error for the same number of elements as shown in Table 5.7 and compare the accuracy and rate of convergence with the results shown in Table 5.7.

Finite Element Method; Sturm–Liouville Equation (Sect. 5.4)

5.10 As indicated in Appendix A.2, the contributions to the various one-dimensional finite element operators are obtained on a uniform grid from the formulae

$$M_x = \frac{1}{\Delta x}\int \phi_j \phi_k \, dx \ ,$$

$$L_x = \frac{1}{\Delta x}\int \frac{d\phi_j}{dx}\phi_k \, dx \ ,$$

$$L_{xx} = -\frac{1}{\Delta x}\int \frac{d\phi_j}{dx}\frac{d\phi_k}{dx} \, dx \ .$$

Show (by evaluation of the integrals) that, for linear elements on a uniform grid, the above operators centered at the kth node have the values

$$M_x = \left(\frac{1}{6}, \frac{2}{3}, \frac{1}{6}\right), \quad L_x = \left(-\frac{0.5}{\Delta x}, 0, \frac{0.5}{\Delta x}\right), \quad L_{xx} = \left(\frac{1}{\Delta x^2}, -\frac{2}{\Delta x^2}, \frac{1}{\Delta x^2}\right).$$

5.11 Modify program STURM so that it will cope with boundary conditions of the form

$$\bar{y}(0) = a, \quad \frac{d\bar{y}}{dx}(1) = b.$$

This will require modification to the evaluation of $b_{m,j}$ and g_m when $m = J$, taking into account all non-zero contributions arising in (5.73). Also some adjustment of the tridiagonal system of equations will be necessary prior to solution by BANFAC/BANSOL. Test the modifications by solving numerically the equation

$$\frac{d^2\bar{y}}{dx^2} + \bar{y} = F = \left(1 - \frac{25\pi^2}{36}\right)\cos\left(\frac{5\pi x}{6}\right),$$

with boundary conditions $\bar{y}(0) = 1$, $d\bar{y}/dx(1) = -5\pi/12$ in the range $0 \le x \le 1.0$. This problem has the exact solution $\bar{y} = \cos(5\pi x/6)$. Obtain solutions using both linear and quadratic interpolation for the grid sizes shown in Table 5.10 and compare the accuracy and convergence rate.

5.12 Apply the Galerkin finite element method, with linear and quadratic interpolation, to the equation

$$\frac{d\bar{y}}{dx} - \bar{y} = 0$$

with the boundary condition $\bar{y}(0) = 1.0$ in the range $0 \le x \le 1.0$. This problem has the exact solution $y = \exp x$. Solve the implicit equations using BANFAC/BANSOL as in STURM. Obtain solutions on a uniform grid of increasing refinement until the rms accuracy is approximately equal to those shown in Table 5.3. Comment on the relative merit (e.g. accuracy and computational efficiency) of using local low-order approximating functions (as in the finite element method) with global higher-order approximating functions as in the traditional MWR (Sect. 5.1).

Finite Element Method—Further Applications (Sect. 5.5)

5.13 Obtain solutions for viscous flow in a rectangular duct, using program DUCT on an 11×11 grid, for various values of b/a until the centreline solution across the smaller dimension is within 1% rms of the one-dimensional parabolic profile, e.g. $u = 1.5(1 - y^2)$.

5.14 Modify program DUCT to solve Laplace's equation in the domain $0 \le x \le 1.0$, $0 \le y \le 1.0$, subject to the boundary conditions

$$\bar{w}(x,0) = \cos(0.5\pi x) , \quad \bar{w}(1,y) = 0 , \quad \bar{w}(x,1) = e^{0.5\pi}\cos(0.5\pi x) ,$$

$$\frac{\partial \bar{w}}{\partial x}(0,y) = 0 .$$

This problem has the exact solution $\bar{w} = \cos(0.5\pi x)\exp(0.5\pi y)$. When applying the finite element method to this problem the evaluation of the Galerkin equations centred on the line $x = 0$ will only involve integrations over two contiguous elements. For homogeneous Neumann boundary conditions (5.101) is still valid since the terms arising from the integration by parts are all zero. Demonstrate that evaluation of the Galerkin equations over the two contiguous elements makes no difference to L_{yy}^T but replaces (5.104, 105) with the expressions

$$L_{xx} = \left(0, -\frac{1}{\Delta x^2}, \frac{1}{\Delta x^2}\right) ; \quad M_x = \left(0, \frac{1}{3}, \frac{1}{6}\right) .$$

Obtain numerical solutions to the above problem on 6×6, 11×11 and 21×21 grids and compare the accuracy of the finite element and finite difference solutions.

5.15 Modify program DUCT to solve the equation

$$\frac{\partial \bar{\phi}}{\partial x} + \frac{\partial \bar{\phi}}{\partial y} - \alpha \frac{\partial^2 \bar{\phi}}{\partial x^2} - \beta \frac{\partial^2 \bar{\phi}}{\partial y^2} = 0$$

in the domain $0 \le x \le 1.0$, $0 \le y \le 1.0$, subject to the boundary conditions

$$\bar{\phi}(x,0) = e^{x/\alpha} , \quad \bar{\phi}(1,y) = e^{1/\alpha}e^{y/\beta} ,$$

$$\bar{\phi}(0,y) = e^{y/\beta} , \quad \bar{\phi}(x,1) = e^{1/\beta}e^{x/\alpha} .$$

This combination has the exact solution $\bar{\phi}(x,y) = \exp(x/\alpha)\exp(y/\beta)$. For the parameter values $\alpha = 1.0$, $\beta = 1.0$, obtain solutions with a linear finite element formulation and a three-point finite difference formulation on 6×6, 11×11 and 21×21 grids. Compare the accuracy and convergence rates of the finite element and finite difference solutions.

Spectral Method (Sect. 5.6)

5.16 Apply the spectral method to the diffusion equation (as described in Sect. 5.6.1) with initial conditions

$$\bar{T}(x,0) = 5x - 4x^2$$

and Dirichlet boundary conditions

$$\bar{T}(0,t) = 0 , \quad \bar{T}(1,t) = 1 .$$

Demonstrate that the results shown in Table 5.13 can be reproduced. Also obtain solutions for $J=7$ and 9 and compare with the approximate solution T_b and exact solution $\bar{T}$.

5.17 Apply the spectral method to the diffusion equation with the initial condition

$$\bar{T}(x, 0) = 3 - 2x - 2x^2 + 2x^3 \ ,$$

and mixed Dirichlet and Neumann boundary conditions (as in Sect. 5.6.2)

$$\frac{\partial T(0, t)}{\partial x} = -2 \ , \qquad \bar{T}(1, t) = 1.0 \ .$$

Obtain numerical solutions with $\alpha = 1.0$ using $J = 3$, 5 and 7 in (5.136) up to $t = 0.1$. Compare the computational solutions with the exact solution

$$\bar{T} = 3 - 2x - \sum_{j=1}^{N} \left(\frac{2}{(2j-1)\pi} \right)^3 \sin[(2j-1)\pi x] e^{-\alpha(2j-1)^2\pi^2 t}$$

to determine the rate of convergence with J. Confirm numerically that the steady-state solution $\bar{T}_{ss} = 3 - 2x$ is achieved by numerically integrating to large time.

6. Steady Problems

Many of the examples considered in Chaps. 3–5 have included time as an independent variable and the construction of the algorithms has taken this into account. However many problems in fluid dynamics are inherently steady, and the governing equations are often elliptic in character (Sect. 2.4).

As indicated in Sect. 3.1.2 any of the major methods, finite difference, finite element, finite volume or spectral can be used to discretise the steady flow equations, and will produce systems of equations that can be written

$$\underline{A}(V)V = B , \qquad (6.1)$$

where V is the vector of unknown nodal values or coefficients in the trial solution (5.2). $\underline{A}$ contains the algebraic coefficients arising from discretisation and, in general, may depend on the solution (V) itself. B is made up of algebraic coefficients associated with discretisation and known values of V, e.g. that are given by the boundary conditions.

The matrix $\underline{A}(V)$ is typically sparse (few nonzero elements) and the nonzero terms are close to the diagonal, for finite difference, finite element and finite volume methods. For the spectral method $\underline{A}(V)$ will be full (few zero elements). In Fig. 6.1 the elements p, q, r, s and t in $\underline{A}$ are associated with the corresponding nodal positions. The elements p and t associated with nodes (j, $k+1$) and (j, $k-1$) are displaced from the diagonal of $\underline{A}$ by N, the number of unknown nodes in the x-direction.

It is possible to construct iterative techniques for the solution of (6.1) as it stands. However, it can be more efficient to set up an outer iteration for which (6.1) is made linear at each step so that direct methods (Sect. 6.2) can be employed. Newton's method is of this type and is considered in Sect. 6.1.

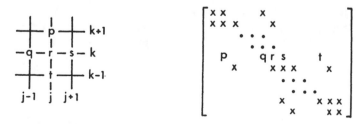

Fig. 6.1. Typical nonzero element pattern in $\underline{A}$ after a finite difference discretisation

If $\underline{A}$ does not depend on V only one step of the outer iteration is required, i.e. (6.1) can be written

$$\underline{A} V = B \tag{6.2}$$

Direct methods (Sect. 6.2) or iterative methods (Sect. 6.3) may be used to solve (6.2). Subroutines FACT and SOLVE are provided in Sect. 6.2.1 to carry out the direct solution of $\underline{A} V = B$, when $\underline{A}$ is full, using Gauss elimination. When $\underline{A}$ is tridiagonal or alternating tridiagonal and pentadiagonal, subroutines BANFAC and BANSOL obtain the solution V also by Gaussian elimination, but much more efficiently than subroutines FACT and SOLVE. For certain structures of $\underline{A}$, such as that produced by a finite difference discretisation of the Poisson equation, special procedures are available (Sect. 6.2.6).

Iterative methods (point Jacobi, Gauss-Seidel and SOR) are illustrated in Sect. 6.3.2 for the Poisson equation arising from the viscous duct flow problem (Sect. 5.5.2). The influence of grid refinement on the optimum relaxation parameter and the convergence rate of point SOR is examined. The extension of point SOR to line SOR and ADI is demonstrated and a comparison with point SOR made. Conjugate gradient (Sect. 6.3.4) and multigrid (Sect. 6.3.5) methods are described primarily as means of accelerating "classical" iterative techniques.

Some iterative methods, when applied directly to (6.1), are conceptually similar to constructing an equivalent unsteady formulation of the original problem. When it is recognised that the transient solution is of no particular value it is possible to modify the time-dependent terms so that convergence to the steady state is achieved with as small an execution time as possible. This is the motivation for pseudo-transient methods (Sect. 6.4).

6.1 Nonlinear Steady Problems

Steady flow problems are often nonlinear due to the nature of the convective terms or, less often, due to the dependence of the flow properties on the solution, e.g. viscosity on the temperature, typically.

Newton's method in several independent variables is a potentially useful technique for solving steady flow problems because of the rapid convergence when close to the solution. The method will be illustrated here for two problems. First a system of highly nonlinear algebraic equations will be solved. Second the two-dimensional steady Burgers' equations (Sect. 10.4) will be considered, since these equations have the same nonlinear structure as the incompressible Navier-Stokes equations (Sect. 11.5).

6.1.1 Newton's Method

Newton's method is a powerful technique for solving nonlinear equations like (6.1). This method is shown schematically in Fig. 6.2.

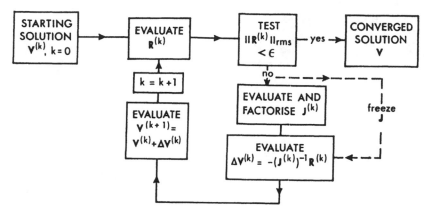

Fig. 6.2. Schematic representation of Newton's method

If (6.1) is written

$$R = \underline{A}(V)V - B = 0 \ , \tag{6.3}$$

then Newton's method can be written as

$$V^{(k+1)} = V^{(k)} - (\underline{J}^{(k)})^{-1}R^{(k)} \ , \tag{6.4}$$

where k is the iteration level and $\underline{J}^{(k)}$ is the Jacobian. An element of $\underline{J}^{(k)}$ is

$$J_{ij} = \frac{\partial R_i^{(k)}}{\partial V_j^{(k)}} \ . \tag{6.5}$$

Equation (6.4) can be rewritten in the form

$$\underline{J}^{(k)} \Delta V^{(k+1)} = -R^{(k)} \ . \tag{6.6}$$

Equation (6.6) is a linear system of equations that is to be solved for the correction vector $\Delta V^{(k+1)} = V^{(k+1)} - V^{(k)}$ at each stage of the iteration.

An attractive feature of Newton's method is that it demonstrates quadratic convergence if the current iteration $V^{(k)}$ is sufficiently close to the converged solution V_c. Quadratic convergence implies

$$\| V^{(k+1)} - V_c \| \approx \| V^{(k)} - V_c \|^2 \ . \tag{6.7}$$

A criterion for the convergence of Newton's method can be developed as follows (after Isaacson and Keller 1966). If $\underline{J}^{(0)}$ has an inverse with its norm bounded by a, i.e.

$$\| \underline{J}^{(0)-1} \| \leq a \ ,$$

and if the first correction vector $\Delta V^{(1)}$ has a norm bounded by b, i.e.

$$\| \Delta V^{(1)} \| = \| -(\underline{J}^{(0)})^{-1} R^{(0)} \| \leq b \ ,$$

and if $R^{(0)}$ has continuous second derivatives which satisfy

$$\sum_{j=1}^{N} \left| \frac{\partial^2 R^{(0)}}{\partial V_i \partial V_j} \right| \le c/N \ ,$$

for all V in $\|\Delta V^{(1)}\| < 2b$, and if the constants satisfy the relation

$$a \ b \ c < 0.5 \ , \tag{6.8}$$

then Newton's method will converge to the solution

$$\lim_{k \to \infty} V^{(k)} = V_c \ ,$$

at which $R(V_c) = 0$ and $\| V^{(k)} - V_c \| \le b/(2^{k-1})$.

The vector norms are maximum norms, e.g. $|V| = \max_i |V_i|$ and the matrix norms are maximum natural norms, e.g.

$$\|\underline{J}\| = \max_i \sum_{j=1}^{N} |J_{ij}| \ .$$

It should be noted that the evaluation of the inequality (6.8) involves the same order of work as attempting to solve the problem itself. The main difficulty with Newton's method is that the radius of convergence, b, decreases as N increases (Rheinboldt 1974), so that $V^{(0)}$ must be close to V_c to ensure convergence.

The main contribution to the execution time in using Newton's method is the factorisation of $\underline{J}^{(k)}$, in the solution of (6.6). It is possible to reduce the execution time by freezing the value of $\underline{J}^{(k)}$ for a number of steps Δk. That is $\underline{J}^{(k)}$ need only be factorised once every Δk steps. However more iterations are required, typically, to reach the converged solution. Methods for solving nonlinear systems of equations, including Newton's method, are discussed by Ortega and Rheinboldt (1970).

6.1.2 NEWTON: Flat-Plate Collector Temperature Analysis

For a flat-plate solar collector, shown schematically in Fig. 6.3, an energy balance for the absorber and two glass covers might produce the following system of nonlinear equations connecting the absolute temperatures of the absorber (T_1) and the two glass covers (T_2 and T_3):

$$R_1 = (T_1^4 + 0.06823T_1) - (T_2^4 + 0.05848T_2) - 0.01509 = 0 \ ,$$

$$R_2 = (T_1^4 + 0.05848T_1) - (2T_2^4 + 0.11696T_2) + (T_3^4 + 0.05848T_3) = 0 \ , \tag{6.9}$$

$$R_3 = (T_2^4 + 0.05848T_2) - (2.05T_3^4 + 0.2534T_3) + 0.06698 = 0.$$

In (6.9) T_1, T_2 and T_3 are absolute temperatures/1000 to make the equations better conditioned.

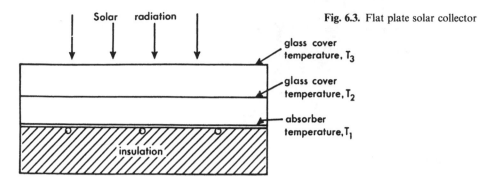

Fig. 6.3. Flat plate solar collector

Multicomponent Newton's method (Sect. 6.1.1) applied to (6.9) requires the solution, at each iteration step (k), of

$$\underline{J}^{(k)}\Delta T^{(k+1)} = -R^{(k)} , \tag{6.10}$$

for $\Delta T^{(k+1)}$. The solution is then updated by

$$T^{(k+1)} = T^{(k)} + \Delta T^{(k+1)} .$$

In (6.10), $J_{ij} = \partial R_i / \partial T_j$.

For this problem,

$$\underline{J} = \begin{pmatrix} (4T_1^3 + 0.06823) & -(4T_2^3 + 0.05848) & 0. \\ (4T_1^3 + 0.05848) & -(8T_2^3 + 0.11696) & (4T_3^3 + 0.05848) \\ 0. & (4T_2^3 + 0.05848) & -(8.2T_3^3 + 0.2534) \end{pmatrix} . \tag{6.11}$$

This problem has been incorporated into a computer program, NEWTON. A flow chart for NEWTON is shown in Fig. 6.4 and the listings of the subroutines

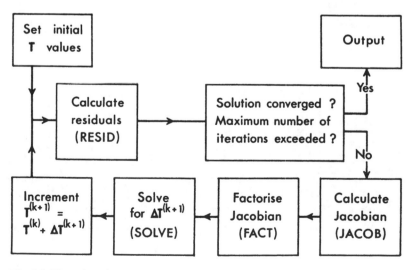

Fig. 6.4. Flow chart for program NEWTON

```
1  C
2  C      NEWTON APPLIES NEWTON'S METHOD TO SOLVE A NONLINEAR SYSTEM OF
3  C      ALGEBRAIC EQUATIONS, R(T).
4  C      RESID EVALUATES THE RESIDUALS
5  C      JACOB EVALUATES THE JACOBIAN
6  C      FACT FACTORISES THE JACOBIAN INTO L.U
7  C      SOLVE SOLVES THE LINEAR SYSTEM FOR DT
8  C
9         DOUBLE PRECISION SUM,RMSR,RMST,DSQRT
10        DIMENSION T(50),AJ(50,50),R(50),JPVT(50),TEX(50)
11 C
12        OPEN(1,FILE='NEWTON.DAT')
13        OPEN(6,FILE='NEWTON.OUT')
14        READ(1,1)N,ITMX,EPS
15        READ(1,2)(T(J),J=1,N)
16        READ(1,3)(TEX(J),J=1,N)
17      1 FORMAT(2I5,E10.3)
18      2 FORMAT(10F5.2)
19      3 FORMAT(7F10.7)
20 C
21        WRITE(6,4)N,ITMX,EPS
22      4 FORMAT(' NEWTONS METHOD FOR N =',I3,'  ITMX= 'I3,'  EPS= ',E10.3)
23        WRITE(6,5)(T(J),J=1,N)
24      5 FORMAT(' INIT SOLN =',3F5.2,//)
25        AN = N
26        IT = 0
27      6 CONTINUE
28 C
29 C      CALCULATE RESIDUALS
30 C
31        CALL RESID(N,T,R)
32 C
33        SUM = 0.
34        DO 7 I = 1,N
35      7 SUM = SUM + R(I)*R(I)
36        RMSR = DSQRT(SUM/AN)
37        SUM = 0.
38        DO 8 I = 1,N
39      8 SUM = SUM + (T(I) - TEX(I))**2
40        RMST = DSQRT(SUM/AN)
41        WRITE(6,9)RMSR,RMST,(R(J),J=1,3)
42      9 FORMAT(' RMS RH=',D11.4,' RMS T =',D11.4,'  R=',3D11.4)
43        IF(RMST .LT. EPS)GOTO 12
44        IT =.IT + 1
45        IF(IT .EQ. ITMX)GOTO 12
46 C
47 C      CALCULATE JACOBIAN
48 C
49        CALL JACOB(N,T,AJ)
50 C
51 C      FACTORISE THE JACOBIAN INTO L.U
52 C
53        CALL FACT(N,AJ,JPVT)
54 C
55        IF(JPVT(N) .EQ. -1)WRITE(6,10)
56     10 FORMAT(' ZERO PIVOT DETECTED')
57        IF(JPVT(N) .EQ. -1)GOTO 15
```

Fig. 6.5. Listing of program NEWTON

```
58 C
59 C      SOLVE FOR THE CORRECTION, DT
60 C
61        CALL SOLVE(N,AJ,JPVT,R)
62 C
63 C      INCREMENT T
64 C
65        DO 11 I = 1,N
66        DT = -R(I)
67     11 T(I) = T(I) + DT
68        GOTO 6
69 C
70 C      GENERATE OUTPUT
71 C
72     12 WRITE(6,13)IT,RMST
73     13 FORMAT(//,' AFTER ',I3,' ITERATIONS THE RMS RESIDUAL IS ',E12.5)
74        WRITE(6,14)(T(I),I=1,N)
75     14 FORMAT(' T = '7F10.7)
76     15 CONTINUE
77        STOP
78        END
```

Fig. 6.5. (cont.) Listing of program NEWTON

indicated in Fig. 6.4 are provided in Figs. 6.5–7. The various parameters used by program NEWTON are shown in Table 6.1. The residuals (6.9) are calculated in RESID (Fig. 6.6) and a test is made (NEWTON, lines 43–45) to stop if the rms residual is below tolerance or if the maximum number of iterations has been exceeded. Subroutine JACOB (Fig. 6.7) evaluates the terms in the Jacobian (6.11). The linear system (6.10) is solved in two stages. First the Jacobian is factorised into LU form by subroutine FACT and subsequently the reduced system is solved by SOLVE. Subroutines FACT and SOLVE are described in Sect. 6.2.1.

The solution of (6.9) by program NEWTON is shown in Fig. 6.8. The solution indicates that the temperatures of the absorber surface and two glass covers are 415 K (142°C), 379 K (106°C) and 334 K (61°C), respectively. The rms error and rms residual are showing approximate quadratic convergence, as in (6.7), as the converged solution is approached. This example demonstrates how effective Newton's method is for *small* systems of highly nonlinear algebraic equations.

```
 1        SUBROUTINE RESID(N,T,R)
 2 C
 3 C      EVALUATES RESIDUALS REQUIRED BY NEWTON'S METHOD
 4 C
 5        DIMENSION T(50),R(50),T4(50)
 6 C
 7        DO 1 J = 1,N
 8        DUM = T(J)
 9      1 T4(J) = DUM*DUM*DUM*DUM
10        C = 0.05848
11        DUM = T4(2) + C*T(2)
12        DAM = T4(3) + C*T(3)
13        R(1) = T4(1) + 0.06823*T(1) - DUM - 0.01509
14        R(2) = T4(1) + C*T(1) - 2.*DUM + DAM
15        R(3) = DUM - 2.05*T4(3) - 0.2534*T(3) + 0.06698
16        RETURN
17        END
```

Fig. 6.6. Listing of subroutine RESID

```
1       SUBROUTINE JACOB(N,T,AJ)
2  C
3  C    EVALUATES JACOBIAN REQUIRED BY NEWTON'S METHOD
4  C
5       DIMENSION T(50),AJ(50,50),T3(50)
6  C
7       DO 2 I = 1,N
8       DO 1 J = 1,N
9     1 AJ(I,J) = 0.
10      DUM = T(I)
11      T3(I) = 4.*DUM*DUM*DUM
12    2 CONTINUE
13      C = 0.05848
14      AJ(1,1) = T3(1) + 0.06823
15      AJ(1,2) = - T3(2) - C
16      AJ(2,1) = T3(1) + C
17      AJ(2,2) = - 2.*T3(2) - 0.11696
18      AJ(2,3) = T3(3) + C
19      AJ(3,2) = T3(2) + C
20      AJ(3,3) = - 2.05*T3(3) - 0.2534
21      RETURN
22      END
```

Fig. 6.7. Listing of subroutine JACOB

Table 6.1. Parameters used by the program NEWTON

Parameter	Description
N	number of equations
IT	iteration counter
ITMX	maximum number of iterations
EPS	tolerance on rms solution error
T	temperature, dependent variable
R	equation residuals, (6.9) holds ΔT on return from SOLVE
RMSR	rms residual
RMST	rms solution error, $\| T - \bar{T} \|_{rms}$
AJ	Jacobian, $\underline{J}$
JPVT	index of kth pivot row (in FACT)
	$JPVT(N) = -1$, if zero pivot occurs
DT	ΔT, correction to T

```
NEWTONS METHOD FOR N =  3  ITMX=  10  EPS=    .100E-04
INIT SOLN =  .30  .30  .30


RMS RH=  .7023D-02 RMS T =  .8307D-01  R= -.1216D-01   .6971D-09 -.1007D-05
RMS RH=  .9400D-02 RMS T =  .2885D-01  R=  .1344D-01   .7722D-02  .4993D-02
RMS RH=  .1317D-02 RMS T =  .3148D-02  R=  .1738D-02   .1449D-02  .2885D-03
RMS RH=  .2372D-04 RMS T =  .4647D-04  R=  .2928D-04   .2881D-04  .4479D-06
RMS RH=  .5353D-08 RMS T =  .4867D-07  R=  .6588D-08   .5570D-08  .3399D-08


AFTER   4 ITERATIONS THE RMS RESIDUAL IS   .48667E-07
T =  .4151283  .3794904  .3335792
```

Fig. 6.8. Typical output produced by program NEWTON

6.1.3 NEWTBU: Two-Dimensional Steady Burgers' Equations

The unsteady form of these equations are discussed more fully in Sect. 10.4. Here the steady two-dimensional Burgers' equations

$$\bar{u}\frac{\partial \bar{u}}{\partial x} + \bar{v}\frac{\partial \bar{u}}{\partial y} - \frac{1}{\text{Re}}\left(\frac{\partial^2 \bar{u}}{\partial x^2} + \frac{\partial^2 \bar{u}}{\partial y^2}\right) = 0 \ ,$$

$$\bar{u}\frac{\partial \bar{v}}{\partial x} + \bar{v}\frac{\partial \bar{v}}{\partial y} - \frac{1}{\text{Re}}\left(\frac{\partial^2 \bar{v}}{\partial x^2} + \frac{\partial^2 \bar{v}}{\partial y^2}\right) = 0 \ ,$$

(6.12)

with Dirichlet boundary conditions on u and v are to be solved using Newton's method after discretisation using three-point centred difference formulae. The discretised form of (6.12) can be written

$$Ru_{j,k} = u_{j,k}L_x u_{j,k} + v_{j,k}L_y u_{j,k} - \frac{1}{\text{Re}}(L_{xx}u_{j,k} + L_{yy}u_{j,k}) = 0 \ ,$$

(6.13)

$$Rv_{j,k} = u_{j,k}L_x v_{j,k} + v_{j,k}L_y v_{j,k} - \frac{1}{\text{Re}}(L_{xx}v_{j,k} + L_{yy}v_{j,k}) = 0 \ ,$$

where

$$L_x u_{j,k} = \frac{u_{j+1,k} - u_{j-1,k}}{2\Delta x} \ ,$$

(6.14)

$$.L_{xx}u_{j,k} = \frac{u_{j-1,k} - 2u_{j,k} + u_{j+1,k}}{\Delta x^2} \ ,$$

$$L_y = \left(\frac{0.5}{\Delta y}, 0, -\frac{0.5}{\Delta y}\right)^T$$

and

$$L_{yy} = \left(\frac{1}{\Delta y^2}, -\frac{2}{\Delta y^2}, \frac{1}{\Delta y^2}\right)^T$$

The solution is sought in the domain shown in Fig. 6.9.

The boundary values of u and v are taken from the exact solution (Fletcher 1983)

$$\bar{u} = -\frac{2}{\text{Re}\phi}\frac{\partial \phi}{\partial x} \quad \text{and} \quad \bar{v} = -\frac{2}{\text{Re}\phi}\frac{\partial \phi}{\partial y} \ ,$$

(6.15)

where

$$\phi = a_1 + a_2 x + a_3 y + a_4 xy + a_5(e^{\lambda(x-x_0)} + e^{-\lambda(x-x_0)})\cos(\lambda y)$$

(6.16)

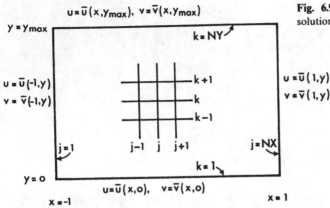

Fig. 6.9. Computation domain for solution of (6.12)

and a_1, a_2, a_3, a_4, a_5, λ and x_0 are chosen to produce different behaviour of the exact solution.

The evaluation of (6.13) at all internal nodes provides $2(NX-2)(NY-2)$ equations that must be solved for $u_{j,k}$ and $v_{j,k}$ for all internal nodes. The solution process, using Newton's method, can be written as

$$\underline{J}^{(n)} \Delta q^{(n+1)} = -R^{(n)} , \tag{6.17}$$

where Δq is made up of $\Delta u_{j,k}$, $\Delta v_{j,k}$ and the vector R contains $Ru_{j,k}$ and $Rv_{j,k}$ evaluated at each internal node. The superscript (n) is used here to denote the iteration level. The Jacobian $\underline{J}$ contains terms like

$$\partial Ru_{j,k}/\partial u_{j-1,k} = -0.5u_{j,k}/\Delta x - 1/(Re\Delta x^2) ,$$

$$\partial Ru_{j,k}/\partial u_{j,k} = (2/Re)(1/\Delta x^2 + 1/\Delta y^2) + (0.5/\Delta x)(u_{j+1,k} - u_{j-1,k}) ,$$

$$\partial Ru_{j,k}/\partial u_{j+1,k} = 0.5u_{j,k}/\Delta x - 1/(Re\Delta x^2) ,$$

$$\partial Ru_{j,k}/\partial u_{j,k-1} = -0.5v_{j,k}/\Delta y - 1/(Re\Delta y^2) , \tag{6.18}$$

$$\partial Ru_{j,k}/\partial u_{j,k+1} = 0.5v_{j,k}/\Delta y - 1/(Re\Delta y^2) ,$$

$$\partial Ru_{j,k}/\partial v_{j,k} = (0.5/\Delta y)(u_{j,k+1} - u_{j,k-1}) .$$

Similar terms arise associated with $Rv_{j,k}$.

The above formulation is implemented in program NEWTBU. Listings of NEWTBU, EXBUR to provide the exact solution, RESBU to evaluate (6.13) and JACBU to evaluate (6.18) are given in Figs. 6.10–13. Table 6.2 lists the parameters used in NEWTBU.

The equations (6.13) are located in R in the order,

$$Ru_{2,2}, Rv_{2,2}, Ru_{2,3}, Rv_{2,3}, \ldots, Ru_{2,NY-1}, Rv_{2,NY-1}, Ru_{3,2}, Rv_{3,2}, \ldots$$

and in q the velocities are ordered in a corresponding manner,

$$u_{2,2}, v_{2,2}, u_{2,3}, v_{2,3}, \ldots, u_{2,NY-1}, v_{2,NY-1}, u_{3,2}, v_{3,2}, \ldots .$$

The exact solution of (6.12), given by (6.15) with parameter values

$$a_1 = a_2 = 110.13 \ , \qquad a_3 = a_4 = 0 \ , \qquad a_5 = 1.0 \ , \qquad \lambda = 5 \ ,$$

$$\text{Re} = 10 \ , \qquad y_{max} = \pi/30 \quad \text{and} \quad x_0 = 1.0 \ ,$$

has been used as the starting solution for $\bar{u}$ and $\bar{v}$. However, program NEWTBU applied on a 5×5 grid, i.e. 18 unknown u and v nodal values, fails to produce a converged solution. To obtain a converged solution it is necessary to

```
 1
 2 C
 3 C      NEWTBU APPLIES NEWTON'S METHOD TO SOLVE BURGERS' EQUATIONS
 4 C      FOR U(X,Y) AND V(X,Y).
 5 C      EXBUR EVALUATES THE EXACT SOLUTION OF BURGERS' EQUATIONS
 6 C      RESBU EVALUATES THE RESIDUALS
 7 C      JACBU EVALUATES THE JACOBIAN
 8 C      FACT FACTORISES THE JACOBIAN INTO L.U
 9 C      SOLVE SOLVES THE LINEAR SYSTEM FOR DT
10 C      IRD CONTROLS READ/WRITE OF STARTING SOLUTION
11 C      IPN IS THE INCREMENT FOR PRINTING ITERATIVE RMS
12 C
13        DOUBLE PRECISION SUM,DSQRT,R,U,V
14        DIMENSION U(21,21),V(21,21),UE(21,21),VE(21,21),A(5),R(50)
15       1,AJ(50,50),JPVT(50),RD(50)
16        COMMON DX,DY,RE,NX,NY,DTI
17 C
18        OPEN(1,FILE='NEWTBU.DAT')
19        OPEN(6,FILE='NEWTBU.OUT')
20        OPEN(7,FILE='NEWTBU.STO')
21        READ(1,1)NX,NY,ITMX,IRD,IPN,EPS,RE,OM,DT
22        READ(1,2)(A(J),J=1,5),AL
23      1 FORMAT(5I5,4E10.3)
24      2 FORMAT(5F8.2,F5.2)
25 C
26        NXP = NX - 1
27        NYP = NY - 1
28        N = (NX -2)*(NY - 2)*2
29        WRITE(6,3)N,ITMX,IRD,IPN
30      3 FORMAT(' NEWTONS METHOD FOR N =',I3,5X,'ITMX=',I3,
31       1' IRD=',I2,'  IPN=',I2)
32        WRITE(6,4)DT,EPS,RE,OM
33      4 FORMAT(' DT=',E12.5,'  EPS=',E10.3,'  RE=',E10.3,
34       1' OM=',E10.3)
35        WRITE(6,5)(A(J),J=1,5),AL
36      5 FORMAT(' A =',5F8.2,'  AL=',F5.2,//)
37 C
38 C      CALCULATE EXACT SOLUTION
39 C
40        CALL EXBUR(UE,VE,A,AL)
41 C
42 C      GENERATE INITIAL SOLUTION
43 C
44        IF(IRD .LT. 2)GOTO 8
45 C
46 C      IRD = 2 OBTAIN STARTING SOLUTION FROM NEWTBU.STO (UNIT 7)
47 C
48        DO 7 K = 1,NY
49        READ(7,6)(U(K,J),J=1,NX)
50        READ(7,6)(V(K,J),J=1,NX)
51      6 FORMAT(10F8.5)
52      7 CONTINUE
53        GOTO 11
54 C
```

Fig. 6.10. Listing of program NEWTBU

```
55     8 DO 10 J = 1,NX
56        DO 9 K = 1,NY
57        U(K,J) = UE(K,J)
58        V(K,J) = VE(K,J)
59      9 CONTINUE
60     10 CONTINUE
61 C
62     11 DO 12 K = 1,NY
63     12 WRITE(6,13)(UE(K,J),J=1,NX)
64     13 FORMAT(' UE=',7F10.4)
65        DO 14 K = 1,NY
66     14 WRITE(6,15)(VE(K,J),J=1,NX)
67     15 FORMAT(' VE=',7F10.4)
68 C
69        AN = N
70        IT = 0
71        ITP = 0
72        OMH = OM
73        DTI = 1./DT
74     16 CONTINUE
75 C
76 C      CALCULATE RESIDUALS
77 C
78        CALL RESBU(U,V,R)
79 C
80        SUM = 0.
81        DO 17 I = 1,N
82        RD(I) = R(I)
83     17 SUM = SUM + R(I)*R(I)
84     18 RMS = DSQRT(SUM/AN)
85        IF(IT .GE. ITP)WRITE(6,19)IT,RMS,(R(J),J=1,3)
86     19 FORMAT(' IT=',I3,'  RMS=',D11.4,'  R=',3D11.4)
87 C
88 C      TEST FOR CONVERGENCE OR MAXIMUM NUMBER
89 C      OF ITERATIONS EXCEEDED
90 C
91        IF(RMS .LT. EPS)GOTO 23
92        IF(IT .GE. ITP)ITP = ITP + IPN
93        IT = IT + 1
94        IF(IT .EQ. ITMX)GOTO 23
95        IF(RMS .GT. 1.0E+03)GOTO 30
96 C
97 C      CALCULATE JACOBIAN
98 C
99        CALL JACBU(U,V,AJ)
100 C
101 C     FACTORISE THE JACOBIAN INTO L.U
102 C
103        CALL FACT(N,AJ,JPVT)
104 C
105        IF(JPVT(N) .EQ. -1)WRITE(6,20)
106     20 FORMAT(' ZERO PIVOT DETECTED')
107        IF(JPVT(N) .EQ. -1)GOTO 30
108 C
109 C     SOLVE FOR THE CORRECTION, DT
110 C
111        CALL SOLVE(N,AJ,JPVT,RD)
112 C
113 C     INCREMENT U AND V
114 C
```

Fig. 6.10. (cont.) Listing of program NEWTBU

```
115        DO 22 J = 2,NXP
116        DO 21 K = 2,NYP
117        MB = 2*(K-2) + 2*(NY-2)*(J-2)
118        DU = - RD(MB+1)
119        DV = - RD(MB+2)
120        U(K,J) = U(K,J) + DU*OM
121        V(K,J) = V(K,J) + DV*OM
122     21 CONTINUE
123     22 CONTINUE
124        GOTO 16
125 C
126 C      GENERATE OUTPUT
127 C
128     23 WRITE(6,24)IT,RMS
129     24 FORMAT(//,' AFTER ',I3,' ITERATIONS THE RMS RESIDUAL IS ',
130        1D12.5)
131        DO 25 K = 1,NY
132     25 WRITE(6,26)(U(K,J),J=1,NX)
133     26 FORMAT(' U=',7F10.4)
134        DO 27 K = 1,NY
135     27 WRITE(6,28)(V(K,J),J=1,NX)
136     28 FORMAT(' V=',7F10.4)
137 C
138        IF(IRD .EQ. 0)GOTO 30
139        DO 29 K = 1,NY
140        WRITE(7,6)(U(K,J),J=1,NX)
141        WRITE(7,6)(V(K,J),J=1,NX)
142     29 CONTINUE
143     30 CONTINUE
144        STOP
145        END
```

Fig. 6.10. (cont.) Listing of program NEWTBU

```
1
2          SUBROUTINE EXBUR(UE,VE,A,AL)
3 C
4 C        CALCULATES THE EXACT SOLUTION OF THE TWO-DIMENSIONAL BURGERS'
5 C        EQUATIONS USING THE COLE-HOPF TRANSFORMATION
6 C
7          DIMENSION A(5),UE(21,21),VE(21,21),PH(21,21)
8          COMMON DX,DY,RE,NX,NY,DTI
9          PI = 3.1415927
10         XZ = 1.0
11         YMAX = PI/6./AL
12         ANY = NY - 1
13         DY = YMAX/ANY
14         ANX = NX - 1
15         DX = 2./ANX
16         DO 2 J = 1,NX
17         AJ = J - 1
18         X = -1. + AJ*DX
19         XD = X - XZ
20         DEX = EXP(AL*XD) + EXP(-AL*XD)
21         DDX = EXP(AL*XD) - EXP(-AL*XD)
22         DO 1 K = 1,NY
23         AK = K - 1
24         Y = AK*DY
25         SY = SIN(AL*Y)
26         CY = COS(AL*Y)
27         PH(K,J) = A(1) + A(2)*X + A(3)*Y + A(4)*X*Y + A(5)*DEX*CY
28         PHX = A(2) + A(4)*Y + A(5)*AL*DDX*CY
29         PHY = A(3) + A(4)*X - A(5)*AL*DEX*SY
30         UE(K,J) = - 2.*PHX/PH(K,J)/RE
31         VE(K,J) = - 2.*PHY/PH(K,J)/RE
32      1 CONTINUE
33      2 CONTINUE
34        RETURN
35        END
```

Fig. 6.11. Listing of subroutine EXBUR

```
1
2          SUBROUTINE RESBU(U,V,R)
3 C
4 C     EVALUATES RESIDUALS OF 2D STEADY BURGERS' EQUATIONS
5 C
6          DOUBLE PRECISION R,U,V,CX,CY,CCX,CCY,DUM
7          DIMENSION U(21,21),V(21,21),R(50)
8          COMMON DX,DY,RE,NX,NY,DTI
9 C
10         NXP = NX - 1
11         NYP = NY - 1
12         CX = 0.5/DX
13         CY = 0.5/DY
14         CCX = 1./RE/DX/DX
15         CCY = 1./RE/DY/DY
16         NCT = 1
17 C
18         DO 2 J = 2,NXP
19         JM = J - 1
20         JP = J + 1
21         DO 1 K = 2,NYP
22         KM = K - 1
23         KP = K + 1
24         DUM = CX*U(K,J)*(U(K,JP)-U(K,JM)) + CY*V(K,J)*(U(KP,J)-U(KM,J))
25         R(NCT) = DUM - CCX*(U(K,JM)-2.*U(K,J)+U(K,JP)) - CCY*(U(KM,J)
26       1 - 2.*U(K,J) + U(KP,J))
27         DUM = CX*U(K,J)*(V(K,JP)-V(K,JM)) + CY*V(K,J)*(V(KP,J)-V(KM,J))
28         R(NCT+1) = DUM - CCX*(V(K,JM)-2.*V(K,J)+V(K,JP)) - CCY*(V(KM,J)
29       1 - 2.*V(K,J) + V(KP,J))
30         NCT = NCT + 2
31       1 CONTINUE
32       2 CONTINUE
33 C
34         RETURN
35         END
```

Fig. 6.12. Listing of subroutine RESBU

underrelax the correction of $q^{(n)}$. Thus the following equation is used:

$$q^{(n+1)} = q^{(n)} + \omega \Delta q^{(n+1)} \tag{6.19}$$

with $\omega = 0.15$. However, the convergence rate is then quite slow (Fig. 6.14). This situation is not unusual in applying Newton's method to the discretised equations governing flow problems. Namely that the corrections calculated by Newton's method are sufficiently large as to throw the current solution outside of the radius of convergence (Sect. 6.1.1), if the corrections are implemented in full.

Equation (6.19) suggests a related strategy. This is to treat $\Delta q^{(n+1)}$ obtained from the solution of (6.17) as a search direction. For three values of ω, new solutions

$$q_m^{(n+1)} = q^{(n)} + \omega_m \Delta q^{(n+1)}$$

are obtained and corresponding residuals $R_m^{(n+1)}$ evaluated. For each value ω_m an rms residual $R_{m,\mathrm{rms}}^{(n+1)}$ is calculated and a quadratic dependence of $R_{m,\mathrm{rms}}^{(n+1)}$ on ω assumed. Then the value of ω, and hence $q^{(n+1)}$, is chosen that leads to the minimum R_{rms} for the given search direction $\Delta q^{(n+1)}$. However, this strategy can be computationally expensive if the residuals Ru and Rv are complicated.

```
1
2            SUBROUTINE JACBU(U,V,AJ)
3    C
4    C       EVALUATES THE JACOBIAN OF THE 2D STEADY BURGERS' EQUATIONS
5    C
6            DOUBLE PRECISION U,V,CX,CY,CCX,CCY
7            DIMENSION U(21,21),V(21,21),AJ(50,50)
8            COMMON DX,DY,RE,NX,NY,DTI
9    C
10           N = (NX - 2)*(NY - 2)*2
11           DO 2 J = 1,N
12           DO 1 K = 1,N
13         1 AJ(K,J) = 0.
14         2 CONTINUE
15   C
16           NXP = NX - 1
17           NYP = NY - 1
18           CX = 0.5/DX
19           CY = 0.5/DY
20           CCX = 1./RE/DX/DX
21           CCY = 1./RE/DY/DY
22   C
23           DO 7 J = 2,NXP
24           JM = J - 1
25           JP = J + 1
26           DO 6 K = 2,NYP
27           KM = K - 1
28           KP = K + 1
29           MB = 2*(K-2) + 2*(NY-2)*(J-2)
30           LU = MB + 1
31           LV = MB + 2
32           ML = MB - 2*(NY - 2)
33           IF(ML .LT. 0)GOTO 3
34           AJ(LU,ML+1) = - CX*U(K,J) - CCX
35           AJ(LV,ML+2) = - CX*U(K,J) - CCX
36         3 AJ(LU,MB+1) = 2.*(CCX+CCY) + CX*(U(K,JP)-U(K,JM)) + DTI
37           AJ(LU,MB+2) = CY*(U(KP,J) - U(KM,J))
38           AJ(LV,MB+1) = CX*(V(K,JP) - V(K,JM))
39           AJ(LV,MB+2) = 2.*(CCX+CCY) + CY*(V(KP,J) - V(KM,J)) + DTI
40           MR = MB + 2*(NY-2)
41           IF(MR .GT. N-2)GOTO 4
42           AJ(LU,MR+1) = CX*U(K,J) - CCX
43           AJ(LV,MR+2) = CX*U(K,J) - CCX
44         4 MK = MB - 2
45           IF(MK .LT. 0)GOTO 5
46           AJ(LU,MK+1) = - CY*V(K,J) - CCY
47           AJ(LV,MK+2) = - CY*V(K,J) - CCY
48         5 MT = MB + 2
49           IF(MT .GT. N-2)GOTO 6
50           AJ(LU,MT+1) = CY*V(K,J) - CCY
51           AJ(LV,MT+2) = CY*V(K,J) - CCY
52         6 CONTINUE
53         7 CONTINUE
54   C
55           RETURN
56           END
```

Fig. 6.13. Listing of subroutine JACBU

Table 6.2. Parameters used by program NEWTBU

Parameter	Description
NX, NY	number of grid points in the x and y directions
ITMX	maximum number of iterations
IRD	$=0, 1$ generate starting solution from subroutine EXBUR
	$=1, 2$ write final solution to file NEWTBU.STA
	$=2$ read starting solution from NEWTBU.STA
IPN	print sample of the residuals every IPN iterations
EPS	tolerance on the rms equation residuals
RE	Reynolds number, Re, (6.13)
OM	under-relaxation factor, ω, (6.19)
DT	time step, used by pseudotransient version (Sect. 6.4)
N	number of equations to be solved
UE, VE	exact solution, $\bar{u}, \bar{v}$
A	$a_1, \ldots , a_5$ (6.16)
U, V	dependent variables (6.13)
R	equation residuals (6.13)
RD	$=R$ after return from RESBU
	contains corrections, Δq, on return from SOLVE
AJ	Jacobian, J, (6.17)
DU, DV	corrections to u and v; Δq in (6.17)

```
NEWTONS METHOD FOR N = 18      ITMX=400 IRD= 1   IPN=20
DT=  .50000E+03  EPS=  .100E-04  RE=  .100E+02  OM=  .150E+00
A =  110.13  110.13     .00     .00    1.00


UE=     .9990    .9586    .4888   -.0559   -.0991
UE=     .9990    .9583    .4863   -.0565   -.0991
UE=     .9990    .9572    .4786   -.0584   -.0991
UE=     .9989    .9553    .4655   -.0614   -.0992
UE=     .9988    .9524    .4462   -.0657   -.0992
VE=     .0000    .0000    .0000    .0000    .0000
VE=     .1317    .1277    .0753    .0090    .0012
VE=     .2679    .2598    .1515    .0179    .0023
VE=     .4142    .4010    .2297    .0266    .0034
VE=     .5774    .5577    .3109    .0349    .0045
IT=  0   RMS=  .1496D+00  R= -.2169D+00 -.2850D-01 -.2188D+00
IT= 20   RMS=  .2854D-01  R= -.8407D-02 -.1105D-02 -.8483D-02
IT= 40   RMS=  .5504D-02  R= -.3259D-03 -.4283D-04 -.3289D-03
IT= 60   RMS=  .1024D-02  R= -.1264D-04 -.1662D-05 -.1277D-04
IT= 80   RMS=  .1892D-03  R= -.4914D-06 -.6470D-07 -.4977D-06
IT=100   RMS=  .3499D-04  R= -.1930D-07 -.2559D-08 -.1981D-07


AFTER 115 ITERATIONS THE RMS RESIDUAL IS   .98726D-05
U=      .9990    .9586    .4888   -.0559   -.0991
U=      .9990    .9605    .4835   -.0567   -.0991
U=      .9990    .9603    .4748   -.0585   -.0991
U=      .9989    .9577    .4625   -.0615   -.0992
U=      .9988    .9524    .4462   -.0657   -.0992
V=      .0000    .0000    .0000    .0000    .0000
V=      .1317    .1282    .0748    .0090    .0012
V=      .2679    .2605    .1507    .0179    .0023
V=      .4142    .4017    .2290    .0266    .0034
V=      .5774    .5577    .3109    .0349    .0045
```

Fig. 6.14. Typical output from program NEWTBU

An alternative, and more effective, strategy to (6.19) is based on combining Newton's method with a pseudotransient formulation (Sect. 6.4). The current problem is solved by the pseudotransient Newton's method in Sect. 6.4.1.

6.1.4 Quasi-Newton Method

An alternative approach (often called the quasi-Newton method) to solving systems of nonlinear algebraic equations, is to replace (6.4) by

$$V^{(k+1)} = V^{(k)} - \omega^{(k)} \underline{H}^{(k)} R^{(k)} , \qquad (6.20)$$

where $\underline{H}^{(k)}$ is an approximation to $(\underline{J}^{(k)})^{-1}$ which is systematically modified at each step of the iteration so that it approaches $(\underline{J}^{(k)})^{-1}$ as the converged solution is approached. The modification of $\underline{H}^{(k)}$ at each step is considerably more economical than the factorisation of $\underline{J}^{(k)}$. Equation (6.20) can be written as

$$V^{(k+1)} = V^{(k)} - \omega^{(k)} \xi^{(k)} , \qquad (6.21)$$

where $\xi^{(k)} = \underline{H}^{(k)} R^{(k)}$, and can be interpreted as a search direction. The scalar $\omega^{(k)}$ is chosen so that $R_{rms}^{(k+1)}$ is a minimum in the search direction, $\xi^{(k)}$. This last feature provides a much larger radius of convergence than the conventional Newton's method, when N is large. A typical algorithm, the BFGS formula (Fletcher 1980, p. 44), would upgrade $V^{(k+1)}$ using (6.21). Subsequently

$$y^{(k)} = R^{(k+1)} - R^{(k)} ,$$

$$P^{(k)} = - \omega^{(k)} \underline{H}^{(k)} R^{(k)} ,$$

$$\beta^{(k)} = 1/P^{(k)T} y^{(k)} , \qquad \text{and}$$

$$\underline{H}^{(k+1)} = \underline{H}^{(k)} - \beta^{(k)} (\underline{H}^{(k)} y^{(k)} P^{(k)T} + P^{(k)} y^{(k)T} \underline{H}^{(k)})$$
$$+ \beta^{(k)} (1 + \beta^{(k)} y^{(k)T} \underline{H}^{(k)} y^{(k)}) P^{(k)} P^{(k)T} . \qquad (6.22)$$

Although J is typically sparse, the structure of (6.22) indicates that $\underline{H}$ is dense. For $k=0$, $\underline{H}^{(0)}$ is set equal to the identity matrix $\underline{I}$.

The effectiveness of quasi-Newton methods often depends on J having special properties, such as positive definiteness (Jennings 1977a). Thus the applicability to fluid flow computational formulations needs to be examined on a case-by-case basis. Much of the literature on quasi-Newton methods relates to unconstrained minimisation (e.g. Powell 1976; Fletcher 1980). Specific discussions of quasi-Newton methods are available in Broyden (1965), Ortega and Rheinboldt (1970) and Shanno (1983). A typical application of a quasi-Newton method to a fluid flow problem is provided by Engelman et al. (1981).

6.2 Direct Methods for Linear Systems

For elliptic problems that are governed by linear equations, e.g. potential flow (Sect. 11.3) or at intermediate stages of Newton's method, it is necessary to solve a linear system of algebraic equations

$$\underline{A}V=B \tag{6.23}$$

for the components of the vector V where all coefficients in the matrix $\underline{A}$ and vector B are known. Gauss elimination (Dahlquist and Bjorck 1974, pp. 147–159) is the preferred technique for solving (6.23) but the efficient implementation of the method depends on the structure of $\underline{A}$. Three categories of $\underline{A}$ are worth identifying:

(a) $\underline{A}$ contains few or no zero coefficients. $\underline{A}$ is said to be full or dense.
(b) $\underline{A}$ contains many zero coefficients. $\underline{A}$ is said to be sparse.
(c) $\underline{A}$ contains many zero coefficients and the non-zero coefficients are clustered close to the main diagonal. $\underline{A}$ is said to be sparse and banded.

Computer programs based on the assumption that $\underline{A}$ is full are more general but also computationally more expensive. Such programs could also be applied to sparse and/or banded matrices, but more economical special procedures are available for these categories. Whatever the structure of $\underline{A}$, Gauss elimination requires two stages. First $\underline{A}$ is factorised into lower triangular, $\underline{L}$, and upper triangular, $\underline{U}$, factors. Second, the factored form of (6.23) is solved as

$$\underline{U}V=\underline{L}^{-1}B \ . \tag{6.24}$$

Because of the structure of $\underline{U}$, (6.24) only involves a back-substitution, once $\underline{L}^{-1}B$ has been formed.

6.2.1 FACT/SOLVE: Solution of Dense Systems

When $\underline{A}$ is dense, all coefficients in $\underline{A}$ are operated on to form $\underline{L}$ and $\underline{U}$. Subroutine FACT (Fig. 6.15) performs the factorisation $\underline{L}\underline{U}=\underline{A}$. The subsequent back-substitution (6.24) is performed by subroutine SOLVE (Fig. 6.16).

At each step of the factorisation a multiple of the current row k is subtracted from each subsequent row i to eliminate v_k. The multiplier is $a_{i,k}/a_{k,k}$ and $a_{k,k}$ is called the pivot. If the pivot is zero or very small then Gauss elimination fails or is very inaccurate. To avoid this situation partial pivoting is introduced. To eliminate v_k a search is made of the kth column in the range $i=k+1, N$ to find the largest pivot $a_{l,k}$ (FACT, lines 17–24). If the largest pivot is zero then an error condition is flagged [JPVT(N)$=-1$] and a return is made to the calling program.

Assuming the largest pivot is not zero, a row interchange is made (FACT, lines 39–41) to bring the largest pivot, $a_{l,k}$, into the position (k, k). For each column j ($j=k+1, N$) coefficients in the ith row ($i=k+1, N$) are modified (FACT, lines 38–46) according to

```
1        SUBROUTINE FACT(N,A,JPVT)
2  C
3  C     FACTORISES A INTO PERMUTED L.U SO THAT PERM*A = L*U
4  C     JPVT(K) GIVES INDEX OF KTH PIVOT ROW
5  C     SETS JPVT(N) = -1 IF ZERO PIVOT OCCURS
6  C
7        DIMENSION A(50,50),JPVT(50)
8        NM = N - 1
9  C
10 C     GAUSS ELIMINATION WITH PARTIAL PIVOTING
11 C
12       DO 5 K = 1,NM
13       KP = K + 1
14 C
15 C     SELECT PIVOT
16 C
17       L = K
18       DO 1 I = KP,N
19       IF(ABS(A(I,K)) .GT. ABS(A(L,K)))L = I
20     1 CONTINUE
21       JPVT(K) = L
22       S = A(L,K)
23       A(L,K) = A(K,K)
24       A(K,K) = S
25 C
26 C     CHECK FOR ZERO PIVOT
27 C
28       IF(ABS(S) .LT. 1.0E-15)GOTO 6
29 C
30 C     CALCULATE MULTIPLIERS
31 C
32       DO 2 I = KP,N
33       A(I,K) = -A(I,K)/S
34     2 CONTINUE
35 C
36 C     INTERCHANGE AND ELIMINATE BY COLUMNS
37 C
38       DO 4 J = KP,N
39       S = A(L,J)
40       A(L,J) = A(K,J)
41       A(K,J) = S
42       IF(ABS(S) .LT. 1.0E-15)GOTO 4
43       DO 3 I = KP,N
44       A(I,J) = A(I,J) + A(I,K)*S
45     3 CONTINUE
46     4 CONTINUE
47     5 CONTINUE
48       RETURN
49     6 JPVT(N) = -1
50       RETURN
51       END
```

Fig. 6.15. Listing of subroutine FACT

$$a_{i,j} = a_{i,j} - \left(\frac{a_{i,k}}{a_{k,k}}\right)a_{k,j} \ . \tag{6.25}$$

Arrays in computers are stored by columns (index i in $a_{i,j}$). Therefore it is more efficient to make the innermost loop in FACT (lines 43–45) operate on the first index.

Subroutine SOLVE first manipulates B (lines 12–21) to interchange with the pivot row and to modify b_i, $i = k + 1$, N by subtracting multiples of the pivot row,

$$b_i = b_i - \left(\frac{a_{i,k}}{a_{k,k}}\right)b_k \ , \tag{6.26}$$

```
1        SUBROUTINE SOLVE(N,A,JPVT,B)
2  C
3  C     SOLVES LINEAR SYSTEM, A*X = B
4  C     ASSUMES A IS FACTORISED INTO L.U FORM (BY FACT)
5  C     RETURNS SOLUTION, X, IN B
6  C
7        DIMENSION A(50,50),JPVT(50),B(50)
8  C
9  C     FORWARD ELIMINATION
10 C
11       NM = N - 1
12       DO 2 K = 1,NM
13       KP = K + 1
14       L = JPVT(K)
15       S = B(L)
16       B(L) = B(K)
17       B(K) = S
18       DO 1 I = KP,N
19       B(I) = B(I) + A(I,K)*S
20     1 CONTINUE
21     2 CONTINUE
22 C
23 C     BACK SUBSTITUTION
24 C
25       DO 4 KA = 1,NM
26       KM = N - KA
27       K = KM + 1
28       B(K) = B(K)/A(K,K)
29       S = - B(K)
30       DO 3 I = 1,KM
31       B(I) = B(I) + A(I,K)*S
32     3 CONTINUE
33     4 CONTINUE
34       B(1) = B(1)/A(1,1)
35       RETURN
36       END
```

Fig. 6.16. Listing of subroutine SOLVE

corresponding to the equivalent operation (6.25) in FACT. The final solution is obtained by SOLVE (lines 25–34) by back-substitution. That is

$$v_N = \frac{b_N}{a_{N,N}} \quad \text{and} \quad v_i = \left(b_i - \sum_{k=i+1}^{N} a_{i,k} v_k \right) \bigg/ a_{i,i} \ .$$

To reduce storage the solution V overwrites and is returned as B.

Implementing Gauss elimination in two parts, subroutines FACT and SOLVE, is useful when solutions for multiple right-hand sides, B in (6.23) are required. This is because FACT requires $O(N^3)$ operations, whereas SOLVE requires $O(N^2)$ operations. Subroutines FACT and SOLVE, in slightly fuller form, are discussed at greater length by Forsythe et al. (1977, Sect. 3.3).

Typically spectral methods (Sect. 5.6) and panel methods (Sect. 14.1) produce matrices $\underline{A}$ with a dense structure. In contrast, the finite element and finite difference methods produce matrices $\underline{A}$ that are sparse and, with the use of splitting (factorisation) techniques (Sect. 8.2), narrowly banded.

When $\underline{A}$ is sparse, but not necessarily banded, Gauss elimination is often implemented by storing A as a single array containing the magnitude of $a_{i,j}$ with an associated pointer array, IA, providing information about the location of (i,j). The major difficulty in coding sparse Gauss elimination is that fill-in of $\underline{A}$ occurs. That is, during the elimination process entries in $\underline{A}$ that were previously zero become non-zero. The different strategies for updating $\underline{A}$ and IA are discussed by Jennings

(1977, Chap. 5), and, at a more advanced level, by Duff (1981), George and Liu (1981) and Duff et al. (1986). Packages for implementing sparse Gauss elimination are available, e.g. Duff (1981, pp. 1–29).

6.2.2 Tridiagonal Systems: Thomas Algorithm

The use of three-point finite difference formulae or finite elements with linear interpolation leads, after splitting (Sect. 8.2), to a tridiagonal structure for $\underline{A}$ in (6.23). The use of higher-order finite difference schemes or higher-order finite elements produces a larger bandwidth in $\underline{A}$. The Thomas algorithm is suitable for solving (6.23) when $\underline{A}$ is tridiagonal. The extension of the Thomas algorithm to the case when $\underline{A}$ is pentadiagonal is described in Sect. 6.2.4. For systems of equations, $\underline{A}$ has a block (tridiagonal) structure typically. The treatment of $\underline{A}$ for this case is considered in Sect. 6.2.5.

When the nonzero elements lie close to the main diagonal it is useful to consider variants of Gauss elimination that take advantage of the banded nature of $\underline{A}$. An example is provided by the convection-diffusion problem of Sect. 9.3. Using centred difference formulae the following algorithm (in the present notation) is obtained

$$-(1+0.5R_{\text{cell}})v_{i-1}+2v_i-(1-0.5R_{\text{cell}})v_{i+1}=0 \ , \tag{6.27}$$

which, when repeated for every node, gives

$$
\begin{bmatrix}
b_1 & c_1 & & & & \\
a_2 & b_2 & c_2 & & & \\
 & \cdot & \cdot & \cdot & & \\
 & & a_i & b_i & c_i & \\
 & & & \cdot & \cdot & \cdot \\
 & & & a_{N-1} & b_{N-1} & c_{N-1} \\
 & & & & a_N & b_N
\end{bmatrix}
\begin{bmatrix}
v_1 \\ v_2 \\ \cdot \\ v_i \\ \cdot \\ v_{N-1} \\ v_N
\end{bmatrix}
=
\begin{bmatrix}
d_1 \\ d_2 \\ \cdot \\ d_i \\ \cdot \\ d_{N-1} \\ d_N
\end{bmatrix} , \tag{6.28}
$$

where $a_i=-(1+0.5R_{\text{cell}})$, $b_i=2$, $c_i=-(1-0.5R_{\text{cell}})$. Nonzero values of d_i are associated with source terms, or, for d_1 and d_N, with boundary conditions. All terms in $\underline{A}$, other than those shown, are zero. The change in notation, particularly in relation to b_i, from that in Sect. 6.2.1, may be noted.

The Thomas algorithm for solving (6.28) consists of two parts (Fig. 6.17).

First (6.28) is manipulated into the form

$$
\begin{bmatrix}
1 & c'_1 & & & & \\
 & 1 & c'_2 & & & \\
 & \cdot & \cdot & \cdot & & \\
 & & 1 & c'_i & & \\
 & & & \cdot & \cdot & \cdot \\
 & & & & 1 & c'_{N-1} \\
 & & & & & 1
\end{bmatrix}
\begin{bmatrix}
v_1 \\ \cdot \\ \cdot \\ v_i \\ \cdot \\ \cdot \\ v_N
\end{bmatrix}
=
\begin{bmatrix}
d'_1 \\ \cdot \\ \cdot \\ d'_1 \\ \cdot \\ \cdot \\ d'_N
\end{bmatrix} ,
$$

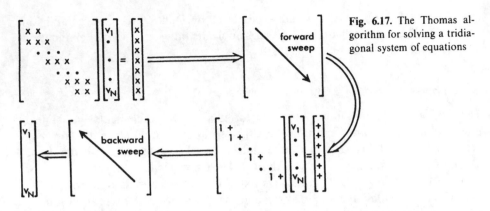

Fig. 6.17. The Thomas algorithm for solving a tridiagonal system of equations

i.e. the a_i coefficients have been eliminated and the b_i coefficients normalised to unity. For the first equation

$$c_1' = \frac{c_1}{b_1} , \qquad d_1' = \frac{d_1}{b_1} , \tag{6.29}$$

and for the general equation

$$c_i' = \frac{c_i}{b_i - a_i c_{i-1}'} , \tag{6.30}$$

$$d_i' = \frac{d_i - a_i d_{i-1}'}{b_i - a_i c_{i-1}'} .$$

The equations are modified, as in (6.30), in a forward sweep (Fig. 6.17). The second stage consists of a back-substitution (backward sweep in Fig. 6.17),

$$v_N = d_N' \qquad \text{and}$$

$$v_i = d_i' - v_{i+1} c_i' . \tag{6.31}$$

The Thomas algorithm is particularly economical; it requires only $5N - 4$ operations (multiplications and divisions). But to prevent ill-conditioning (and hence round-off contamination) it is necessary that

$$|b_i| > |a_i| + |c_i| .$$

The use of splitting for multidimensional problems (Sect. 8.2) typically generates tridiagonal systems of equations that can be solved efficiently using the Thomas algorithm.

6.2.3 BANFAC/BANSOL: Narrowly Banded Gauss Elimination

When A is narrowly banded subroutines BANFAC and BANSOL are suitable for performing Gauss elimination. Subroutine BANFAC (Fig. 6.18) carries out the forward sweep or repeated forward sweeps to render A upper triangular. Sub-

```
1
2          SUBROUTINE BANFAC(B,N,INT)
3  C
4  C       FACTORISES BAND MATRIX ARISING FROM LINEAR OR QUADRATIC ELEMENTS
5  C       INTO L.U
6  C
7          DIMENSION B(5,65)
8          IF(INT .EQ. 2)GOTO 2
9  C
10 C       INT = 1,  LINEAR ELEMENTS = TRIDIAGONAL SYSTEM
11 C
12         NP = N - 1
13         DO 1 J = 1,NP
14         JP = J + 1
15         B(2,JP) = B(2,JP)/B(3,J)
16         B(3,JP) = B(3,JP) - B(2,JP)*B(4,J)
17       1 CONTINUE
18         RETURN
19 C
20 C       INT = 2,  QUADRATIC ELEMENTS = PENTADIAGONAL SYSTEM
21 C       ASSUMES FIRST EQUATION FORMED AT MIDSIDE NODE
22 C
23       2 NH = N/2
24         DO 5 I = 1,2
25         JS = 3 - I
26         DO 4 J = JS,NH
27         JA = 2*(J-1)
28         IF(I .EQ. 2)GOTO 3
29 C
30 C       I = 1,  FIRST PASS, REDUCE TO TRIDIAGONAL
31 C
32         JB = JA + 2
33         B(1,JB) = B(1,JB)/B(2,JB-1)
34         B(2,JB) = B(2,JB) - B(1,JB)*B(3,JB-1)
35         B(3,JB) = B(3,JB) - B(1,JB)*B(4,JB-1)
36         GOTO 4
37 C
38 C       I = 2,  SECOND PASS, REDUCE TO UPPER TRIANGULAR
39 C
40       3 JB = JA + 3
41         B(2,JB-1) = B(2,JB-1)/B(3,JB-2)
42         B(3,JB-1) = B(3,JB-1) - B(2,JB-1)*B(4,JB-2)
43         IF(JB .GT. N)GOTO 4
44         B(2,JB) = B(2,JB)/B(3,JB-1)
45         B(3,JB) = B(3,JB) - B(2,JB)*B(4,JB-1)
46         B(4,JB) = B(4,JB) - B(2,JB)*B(5,JB-1)
47       4 CONTINUE
48       5 CONTINUE
49         RETURN
50         END
```

Fig. 6.18. Listing of subroutine BANFAC

routine BANSOL (Fig. 6.19) modifies the right-hand side, B in (6.23), and solves the resultant system of equations by back-substitution.

For INT = 2, BANFAC and BANSOL have been coded to make use of the particular structure of $\underline{A}$ that is produced by one-dimensional quadratic elements. That is, $\underline{A}$ is alternately "tridiagonal" (three contiguous non-zeros per row) and "pentadiagonal" (five contiguous non-zeros per row), corresponding to equations associated with midside or corner nodes (Fig. 5.10), respectively.

For this case the factorisation process (BANFAC) consists of two sweeps. The first sweep operates only on the "pentadiagonal" rows to eliminate the extreme coefficient $a_{k,k-2}$. The second sweep reduces the matrix to upper triangular form. The precise implementation depends on whether the first and last equations in

```
1
2          SUBROUTINE BANSOL(R,X,B,N,INT)
3  C
4  C       USES L.U FACTORISATION TO SOLVE FOR X, GIVEN R
5  C
6          DIMENSION R(65),X(65),B(5,65)
7          IF(INT .EQ. 2)GOTO 3
8  C
9  C       INT = 1,  TRIDIAGONAL SYSTEM
10 C
11         NP = N - 1
12         DO 1 J = 1,NP
13         JP = J + 1
14       1 R(JP) = R(JP) - B(2,JP)*R(J)
15         X(N) = R(N)/B(3,N)
16         DO 2 J = 1,NP
17         JA = N - J
18         X(JA) = (R(JA) - B(4,JA)*X(JA+1))/B(3,JA)
19       2 CONTINUE
20         RETURN
21 C
22 C       INT = 2,  PENTADIAGONAL SYSTEM
23 C       ASSUMES FIRST EQUATION FORMED AT MIDSIDE NODE
24 C       SET IBC = 0 IF LAST EQUATION FORMED AT MIDSIDE NODE
25 C       SET IBC = 1 IF LAST EQUATION FORMED AT CORNER NODE
26 C
27       3 IBC = 1
28         NH = N/2
29         IF(2*NH .EQ. N)R(N+1) = 0.
30         IF(2*NH .EQ. N)B(2,N+1) = 0.
31         DO 6 I = 1,2
32         DO 5 J = 1,NH
33         JA = 2*J
34         DO 4 K = 1,I
35         JB = JA - 1 + K
36       4 R(JB) = R(JB) - B(I,JB)*R(JB-1)
37       5 CONTINUE
38       6 CONTINUE
39         NEN = NH - IBC
40         X(N) = R(N)/B(3,N)
41         IF(IBC .EQ. 1)X(N-1) = (R(N-1) - B(4,N-1)*X(N))/B(3,N-1)
42         DO 7 J = 1,NEN
43         JA = N - 2*J + 1 - IBC
44         X(JA) = (R(JA) - B(4,JA)*X(JA+1) - B(5,JA)*X(JA+2))/B(3,JA)
45         X(JA-1) = (R(JA-1) - B(4,JA-1)*X(JA))/B(3,JA-1)
46       7 CONTINUE
47         RETURN
48         END
```

Fig. 6.19. Listing of subroutine BANSOL

(6.23) are associated with midside or corner nodes. Dirichlet boundary conditions correspond to midside nodes; Neumann boundary conditions correspond to corner nodes. The reader may find it helpful to examine the code relating to the option, INT = 2, after looking at Sect. 6.2.4.

An example of the use of BANFAC/BANSOL for both tridiagonal and alternating tridiagonal/pentadiagonal matrices is provided by the solution of the Sturm-Liouville problem (Sect. 5.4.2).

A useful test that the subroutines solving (6.23) have been coded correctly is to multiply up the left-hand side of (6.23), after V has been obtained, and to ensure that the left-hand side is equal to the right-hand side. However, this does not guarantee an accurate solution V if A is ill-conditioned. That is if the equations forming the system (6.23) are close to linear dependence. Ill-conditioned systems, and the effect on solution accuracy, are discussed by Gerald (1978, pp. 106–113) and Dahlquist and Bjorck (1974, pp. 174–185).

6.2.4 Generalised Thomas Algorithm

The use of higher-order finite difference and finite element schemes leads to bandwidths larger than tridiagonal. However, the Thomas algorithm can be generalised without difficulty. Suppose that the following pentadiagonal system is to be solved:

$$
\begin{bmatrix}
b_1 & c_1 & f_1 \\
a_2 & b_2 & c_2 & f_2 \\
e_3 & a_3 & b_3 & c_3 & f_3 \\
& & \cdot & \cdot & \cdot \\
& e_i & a_i & b_i & & c_i & & f_i \\
& & \cdot & & \cdot & & \cdot & & \cdot \\
& & & e_{N-1} & a_{N-1} & b_{N-1} & c_{N-1} \\
& & & & e_N & a_N & b_N
\end{bmatrix}
\begin{bmatrix}
v_1 \\ \cdot \\ \cdot \\ \cdot \\ v_i \\ \cdot \\ \cdot \\ v_N
\end{bmatrix}
=
\begin{bmatrix}
d_1 \\ \cdot \\ \cdot \\ \cdot \\ d_i \\ \cdot \\ \cdot \\ d_N
\end{bmatrix}
$$

The first stage consists of eliminating all the e_i terms to give

$$
\begin{bmatrix}
b_1 & c_1 & f_1 \\
a_2 & b_2 & c_2 & f_2 \\
& a_3' & b_3' & c_3' & f_3' \\
& & \cdot & \cdot & \cdot & \cdot \\
& & & a_i' & b_i' & c_i' & f_i' \\
& & & & \cdot & \cdot & \cdot & & \cdot \\
& & & & & a_{N-1}' & b_{N-1}' & c_{N-1}' \\
& & & & & & a_N' & b_N'
\end{bmatrix}
\begin{bmatrix}
v_1 \\ \cdot \\ \cdot \\ \cdot \\ v_i \\ \cdot \\ \cdot \\ v_N
\end{bmatrix}
=
\begin{bmatrix}
d_1 \\ \cdot \\ \cdot \\ \cdot \\ d_i' \\ \cdot \\ \cdot \\ d_N
\end{bmatrix}
,
$$

where

$$
a_i' = a_i - e_i b_{i-1}'/a_{i-1}' \ ,
$$
$$
b_i' = b_i - e_i c_{i-1}'/a_{i-1}' \ ,
$$
$$
c_i' = c_i - e_i f_{i-1}'/a_{i-1}' \ , \tag{6.32}
$$
$$
f_i' = f_i \ ,
$$
$$
d_i' = d_i - e_i d_{i-1}'/a_{i-1}' \ .
$$

The next stage is essentially the same as the first stage of the conventional Thomas algorithm. That is, all the a_i' terms are eliminated to give

$$
\begin{bmatrix}
1 & c_1'' & f_1 & & & & \\
& 1 & c_2'' & f_2'' & & & \\
& & \cdot & \cdot & \cdot & & \\
& & & \cdot & \cdot & \cdot & \\
& & & 1 & c_i'' & f_i'' & \\
& & & & \cdot & \cdot & \cdot \\
& & & & & \cdot & \cdot \\
& & & & & 1 & c_{N-1}'' \\
& & & & & & 1
\end{bmatrix}
\begin{bmatrix}
v_1 \\ \cdot \\ \cdot \\ \cdot \\ v_i \\ \cdot \\ \cdot \\ \cdot \\ v_N
\end{bmatrix}
=
\begin{bmatrix}
d_1 \\ \cdot \\ \cdot \\ \cdot \\ d_i'' \\ \cdot \\ \cdot \\ \cdot \\ d_N''
\end{bmatrix} , \tag{6.33}
$$

where

$$
c_i'' = \frac{c_i' - a_i' f_{i-1}''}{b_i' - a_i' c_{i-1}''} ,
$$

$$
f_i'' = \frac{f_i'}{b_i' - a_i' c_{i-1}''} , \tag{6.34}
$$

$$
d_i'' = \frac{d_i' - a_i' d_{i-1}''}{b_i' - a_i' c_{i-1}''} .
$$

The solution of (6.33) then requires a back-substitution of the general form

$$
v_i = d_i'' - c_i'' v_{i+1} - f_i'' v_{i+2} . \tag{6.35}
$$

It is apparent that the pentadiagonal system requires two forward sweeps and one backward sweep.

The various stages of the generalised Thomas algorithm can be interpreted as a sequence of operations (forward sweeps) to reduce the matrix $\underline{A}$ to upper-triangular form at which point a back-substitution can be performed to obtain V. Clearly the greater the bandwidth of $\underline{A}$ the less economical will be the above algorithm.

If the bandwidth itself contains a large number of embedded zeros then alternative sparse matrix techniques (Jennings 1977a, Chap. 5) are usually more economical, although more difficult to code.

6.2.5 Block Tridiagonal Systems

The Thomas algorithm for solving tridiagonal systems of equations is given in Sect. 6.2.2. This algorithm is required, typically, when a *single* governing equation is discretised using an implicit algorithm (Sect. 7.2). However, many flow problems are governed by systems of equations (Sect. 10.2 and Chap. 11). Attempts to implement implicit algorithms produce, typically, block tridiagonal rather than scalar tridiagonal systems of equations. However the Thomas algorithm can be extended to handle this situation without difficulty.

A block tridiagonal system of equations, equivalent to (6.28) can be written as

$$
\begin{bmatrix}
\underline{b}_1 & \underline{c}_1 & & & & \\
\underline{a}_2 & \underline{b}_2 & \underline{c}_2 & & & \\
 & & \ddots & \ddots & & \\
 & & \underline{a}_i & \underline{b}_i & \underline{c}_i & \\
 & & & \ddots & \ddots & \ddots \\
 & & & & \underline{a}_{N-1} & \underline{b}_{N-1} & \underline{c}_{N-1} \\
 & & & & & \underline{a}_N & \underline{b}_N
\end{bmatrix}
\begin{bmatrix}
V_1 \\ V_2 \\ \vdots \\ V_i \\ \vdots \\ V_{N-1} \\ V_N
\end{bmatrix}
=
\begin{bmatrix}
d_1 \\ d_2 \\ \vdots \\ d_i \\ \vdots \\ d_{N-1} \\ d_N
\end{bmatrix},
$$

$$(6.36)$$

where $\underline{a}_i, \underline{b}_i, \underline{c}_i$ are $M \times M$ submatrices and V_i and d_i are M-component subvectors. The number M corresponds to the number of equations formed at each grid point, e.g. $M = 5$ for three-dimensional viscous compressible flow (Sect. 11.6.3). Consequently V_i is the solution subvector associated with a particular grid point. Equation (6.36) represents a system of N blocks of equations, where each block associated with a particular grid point contains M equations.

The solution of (6.36) follows the solution of (6.28) closely. First, the block tridiagonal matrix, (6.36), is converted into upper triangular form by eliminating the submatrices $\underline{a}_i$. By analogy with (6.29) the first block of equations gives

$$\underline{c}_1' = (\underline{b}_1)^{-1}\underline{c}_1 \quad \text{and} \quad d_1' = (\underline{b}_1)^{-1}d_1 \;, \tag{6.37}$$

and for the general block

$$\underline{b}_i' = \underline{b}_i - \underline{a}_i\underline{c}_{i-1}' \;, \qquad \underline{c}_i' = (\underline{b}_i')^{-1}\underline{c}_i \;, \qquad d_i' = (\underline{b}_i')^{-1}\{d_i - \underline{a}_i d_{i-1}'\} \;. \tag{6.38}$$

Explicit matrix inverses are shown in (6.37 and 38). In practice it is more economical to solve the M component subsystem, e.g. $\underline{b}_i'\underline{c}_i' = \underline{c}_i$ is solved for $\underline{c}_i'$. After execution of (6.37 and 38), (6.36) is of upper triangular form with $\underline{c}_i'$ and d_i' replacing $\underline{c}_i$ and d_i and the unit matrix $\underline{I}$ replacing $\underline{b}_i$.

The second stage, equivalent to (6.31) requires a back-substitution

$$V_N = d_N' \quad \text{and}$$
$$V_i = d_i' - \underline{c}_i' V_{i+1} \;. \tag{6.39}$$

The operation count for the block Thomas algorithm is $O(5NM^3/3)$ operations, which is clearly preferable to $O((NM)^3/3)$ for full Gauss elimination. However, if it were possible to split the block tridiagonal system (6.36) into M scalar tridiagonal systems the total operation count would be $O(5NM)$ operations, i.e. roughly an $M^2/3$ saving for $N \gg M$. Consequently there is an incentive to construct implicit algorithms that allow some of the equations to decouple, e.g. Sect. 14.2.8. The block Thomas algorithm described above follows Isaacson and Keller (1966, pp. 58–61).

6.2.6 Direct Poisson Solvers

The Poisson equation (and the related Laplace's equation) occurs sufficiently often in fluid dynamics to make special procedures for solving the discretised Poisson equation worth investigating. A three-point finite difference discretisation of the two-dimensional Poisson equation produces a system of equations (6.23) for which $\underline{A}$ has the form shown in Fig. 6.1.

For a uniform grid ($\Delta x = \Delta y$) in a rectangular domain $0 \leq x \leq 1$, $0 \leq y \leq 1$, one block of the system of equations can be written

$$\underline{I}V_{k-1} + \underline{g}V_k + \underline{I}V_{k+1} = h_k , \qquad k = 1, \ldots, M , \tag{6.40}$$

where M is the number of active points in the y direction. Each vector V_k contains N unknown $v_{j,k}$ values associated with one grid line ($y = y_k$) in the x-direction. The matrix $\underline{g}$ is tridiagonal of order N with non-zero entries $1, -4, 1$ associated with the points $(j-1, k)$, (j, k) and $(j+1, k)$, respectively. The vector h_k contains all the right-hand side contributions to B in (6.23) coming from a single x grid line ($y = y_k$).

Two techniques, cyclic reduction and Fourier series representation, are particularly efficient in solving (6.40). The preferred strategy (Swartztrauber 1977) is to start with cyclic reduction to reduce the size M of (6.40) and then to solve the reduced system using a Fourier series representation.

In cyclic reduction the block for the kth (even) line of (6.40), multiplied by $-\underline{g}$, is added to the blocks for the $(k-1)$-th and $(k+1)$-th lines to eliminate V_{k-1} and V_{k+1}. The result is

$$\underline{I}V_{k-2} + \underline{g}^{(1)}V_k + \underline{I}V_{k+2} = h_k^{(1)} , \qquad \text{where}$$

$$\underline{g}^{(1)} = 2\underline{I} - \underline{gg} = (\sqrt{2}\,\underline{I} + \underline{g})(\sqrt{2}\,\underline{I} - \underline{g}) \quad \text{and} \tag{6.41}$$

$$h_k^{(1)} = h_{k-1} - \underline{g}h_i + h_{k+1} .$$

This process can be repeated as many times as required. After l reductions, (6.40) becomes

$$\underline{I}V_{k-2^l} + \underline{g}^{(l)}\underline{V} + \underline{I}V_{k+2^l} = h^{(l)} , \qquad \text{where} \tag{6.42}$$

$$\underline{g}^{(l)} = -\prod_{i=1}^{2^l}(\underline{g} - \beta_i\underline{I})$$

and $\beta^i = 2\,\cos[(2k-1)\pi/2^{l+1}]$.

In principle the reduction can be continued until the boundary conditions determine V_{k-2^l} and V_{k+2^l} (Dorr 1970). This permits back-substitution to determine all the intermediate V_k vectors. However, in practice, it is more efficient to terminate the process after l reductions and to switch to a Fourier series representation.

If Dirichlet boundary conditions are applied on all boundaries the following would be an appropriate Fourier series representation:

$$v_{j,k} = \sum_{s=1}^{N} U_{s,k} \sin\left(\frac{sj\pi}{(N+1)}\right) \quad \text{for } k=1, \ldots, M_l , \tag{6.43}$$

where M_l is the number of x gridlines left in (6.42) after l reductions. The coefficients $U_{s,k}$ are obtained by solving the tridiagonal systems

$$U_{s,k-1} + \lambda_s U_{s,k} + U_{s,k+1} = H_{s,k} \quad \text{for } s=1, \ldots, N , \tag{6.44}$$

where

$$H_{s,k} = \left(\frac{2}{N+1}\right)\sum_{j=1}^{N} h(x_j, y_k) \sin\left(\frac{sj\pi}{(N+1)}\right) , \tag{6.45}$$

$h(x_j, y_k)$ is an element of $\boldsymbol{h}^{(l)}$ in (6.42) and

$$\lambda_s = -2\left[1 + 2\sin^2\left(\frac{0.5s\pi}{N+1}\right)\right] . \tag{6.46}$$

The fast Fourier transform (Cooley et al. 1970; Brigham 1974) is used to solve (6.43 and 45). Slightly different forms of these equations and (6.46) occur with different choices of the boundary conditions (Swartztrauber 1977).

For an $N = M$ grid the optimum number of cyclic reductions is given by Swartztrauber (1977) as $l = \log_2(\log_2 N) - 1$. The combined algorithm FACR(l), which was first described by Hockney (1970), has an operation count of $N^2\log_2(\log_2 N)$. For $N = 1000$, $l = 2$, the FACR(2) algorithm is approximately 30 times faster than the ADI algorithm (Sect. 6.3.2) applied to a Poisson equation. The implementation of FACR(l) on supercomputers is discussed by Hockney and Jesshope (1981, pp. 346–351) as are strategies for solving discretised Poisson equations in three dimensions (Hockney and Jesshope 1981, pp. 351–352).

For detailed discussions of direct methods for the Poisson equation and the fast Fourier transform the reader is referred to Hockney and Jesshope (1981), Swartztrauber (1977) and Dorr (1970) and the references cited therein.

It should be made clear that the speed of cyclic reduction and the Fourier series representation comes from the symmetric, constant coefficient nature of the discrete Laplacian on a uniform rectangular grid. For an arbitrary grid, discretisation of the Poisson equation, e.g. via the finite volume method (Sect. 5.2), an isoparametric finite element formulation (Sect. 5.5.3) or the use of generalised coordinates (Chap. 12), leads to an algebraic equation with variable coefficients. For example, with generalised coordinates this could be written

$$a_{j,k}V_{j-1,k} + b_{j,k}V_{j+1,k} + c_{j,k}V_{j,k-1} + d_{j,k}V_{j,k+1} + e_{j,k}V_{j,k} = h_{j,k} , \tag{6.47}$$

or, equivalent to (6.40),

$$C_k V_{k-1} + \underline{G}_k V_k + \underline{D}_k V_{k+1} = \boldsymbol{h}_k . \tag{6.48}$$

As before, $\underline{G}_k$ is tridiagonal and C_k and D_k are diagonal. However, the cancellation that led to (6.41) does not occur with (6.48).

In applying the Fourier series representation to (6.47) a dense system of equations for $U_{s,k}$ replaces the tridiagonal system (6.44). Consequently the Fourier series representation applied to (6.47) is no more economical than Gauss elimination (Sect. 6.2.1) applied directly to (6.47). Depending on the grid, (6.47) may also include non-zero contributions associated with grid points $(j-1, k+1)$, $(j-1, k-1)$, $(j+1, k+1)$ and $(j+1, k-1)$.

6.3 Iterative Methods

Iterative methods can be applied directly to the nonlinear system of equations (6.1); however, it is easier to generalise iterative techniques by considering the linear form (6.2). All iterative techniques can be viewed as procedures for successively modifying an initial guess so that the solution is systematically approached. Generally, simple iterative methods will not converge unless $\underline{A}$, in (6.2), has large elements on the main diagonal (6.52).

6.3.1 General Structure

The general structure of stationary iterative techniques is illustrated by rewriting (6.2) as

$$(\underline{N} - \underline{P})V = B ,\tag{6.49}$$

where $\underline{N}$ is close to $\underline{A}$ in some sense, i.e. $\|\underline{N}\| \approx \|\underline{A}\|$, but computationally efficient to factorise, e.g. $\underline{N}$ might be tridiagonal. Equation (6.49) can be rewritten as

$$\underline{N}V^{(n+1)} = \underline{P}V^{(n)} + B \quad \text{or}$$

$$V^{(n+1)} = \underline{N}^{-1}\underline{P}V^{(n)} + \underline{N}^{-1}B \quad \text{or} \tag{6.50}$$

$$V^{(n+1)} = V^{(n)} - \underline{N}^{-1}R^{(n)} ,\tag{6.51}$$

where $R^{(n)}$ is the vector of equation residuals at the nth step of the iteration, i.e.

$$R^{(n)} = \underline{A}V^{(n)} - B .$$

As the exact solution is approached $\|R^{(n)}\|$ tends to zero, so that monitoring $\|R^{(n)}\|$ indicates the progress towards convergence. The general iterative method consists of an initial guess, $V^{(1)}$, and the successive improvement using (6.51). The various methods differ, primarily, in the choice for $\underline{N}$.

The scheme (6.51) will converge if the spectral radius, (i.e. the magnitude of the maximum eigenvalue) of $\underline{N}^{-1}\underline{P}$ is less than unity. This condition often corresponds to the more restrictive condition that $\underline{A}$ is diagonally dominant, i.e.

$$|A_{jj}| > \sum_{\substack{i \\ i \neq j}} |A_{ij}| \ . \tag{6.52}$$

The Jacobi method is one of the simpler iterative schemes for solving (6.2). In this method

$$\underline{N} = \underline{DI} \quad \text{and} \quad \underline{P} = (\underline{L} + \underline{U}) \ ,$$

where $\underline{DI}$ is the diagonal matrix and $\underline{L}$ and $\underline{U}$ are the strictly lower and upper triangular matrices respectively. Thus (6.50) becomes

$$v_i^{(n+1)} = \left(B_i - \sum_{\substack{j \\ j \neq i}}^{N} A_{ij} v_j^{(n)} \right) \bigg/ A_{ii} \ . \tag{6.53}$$

The Jacobi method in the above form is uneconomic, requiring too many iterations to reach convergence. However, it can be significantly improved by either Chebyshev acceleration (Hageman and Young 1981, Chap. 4) or conjugate gradient acceleration (Sect. 6.3.4).

A more immediate improvement to the Jacobi method is provided by the Gauss-Seidel method in which values of $v_j^{(n+1)}$ are used on the right-hand side of (6.53) as soon as they are available.

In this case

$$\underline{N} = \underline{DI} - \underline{L} \ , \qquad \underline{P} = \underline{U} \tag{6.54}$$

and the equivalent of (6.53) is

$$v_i^{(n+1)} = \left(B_i - \sum_{j=1}^{i-1} A_{ij} v_j^{(n+1)} - \sum_{j=i+1}^{N} A_{ij} v_j^{(n)} \right) \bigg/ A_{ii} \ . \tag{6.55}$$

Gauss-Seidel iteration is typically twice as fast as Jacobi iteration but cannot be accelerated. If $\underline{A}$ is diagonally dominant, (6.52), the convergence of Jacobi and Gauss-Seidel methods is guaranteed.

Successive overrelaxation (SOR) provides a significant improvement over Gauss-Seidel by evaluating $v^{(n+1)}$ as a weighted average of $v_i^{(n)}$ and $(v_i^{(n+1)})_{GS}$. Thus the SOR scheme can be written as

$$v_i^{(n+1)} = \lambda \left[\left(B_i - \sum_{j=1}^{i-1} A_{ij} v_j^{(n+1)} - \sum_{j=i+1}^{N} A_{ij} v_j^{(n)} \right) \bigg/ A_{ii} \right] + (1 - \lambda) v_i^{(n)} \ . \tag{6.56}$$

The term λ is the relaxation parameter. The restriction $0 < \lambda < 2$, is necessary for convergence of SOR. In terms of (6.51) the SOR scheme corresponds to

$$\underline{N} = \frac{\underline{DI}}{\lambda} - \underline{L} \ . \tag{6.57}$$

The number of iterations for convergence is sensitive to the choice of λ. An optimum choice is given by

$$\lambda_{\text{opt}} = \frac{2}{1 + (1 - \mu^2)^{1/2}} \, , \tag{6.58}$$

where μ is the largest eigenvalue of $\underline{I} - \underline{DI}^{-1}\underline{A}$. However, finding μ explicitly can be as expensive as solving the original problem. Therefore a preferred strategy is to obtain an estimate of μ as the iteration (6.56) proceeds. Equation (6.58) then provides an improved value for λ. Details are provided by Hageman and Young (1981, Chap. 9).

The SOR scheme with a good estimate of λ_{opt} is considerably more efficient than either the Jacobi or Gauss-Seidel methods. However, in the original form it is not possible to apply acceleration techniques, Chebyshev or conjugate gradient, to SOR. But by making a small modification to give the symmetric successive overrelaxation (SSOR) method it is possible to introduce acceleration techniques. The SSOR method consists of two stages. In the first the SOR scheme is used. In the second stage the unknowns are iterated in the reverse order using the SOR scheme with the same value of λ as in the first stage. The SSOR method corresponds to the following choice for $\underline{N}$ in (6.51):

$$\underline{N} = \frac{(\underline{DI} + \lambda\underline{L})\underline{DI}^{-1}(\underline{DI} + \underline{U})}{\lambda(2 - \lambda)} \, . \tag{6.59}$$

It should be emphasised that the SSOR method is less efficient than the SOR method, unless acceleration techniques are included.

6.3.2 Duct Flow by Iterative Methods

In this section we apply the three iterative methods, Jacobi, Gauss-Seidel and successive over-relaxation (SOR), to the problem of fully developed laminar flow in a square duct. These three iterative methods are discussed in Sect. 6.3.1, as explicit or point algorithms. Using the Thomas algorithm (subroutines BANFAC and BANSOL, Sect. 6.2.3), these iterative methods can also be applied as implicit or line algorithms; line SOR and ADI will also be described in this section.

The current example, the duct-flow problem, is also used to illustrate the finite element method in Sect. 5.5.2 and a computer program, DUCT (Fig. 5.22), is provided. Viscous flow in a duct is governed by the nondimensional equation (5.97)

$$\left(\frac{b}{a}\right)^2 \frac{\partial^2 \bar{w}}{\partial x^2} + \frac{\partial^2 \bar{w}}{\partial y^2} + 1 = 0 \, , \tag{6.60}$$

with boundary conditions $\bar{w} = 0$ at $x = \pm 1$, $y = \pm 1$. The parameter b/a is the duct aspect ratio, Fig. 5.21. A three-point finite difference discretisation of (6.60) is

$$\left(\frac{b}{a}\right)^2 \frac{w_{j-1,k} - 2w_{j,k} + w_{j+1,k}}{\Delta x^2} + \frac{w_{j,k-1} - 2w_{j,k} + w_{j,k+1}}{\Delta y^2} + 1 = 0 \, . \tag{6.61}$$

Applying the Jacobi scheme (Sect. 6.3.1) to (6.61) produces the algorithm

$$w_{j,k}^{(n+1)} = 0.5\left[1+\left(\frac{b}{a}\right)^2 \frac{w_{j-1,k}^{(n)}+w_{j+1,k}^{(n)}}{\Delta x^2} + \frac{w_{j,k-1}^{(n)}+w_{j,k+1}^{(n)}}{\Delta y^2}\right]\bigg/ PAR1 \ , \qquad (6.62)$$

where (n) is the iteration index and $PAR1 = (b/a)^2/\Delta x^2 + 1/\Delta y^2$. In the Jacobi scheme all grid-point values on the right-hand side of (6.62) are evaluated at the nth iteration level.

In the Gauss-Seidel scheme grid-point values are evaluated at the latest iteration level available. Thus if the iteration sweeps repeatedly in the y direction (increasing k) for successive x values (increasing j) the Gauss-Seidel equivalent of (6.62) would be

$$w_{j,k}^{(n+1)} = 0.5\left[1+\left(\frac{b}{a}\right)^2 \frac{w_{j-1,k}^{(n+1)}+w_{j+1,k}^{(n)}}{\Delta x^2} + \frac{w_{j,k-1}^{(n+1)}+w_{j,k+1}^{(n)}}{\Delta y^2}\right]\bigg/ PAR1 \ . \qquad (6.63)$$

In the SOR scheme the solution of (6.63), now called $w_{j,k}^{(*)}$, is combined with the previous solution, $w_{j,k}^{(n)}$, as

$$w_{j,k}^{(n+1)} = \lambda w_{j,k}^{(*)} + (1-\lambda)w_{j,k}^{(n)} = w_{j,k}^{(n)} + \lambda(w_{j,k}^{(*)} - w_{j,k}^{(n)}) \ , \qquad (6.64)$$

where λ is the relaxation factor. The Gauss-Seidel scheme is obtained when $\lambda = 1$. The SOR scheme applied to (6.61) in the range $0 < \lambda < 2$ will produce a converged solution.

The three iterative methods are also applied with the finite element discretisation of (6.60), which is described in Sect. 5.5.2. For example, the equivalent of (6.63) is given by (5.108).

Table 6.3 provides a comparison of the number of iterations to reach convergence for the duct-flow problem, solved by the program DUCT, on an 11×11 grid.

Table 6.3. Number of iterations to convergence ($b/a = 1.0$)

λ	Point FDM $(TOL = 1 \times 10^{-5})$	Point FDM $(TOL = 1 \times 10^{-6})$	Point FEM $(TOL = 1 \times 10^{-6})$	ADI-FDM $(TOL = 1 \times 10^{-6})$	ADI-FEM $(TOL = 1 \times 10^{-6})$	λ_y
Jacobi	41	87	64	19	19	0.9
0.8	39	74	55	16	18	1.1
0.9	33	62	45	14	17	1.3
1.0 (GS)	28	51	38	12	16	1.7
1.1	24	43	32	12	15	2.1
1.2	21	36	26	12	14	2.5
1.3	18	29	21	12	13	2.8
1.4	15	23	17	11	14	3.3
1.5	12	18	12	11	15	3.7
1.55	11	15	13	11	16	4.1
1.6	11	17	15	12	17	4.5
1.7	14	21	20			
1.8	21	28	29			

Convergence is assumed when $\|R^{(n)}\|_{rms} <$ TOL. Results for two tolerances are presented. The starting solution for the iteration is the exact solution of (6.60). For a finite grid this solution is not the same as the solution of the discretised equations (6.61). Starting from the solution $w_{j,k}^{(0)}=0$ requires more iterations, as can be appreciated by comparing the results shown in Tables 6.3 and 6.4.

The results shown in Table 6.3 indicate that the SOR scheme, with the optimum value of λ, converges more quickly than the Gauss-Seidel scheme, which converges more quickly than the Jacobi scheme. For the finite difference method the reduction of the rms value of the change in the solution, $w_{j,k}^{(n+1)} - w_{j,k}^{(n)}$, is plotted against iteration number n in Fig. 6.20.

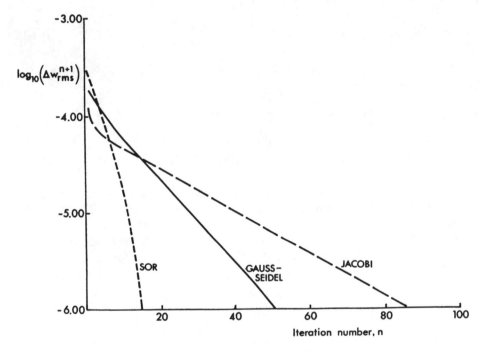

Fig. 6.20. Convergence history for point FDM (TOL $= 1 \times 10^{-6}$)

The results shown in Table 6.3 indicate that the point finite element method (5.108, 109) converges more rapidly than the point finite difference method for values of $\lambda \leq 1.7$. However the computational work per iteration is greater for the finite element method, so that it is not necessarily more efficient than the finite difference method for this problem.

All the algorithms given above, that is (6.62–64) have been explicit. It is possible, and desirable, to consider groups of nodal unknowns implicitly, as long as the resultant system of equations can be solved efficiently. This can be done by forming tridiagonal systems of equations associated with each line (constant k) in the x direction. Thus (6.61) is then written as an algorithm as

$$\left(-\frac{(b/a)^2}{\Delta x^2}\right)w^{(*)}_{j-1,k} + \left(\frac{2(b/a)^2}{\Delta x^2}\right)w^{(*)}_{j,k} - \left(\frac{(b/a)^2}{\Delta x^2}\right)w^{(*)}_{j+1,k}$$

$$= 1 + \frac{w^{(n)}_{j,k-1} - 2w^{(n)}_{j,k} + w^{(n)}_{j,k+1}}{\Delta y^2}. \tag{6.65}$$

This system can be solved conveniently using BANFAC/BANSOL (Sect. 6.2.3). Since the coefficients multiplying $w^{(*)}$ are constant on the left-hand side of (6.65), BANFAC is only required for the grid-line $k=2$. The same factorisation is used by BANSOL for all grid-lines. The combination of (6.65) and (6.64) is known as successive line overrelaxation (SLOR).

The ADI formulation (Varga 1962, Chap. 7) is similar to SLOR but with the implicit direction alternating at each iteration. Thus in place of (6.65) the following two-stage algorithm is used. For the first stage the following tridiagonal system associated with each line (constant k) in the x direction is solved:

$$-\lambda_x w^{(*)}_{j-1,k} + (1+2\lambda_x)w^{(*)}_{j,k} - \lambda_x w^{(*)}_{j+1,k} = \lambda_y \Delta y^2 + \lambda_y w^{(n)}_{j,k-1}$$

$$+ (1-2\lambda_y)w^{(n)}_{j,k} + \lambda_y w^{(n)}_{j,k+1}. \tag{6.66}$$

For the second stage the following tridiagonal system associated with each line (constant j) in the y direction is solved:

$$-\lambda_y w^{(n+1)}_{j,k-1} + (1+2\lambda_y)w^{(n+1)}_{j,k} - \lambda_y w^{(n+1)}_{j,k+1} = \lambda_y \Delta y^2 + \lambda_x w^{(*)}_{j-1,k}$$

$$+ (1-2\lambda_x)w^{(*)}_{j,k} + \lambda_x w^{(*)}_{j+1,k}, \tag{6.67}$$

where λ_y is chosen to accelerate convergence and $\lambda_x = (b\Delta y/a\Delta x)^2 \lambda_y$. Strategies for choosing λ_y are discussed by Varga (1962) and Wachpress (1966).

The above ADI scheme is equivalent to the ADI scheme described in Sect. 8.2.1. Comparable formulae to (6.66, 67) can be obtained for the finite element method by adapting (8.35, 36). The number of iterations for the ADI finite difference method and ADI finite element method are shown in Table 6.3. Both methods give better convergence than the corresponding point algorithms for λ not at the optimum value.

The convergence rate is a strong function of grid refinement. As the grid is refined the number of iterations to convergence increases and the optimum λ becomes larger. This situation is illustrated in Table 6.4, which is based on a starting solution $w^{(0)}_{j,k} = 0$.

The iterative algorithms considered above can be written as

$$w^{(n+1)} = \underline{G} w^{(n)}.$$

The reduction in the rate of convergence as the grid is refined corresponds to the maximum eigenvalue of $\underline{G}$ approaching unity. Some means of accelerating the convergence rate is clearly desirable. Hageman and Young (1981) discuss conjugate gradient and Chebyshev acceleration techniques. However, applied to Jacobi or

Table 6.4. Effect of grid refinement on convergence: Duct problem by point SOR (FDM)

Grid	Optimum λ	Number of iterations to convergence (TOL $= 1 \times 10^{-6}$)
6×6	1.30	12
11×11	1.55	23
21×21	1.74	41
41×41	1.86	79

Gauss-Seidel schemes neither Chebyshev nor conjugate gradient acceleration produce a faster rate of convergence than SOR with an optimum value of the relaxation factor λ. As a very robust algorithm, Hageman and Young recommend SOR with an adaptive algorithm for choosing λ. A subroutine for this purpose is provided by Hageman and Young (1981, pp. 368–372). Jennings (1977a, pp. 210–212) discusses the use of Aitken acceleration.

It should be pointed out that for iterative techniques, such as those described above, to be effective the problem being solved should be strongly elliptic, e.g. second-order derivatives are present but not first-order derivatives. Often this leads to the matrix of equations, e.g. those formed from (6.61), being symmetric and positive definite. The matrix $\underline{A}$ is symmetric and positive definite if $\underline{A}^T = \underline{A}$ and $x^T \underline{A} x > 0$ for all $x \neq 0$.

Examples of equations that are effectively solved by the above iterative techniques are the Poisson equation (for the stream function or pressure) that arises in viscous flow problems based on the stream function vorticity formulation (Sect. 17.3) and the potential equation that often arises in inviscid flow problems (Chap. 14), particularly transonic flow (Sect. 14.3).

If first derivatives of significant magnitude are present the above techniques are very ineffective and may fail to converge. For example, applying the point SOR (FDM) technique to the two-dimensional Burgers' equations considered in Sect. 6.1.3, on a 21×11 grid, requires $\lambda \leq 1.0$ for convergence. After 270 iterations the equation residuals are reduced to about 1.4×10^{-5}. For comparison, an approximate factorisation of the unsteady Newton's method (Sect. 6.4.1) achieves the same degree of convergence in 15 iterations.

Significant first derivatives occur in the momentum equations for the primitive variables (Sect. 17.1), the transport equation for the vorticity (Sect. 17.3) and the energy equation (Sect. 11.2.4).

6.3.3 Strongly Implicit Procedure

This strongly implicit procedure (SIP) scheme starts from the general stationary iterative technique (6.49) and factorises $\underline{N}$ into $\underline{L} \underline{U}$ for the situation where $\underline{A}$ arises from three-point centred difference discretisations in two dimensions. In this case $\underline{A}$ has the structure shown in Fig. 6.1.

Stone (1968) showed that this could be done in a way that produces three diagonals in $\underline{L}$ and three diagonals in $\underline{U}$ of which the main diagonal elements are unity. Forming $\underline{N} = \underline{L}\,\underline{U}$ indicates that $\underline{P}$ consists of two diagonals multiplying $v_{i+1,j-1}$ and $v_{i-1,j+1}$. For a linear system of equations (6.2) the elements of $\underline{L}$, $\underline{U}$ and $\underline{P}$ can be evaluated once and for all.

The algorithm is implemented in two stages. First a forward sweep solves

$$\underline{L} w^{(n+1)} = B + \underline{P}\, V^{(n)} \ . \tag{6.68}$$

This is followed by a back-substitution

$$\underline{U}\, V^{(n+1)} = w^{(n+1)} \ . \tag{6.69}$$

The algorithm is robust and is often more efficient than the ADI scheme (6.66, 67).

Schneider and Zedan (1981) have developed a modified strongly implicit (MSI) algorithm that is applicable to either five-point or nine-point two-dimensional discretisations, i.e. it can be applied to three-point finite difference discretisations or linear finite element discretisations. Here the five-point version of MSI will be described; the reader is referred to Schneider and Zedan for the nine-point version.

The MSI algorithm is executed using (6.68 and 69). However $\underline{L}$, $\underline{U}$ and $\underline{P}$ differ from the form used by Stone. In the MSI algorithm $\underline{L}$ has four non-zero entries per row, $\underline{U}$ has three non-zero off-diagonal entries plus unity on the diagonal and $\underline{P}$ has two entries. If V is numbered in the increasing j and then increasing k directions, (6.68) becomes

$$
\begin{aligned}
f_{j,k} w_{j,k}^{(n+1)} = {} & b_{j,k} + p_{j,k}^1 [v_{j+2,k-1} - \alpha(-2v_{j,k} + 2v_{j+1,k} + v_{j,k-1})]^{(n)} \\
& + p_{j,k}^2 [v_{j-2,k+1} - \alpha(-2v_{j,k} + 2v_{j-1,k} + v_{j,k+1})]^{(n)} \\
& - [c_{j,k} v_{j,k-1} + d_{j,k} v_{j+1,k-1} + e_{j,k} v_{j-1,k}]^{(n+1)} \ .
\end{aligned} \tag{6.70}
$$

The back-substitution (6.69) is implemented as

$$v_{j,k}^{(n+1)} = w_{j,k}^{(n+1)} - (g_{j,k} v_{j+1,k} + h_{j,k} v_{j-1,k+1} + t_{j,k} v_{j,k+1})^{(n+1)} \ . \tag{6.71}$$

In (6.70 and 71) coefficients c, d, e and f are part of $\underline{L}$, coefficients g, h and t are the part of $\underline{U}$ and coefficients p^1 and p^2 are part of $\underline{P}$. The coefficients are related to the elements of $\underline{A}$ by

$$c_{j,k} = \frac{a_{j,k}}{1 - \alpha g_{j,k-1} g_{j+1,k-1}} \ , \qquad d_{j,k} = -c_{j,k} g_{j,k-1} \ , \qquad e_{j,k} = a_{j,k} - c_{j,k} h_{j,k-1} \ ,$$

$$p_{j,k}^1 = d_{j,k} g_{j+1,k-1} \ , \qquad p_{j,k}^2 = e_{j,k} h_{j-1,k} \ ,$$

$$f_{j,k} = a_{j,k} - c_{j,k} t_{j,k-1} - d_{j,k} h_{j+1,k-1} - e_{j,k} g_{j-1,k} + 2\alpha(p_{j,k}^1 + p_{j,k}^2) \ , \tag{6.72}$$

$$g_{j,k} = a_{j+1,k} - d_{j,k} t_{j+1,k-1} - 2\alpha p_{j,k}^1 \ , \qquad h_{j,k} = \frac{-e_{j,k} t_{j-1,k}}{f_{j,k}} \ ,$$

$$t_{j,k} = \frac{a_{j,k+1} - \alpha p_{j,k}}{f_{j,k}} \ .$$

The parameter α is used to accelerate convergence; $\alpha = 0.5$ is close to optimal. Schneider and Zedan (1981) indicate that MSI is typically two to four times more economical than SIP and does not require grid renumbering after every iteration as does SIP.

Application of strongly implicit procedures, related to the above formulations, are reported by Rubin and Khosla (1981), Zedan and Schneider (1985) and Lin (1985).

6.3.4 Acceleration Techniques

The various iterative techniques described in Sects. 6.3.1–6.3.3 are effective but for some problems demonstrate rather slow convergence after the residual R has undergone an initial rapid reduction. If the error in the iterative solution, $e = V^{(n)} - V^c$, is expanded as a Fourier series, the initial rapid reduction in the residual corresponds to the reduction in the amplitude of the short wavelength modes in the Fourier series. The subsequent reduction in the amplitude of the long wavelength modes often proceeds at a much slower rate.

However, it is possible to modify the iterative techniques already considered to accelerate convergence. Two of the more effective techniques are Chebyshev acceleration and conjugate gradient acceleration (Hageman and Young 1981). Here we will describe the key features of conjugate gradient acceleration.

Initially it will be assumed that the matrix $\underline{A}$ in (6.2) is symmetric positive definite. That is, $x^T \underline{A} x > 0$ for any choice of x. The discretisation of Laplace's equation with centred finite difference expressions produces a linear system of equations for which $\underline{A}$ is symmetric positive definite.

The steps in the traditional conjugate gradient method (Hestenes and Stiefel 1952) can be described as follows:

$$\text{i)} \quad V^{(n+1)} = V^{(n)} + \lambda^{(n)} P^{(n)} \ ,$$

$$\text{ii)} \quad R^{(n+1)} = R^{(n)} - \lambda^{(n)} U^{(n)} \ ,$$

$$\text{iii)} \quad \varrho^{(n+1)} = (R^{(n+1)})^T R^{(n+1)} \ ,$$

$$\text{iv)} \quad \alpha^{(n+1)} = \frac{\rho^{(n+1)}}{\rho^{(n)}} \ , \tag{6.73}$$

$$\text{v)} \quad P^{(n+1)} = R^{(n+1)} + \alpha^{(n+1)} P^{(n)} \ ,$$

$$\text{vi)} \quad U^{(n+1)} = \underline{A} P^{(n+1)}$$

$$\text{vii)} \quad \lambda^{(n+1)} = \frac{\varrho^{(n+1)}}{(P^{(n+1)})^T U^{(n+1)}} \ .$$

The algorithm is relatively economical since only one matrix-vector multiplication, step (vi), is required. It may be noted that the residual, $R^{(n+1)}$, is evaluated recursively, from step (ii), instead of from $R^{(n+1)} = B - \underline{A} V^{(n+1)}$. The parameter $\varrho^{(n+1)}$ is monitored to indicate proximity to convergence. In contrast to the SOR scheme (Sect. 6.3.1) there are no empirically chosen parameters in the conjugate gradient method.

In (6.73) $P^{(n)}$ constitutes a search direction which at the beginning of the iteration is given by $P^{(0)} = R^{(0)}$. The choice for $\alpha^{(n+1)}$ in (6.73) ensures that the search directions $P^{(i)}$ are mutually A-conjugate (orthogonal) in the sense that

$$P^{(i)T} \underline{A} P^{(j)} = 0 \quad \text{for} \quad i \neq j . \tag{6.74}$$

In addition the overall algorithm ensures that $R^{(i)T} R^{(j)} = 0$ for $i \neq j$. These feature produce a robust algorithm that, in the absence of round-off errors, would obtain the solution of a system of N linear algebraic equations (6.2) in no more than N iterations. The choice of $\lambda^{(n)}$ in (6.73) ensures that the equivalent quadratic function

$$F(V) = 0.5 V^T \underline{A} V - V^T B \tag{6.75}$$

has a minimum in the $P^{(n)}$ direction at $V^{(n+1)}$.

If the residual R is expanded as a series based on the eigenfunctions of $\underline{A}$ it is found that each step of the conjugate gradient method has the effect of approximately eliminating the contribution from each eigenvector in turn (Jennings 1977b). Consequently if some of the eigenvalues of $\underline{A}$ are bunched, convergence is achieved in less than N iterations.

In addition if it is possible to premultiply (6.2) so that the eigenvalues of the resulting matrix are more bunched than for $\underline{A}$ and the spread of the eigenvalues $(\lambda_{min}, \lambda_{max})$ is reduced, the subsequent application of the conjugate gradient method to the "preconditioned" system converges faster. The best choice for the preconditioner would be $\underline{A}^{-1}$ since then the solution follows trivially. A good choice is $\underline{N}^{-1}$, which appears in the general form of a stationary iterative technique (6.51). Thus (6.2) is replaced by

$$\underline{N}^{-1} \underline{A} V = \underline{N}^{-1} B . \tag{6.76}$$

To use the above conjugate gradient method it is necessary to make $\underline{N}^{-1} \underline{A}$ symmetric. This can be done by forming $\underline{W} \underline{N}^{-1} \underline{A} \underline{W}^{-1}$ where $\underline{W}$ is chosen, typically, so that $\underline{W}^T \underline{W} = \underline{N}$. Consequently (6.76) is replaced by

$$\hat{\underline{A}} \hat{V} = \hat{B} , \quad \text{where} \tag{6.77}$$

$$\hat{\underline{A}} = \underline{W} \underline{N}^{-1} \underline{A} \underline{W}^{-1} , \quad \hat{V} = \underline{W} V \quad \text{and} \quad \hat{B} = \underline{W} \underline{N}^{-1} B .$$

The application of the conjugate gradient method to (6.77) is called a preconditioned conjugate method and, with a good choice for N, will produce more rapid convergence than the direct application to (6.2).

In practice (6.77) is not formed explicitly. Instead (6.73) is replaced by

i) $V^{(n+1)} = V^{(n)} + \lambda^{(n)} P^{(n)}$,

ii) $\delta^{(n+1)} = \delta^{(n)} - \lambda^{(n)} U^{(n)}$; $\hat{R}^{(n+1)} = \underline{W} \delta^{(n+1)} = \underline{W} \underline{N}^{-1} R^{(n+1)}$,

iii) $\varrho^{(n+1)} = (\hat{R}^{(n+1)})^T \hat{R}^{(n+1)}$,

iv) $\alpha^{(n+1)} = \varrho^{(n+1)} / \varrho^{(n)}$,   (6.78)

v) $P^{(n+1)} = \delta^{(n+1)} + \alpha^{(n+1)} P^{(n)}$; $\hat{P}^{(n+1)} = \underline{W} P^{(n+1)}$,

vi) $U^{(n+1)} = \underline{N}^{-1} \underline{A} P^{(n+1)}$; $\hat{U}^{(n+1)} = \underline{W} U^{(n+1)}$,

vii) $\lambda^{(n+1)} = \dfrac{\varrho^{(n+1)}}{(\hat{P}^{(n+1)})^T \hat{U}^{(n+1)}}$.

In the above sequence, δ and $\hat{R}$ play the roles of pseudoresiduals and the value of $\varrho^{(n+1)}$ indicates the closeness to convergence.

Different choices of $\underline{N}$, and hence $\underline{W}$, correspond to different iteration schemes in Sect. 6.3.1. Thus (6.78) can be interpreted as a conjugate gradient acceleration of the basic scheme. An alternative, but very effective, preconditioner can be constructed by approximately factorising $\underline{A}$ into lower triangular, $\underline{L}$, and upper triangular, $\underline{U}$, matrices (Jackson and Robinson 1985).

For a finite difference discretisation of the steady transport equation, (9.81) without the time-dependent term, the resulting system of equations, equivalent to (6.2), will not be symmetric due to the influence of the discretised first derivative terms, $u \partial T / \partial x$ and $v \partial T / \partial y$. The conjugate gradient method can still be used with modifications (Axelsson and Gustafsson 1979).

One such scheme, ORTHOMIN (due to Vinsome 1976), is described here. In the presence notation (6.78) is replaced by

$$V^{(n+1)} = V^{(n)} + \lambda^{(n)} P^{(n)} \ ,$$

$$P^{(n)} = \delta^{(n)} + \sum_{i=0}^{n-1} \alpha^{(n,i)} P^{(i)} \ , \qquad \text{where}$$ (6.79)

$$\delta^{(n)} = \underline{N}^{-1} R^{(n)} \ ,$$

$$\alpha^{(n,i)} = \frac{(\underline{Z} \underline{A} \delta^{(n)})^T P^{(i)}}{(\underline{Z} \underline{A} P^{(i)})^T P^{(i)}} \ ,$$

$$\lambda^{(n)} = \frac{(\underline{Z} \underline{N} \delta^{(n)})^T P^{(n)}}{(\underline{Z} \underline{A} P^{(n)})^T P^{(n)}} \ , \qquad \text{and}$$

$$\underline{Z} = (\underline{N}^{-1} \underline{A})^T \ .$$

If all terms are retained in the $\sum\limits_{i}$ in the expression for $P^{(n)}$ the method requires considerable storage and is rather uneconomic. In practice good performance is obtained with $i = 0, \ldots, 4$.

Markham (1984) has compared various preconditioned conjugate gradient-type methods for asymmetric systems of equations. ORTHOMIN is effective but

Markham finds that the biconjugate gradient method of Fletcher (1976) and variants of it are more efficient. However Jackson and Robinson (1985) find that minimum residual methods are more effective than the biconjugate gradient method if the test problem is very asymmetric. Thus there is no universally preferred conjugate gradient method if $\underline{A}$ is of general form.

It is possible to symmetrise a general matrix equation (6.2) by replacing it with

$$\underline{A}^T \underline{A} V = A^T B . \tag{6.80}$$

This approach is not generally recommended because if $\underline{A}$ is at all ill-conditioned $\underline{A}^T\underline{A}$ will be even more so, with a consequent loss of accuracy when (6.80) is solved for V. However Khosla and Rubin (1981) have solved the equations governing incompressible viscous flow past a circular cylinder at $Re = 100$ by symmetrising the governing equations as in (6.80) and applying a SIP (Sect. 6.3.3) preconditioned conjugate gradient method. The SIP method alone requires approximately 240 iterations for convergence on a 61×24 grid, whereas the SIP preconditioned conjugate gradient method requires only about 40 iterations for convergence.

6.3.5 Multigrid Methods

These methods are applicable to both linear systems (6.2) and nonlinear systems (6.1). A typical method will be described initially for linear systems, the extension to nonlinear systems being described subsequently.

Multigrid methods work with a sequence of grids $m = 1, \ldots, M$, where the grid size ratio $\Delta_{m+1}/\Delta_m = 0.5$. The linear system to be solved on the finest grid is written

$$\underline{A}^M V^M = B^M . \tag{6.81}$$

An approximation to V^M is provided by the solution on the next coarser grid, V^{M-1}. Similarly for an intermediate grid the solution, V^m, is a good approximation to the solution on the next finer grid, V^{m+1}.

If an approximation to the solution of $\underline{A}^{m+1} V^{m+1} = B^{m+1}$ is denoted by $V^{m+1,a}$ so that

$$V^{m+1} = V^{m+1,a} + W^{m+1} , \quad \text{then} \tag{6.82}$$

$$\underline{A}^{m+1} W^{m+1} = B^{m+1} - \underline{A}^{m+1} V^{m+1,a} = R^{m+1} . \tag{6.83}$$

The correction, W^{m+1}, and the residual (or defect), R^{m+1}, are closely approximated by the correction and residual on the next coarser grid, W^m and R^m, if they are smooth enough, i.e. if the amplitudes of high frequency components are small. The highest frequency that can be represented on a discrete grid, Δ_m, is $2\Delta_m$ (Sect. 3.4.1).

Relaxation (iterative) procedures, such as Jacobi, Gauss-Seidel and SOR described in Sect. 6.3.1, remove high frequency components in a few iterations. It is the removal of the low frequency components of the error, and equivalently of the residual, that causes the slow convergence of relaxation methods on a fixed grid.

However, a low frequency component on a fine grid becomes a high frequency component on a coarse grid. Consequently multigrid methods seek to exploit the high frequency smoothing of relaxation methods in the following way.

For a given $(m+1)$-th grid a few (v^1) relaxation steps are made to smooth the high frequency components in the correction and in the residual. From (6.83) this can be written symbolically as

$$W^{m+1,v^1} = \text{RELAX}^{v^1}(W^{m+1}, \underline{A}^{m+1}, R^{m+1}) \quad \text{and}$$

$$R^{m+1,v^1} = R^{m+1} - \underline{A}^{m+1} W^{m+1,v^1} . \tag{6.84}$$

For a linear system of equations the relaxation can be applied to either the original equation (6.2) or to the residual/correction equation (6.83). For nonlinear systems of equations the relaxation must be applied to the original equation (6.1).

The equivalent residual distribution on the mth grid, R^m, is computed from R^{m+1,v^1} by a restriction operator I_{m+1}^m. This can be written symbolically as

$$R^m = I_{m+1}^m R^{m+1,v^1} . \tag{6.85}$$

Appropriate forms of the restriction operator will be indicated later. The process of relaxation and restriction is continued until the coarsest grid, $m=1$ is reached.

On the coarsest grid it is economical to continue the relaxation (or any other iterative method) until the converged correction, W^1, is obtained. Alternatively $\underline{A}^1 W^1 = R^1$ or (6.2) can be solved by a direct method (Sect. 6.2). It may be noted that W^1 is non-zero if R^1 is non-zero, in general.

The correction W^2, can be obtained from W^1 by prolongation (interpolation). This is written symbolically as $W^2 = I_1^2 W^1$ or, more generally,

$$W^{m+1} = I_m^{m+1} W^m . \tag{6.86}$$

Appropriate forms of the prolongation operator I_m^{m+1} will be indicated later. On the finer grid, $m+1$, a few (v^2) relaxation steps are made, i.e.

$$W^{m+1,v^2} = \text{RELAX}^{v^2}(W^{m+1}, \underline{A}^{m+1}, R^{m+1}) , \tag{6.87}$$

before prolongating to the next finer grid.

A single multigrid cycle starting from the finest grid, $m=M$, consists of the repeated application of (6.84 and 85) until the coarsest grid is reached, the exact or complete iterative solution of $\underline{A}^1 W^1 = R^1$ and the repeated application of (6.86 and 87) until the finest grid, $m=M$, is again reached. This is called a V-cycle and a flow chart is shown in Fig. 6.21.

The cycle is repeated until satisfactory convergence on the finest grid is achieved. Ten cycles to reduce the change in the solution to 10^{-5} for successive cycles might be typical if (6.2) were a Poisson equation on a uniform grid. For the first cycle W^m is set equal to the initial guess of the fine-grid solution, $V^{M,a}$.

The restriction operator, I_{m+1}^m in (6.85), permits the residual distribution on the coarser grid R^m to be constructed from the residual distribution on the finer grid

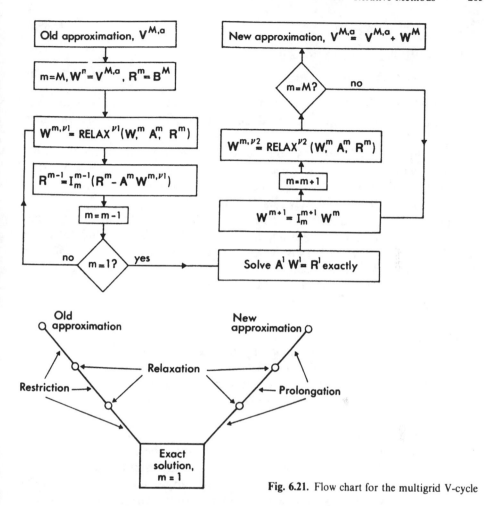

Fig. 6.21. Flow chart for the multigrid V-cycle

R^{m+1}. On a physical grid, with j and k denoting grid point locations in x and y on the coarser grid, the simplest choice is given by direct injection, $R^m_{j,k} = R^{m+1}_{j,k}$. This is stable and computationally efficient for strongly elliptic problems, e.g. Laplace's equation in Sect. 5.2.3. However for weakly elliptic nonlinear problems, i.e. high Re viscous flow (Chap. 17), a more stable choice is given by the five- and nine-point operators,

$$I^m_{m+1} \equiv \frac{1}{16} \begin{bmatrix} 0 & 2 & 0 \\ 2 & 8 & 2 \\ 0 & 2 & 0 \end{bmatrix} \quad \text{and} \quad \frac{1}{16} \begin{bmatrix} 1 & 2 & 1 \\ 2 & 4 & 2 \\ 1 & 2 & 1 \end{bmatrix} \tag{6.88}$$

The prolongation (interpolation) operator in (6.86) is usually based on bilinear interpolation in two dimensions. Thus if the j, k grid point labels are associated with the coarser, m, grid then

$$w_{j,k}^{m+1} = w_{j,k}^m ,$$ (6.89a)

$$w_{j\pm\frac{1}{2},k}^{m+1} = 0.5(w_{j,k}^m + w_{j\pm 1,k}^m) ,$$ (6.89b)

$$w_{j+\frac{1}{2},k+\frac{1}{2}}^{m+1} = 0.25(w_{j,k}^m + w_{j+1,k}^m + w_{j,k+1}^m + w_{j+1,k+1}^m) .$$ (6.89c)

The w^{m+1} at points $(j, k\pm\frac{1}{2})$ are handled as in (6.89b). It should be noted that grid points (j, k) on the coarser grid will be relabelled $(2j+1, 2k+1)$ on the finer grid.

For nonlinear problems it is possible to combine a multigrid method with Newton's method (Sect. 6.1.1). A linear system is solved for the correction to the solution at every step of the Newton iteration (6.6). The multigrid algorithm in Fig. 6.21 can be applied iteratively to solve (6.6).

However, a multigrid method can be applied directly to the nonlinear system (6.1) without the need for introducing Newton's method. The essential difference from the algorithm described in Fig. 6.21 is that at the restriction stage, (6.85), both R^{m+1,v^1} and the approximate solution, $V^{m+1,a}$, must be restricted onto the coarser grid m. At the mth level a solution of

$$\underline{A}^m V^{m,a} = R^m + \underline{A}^m I_{m+1}^m V^{m+1,a}$$ (6.90)

is sought in place of (6.83).

In practice the solution $V^{m,a}$ is obtained by relaxation and further restriction to the coarsest grid, exact solution on the coarsest grid and then relaxation and prolongation to successively finer grids, in a manner equivalent to that shown in Fig. 6.21. Once the solution $V^{m,a}$ is available, the correction

$$W^m = V^{m,a} - I_{m+1}^m V^{m+1,a}$$ (6.91)

is computed and prolongated to the $(m+1)$-th grid. There a revised solution, $V^{m+1,a} + I_m^{m+1} W^m$, is relaxed to give V^{m+1,v^2} essentially as in the linear case.

The nonlinear scheme is referred to as the full approximation storage (FAS) method by Brandt (1977) to distinguish it from the correction storage (CS) method given in Fig. 6.21. Since the FAS method passes both the solution and the residual to coarser grids it is slightly more expensive (about 5%–10%) than the CS scheme; generally it should not be used for solving linear problems. Like the CS method, the FAS method relies on the smoothing of the solution correction W and the residual R for part of its effectiveness.

The description of both the CS and the FAS methods has started with an approximation to the solution on the finest grid, $m = M$, and has followed a V-cycle to the coarsest grid and back again to improve that solution. This is an appropriate strategy if a good approximation to the solution, V^M, is available.

If nothing is known about the solution it is better to start with the solution on the coarsest grid, obtained either directly or by (conventional) iteration. A high-order interpolation is made to the next finer grid and r multigrid V-cycles are applied to improve the solution. The high-order interpolation plus r multigrid V-cycle process is repeated on successively finer grids until $m = M$ is reached, and the multigrid V-cycle is repeated until convergence.

This is the basis of the full multigrid (FMG) method, which is a generalisation of both the FAS and CS methods. The FMG method is shown schematically in Fig. 6.22 for a linear system of equations. The function $INT(V^m, m)$ represents a cubic interpolation of V^m first in the x direction and then, for all $x^{(m+1)}$ grid points, in the y direction. The function $MGI(V^{m,r-1}, \underline{A}^m, B^m)$ denotes one pass through the multigrid V-cycle shown in Fig. 6.21.

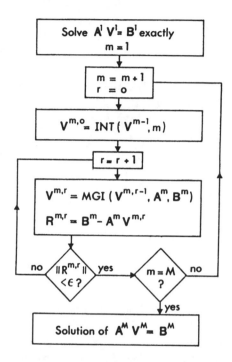

Fig. 6.22. Flow chart for the FMG method

The multigrid strategy can be viewed as an acceleration technique like the conjugate gradient method (Sect. 6.3.4) applied to a basic iterative scheme (Sects. 6.3.1–6.3.3). Alternatively the basic iterative scheme is interpreted as a smoother for the multigrid method just as it was treated as a preconditioner for the conjugate gradient method. As with conjugate gradient acceleration, the use of an incomplete (approximate) line $\underline{L}.\underline{U}$ factorisation of $\underline{A}$ for the smoothing process (6.84, 87) is found to be very effective (Sonneveld et al. 1985). This is particularly the case when $\underline{A}$ is very asymmetric, i.e. for convection-dominated problems (Chaps. 9 and 10).

Multigrid methods are described in detail by Brandt (1977), Stuben and Trottenberg (1982) and Hackbusch (1985), amongst others. Brandt, Stuben and Trottenberg and Briggs (1987a) provide FORTRAN programs for a simple multigrid method applied to a Poisson equation. The description given here follows Stuben and Trottenberg primarily; but Briggs (1987b) probably provides an easier orientation. Applications of multigrid methods to inviscid flow are indicated in Sects. 14.2.9 and 14.3.5 and to viscous incompressible flow in Sects. 17.2.3 and 17.3.1.

6.4 Pseudotransient Method

An alternative to the solution of the algebraic equations produced by discretising the steady problem is to construct an equivalent unsteady problem and to march the transient solution until the steady state is reached. Time then plays the role of an iteration parameter.

Thus if Laplace's equation

$$\frac{\partial^2 \bar{v}}{\partial x^2} + \frac{\partial^2 \bar{v}}{\partial y^2} = 0 \ , \tag{6.92}$$

is discretised using centred finite difference formulae with $\Delta x = \Delta y$, the Jacobi method gives, for internal nodes (on a j, k grid as in Fig. 6.1),

$$v_{j,k}^{(n+1)} = 0.25(v_{j,k+1}^{(n)} + v_{j,k-1}^{(n)} + v_{j+1,k}^{(n)} + v_{j-1,k}^{(n)}) \ . \tag{6.93}$$

To generate an equivalent unsteady method, (6.92) would be replaced by

$$\frac{\partial \bar{v}}{\partial t} = \alpha \left(\frac{\partial^2 \bar{v}}{\partial x^2} + \frac{\partial^2 \bar{v}}{\partial y^2} \right) \ . \tag{6.94}$$

Discretisation using finite difference formulae and $\Delta x = \Delta y$ gives the algorithm

$$v_{j,k}^{n+1} = (1 - 4s)v_{j,k}^n + s(v_{j,k+1}^n + v_{j,k-1}^n + v_{j+1,k}^n + v_{j-1,k}^n) \ , \tag{6.95}$$

where $s = \alpha \Delta t / \Delta x^2$. The choice $s = 0.25$ reproduces (6.93). Thus there is a clear link between the unsteady formulation and the iterative techniques considered in Sect. 6.3.

A major advantage of the unsteady formulation is that it makes available the various splitting techniques discussed in Sect. 8.2. For example (6.94) solved by the ADI method would generate equations (8.14 and 15). The solution of these equations is obtained by solving the subsystems of equations associated with each grid line using the Thomas algorithm (Sect. 6.2.2). This is illustrated in Fig. 6.23.

Since the transient solution is not of interest it is customary to choose a sequence of time-steps to minimise the number of time-steps to convergence. The sequence is problem dependent but is typically geometric with the range of time-steps corresponding to the range of eigenvalues associated with the x-gridline and y-gridline tridiagonal matrices (Wachpress 1966, Chap. 6).

For flow problems it is often found that the approach to the steady state is considerably more rapid in some parts of the computational domain than in others. For the flow over a backward-facing step (Sect. 17.3.3) the solution in the separated flow region behind the step is much slower to converge than in the regions far from the step. The spatial dependence of the convergence rate suggests the following modification to (6.94):

$$c(x, y)\frac{\partial v}{\partial t} = \alpha \left(\frac{\partial^2 \bar{v}}{\partial x^2} + \frac{\partial^2 \bar{v}}{\partial y^2} \right) \ , \tag{6.96}$$

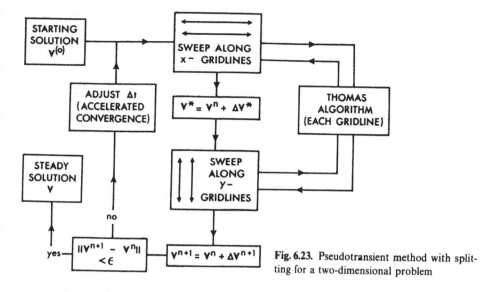

Fig. 6.23. Pseudotransient method with splitting for a two-dimensional problem

where $c(x, y)$ is adjusted empirically to balance the convergence rate in the different parts of the computational domain. The splitting scheme is then applied in the conventional manner (Sect. 8.2).

6.4.1 Two-Dimensional, Steady Burgers' Equations

Here we illustrate the effectiveness of the pseudotransient method by combining it with Newton's method applied to the two-dimensional steady Burgers' equations (Sect. 6.1.3). The unsteady equivalents of (6.12) are

$$\frac{\partial \bar{u}}{\partial t} + \bar{u}\frac{\partial \bar{u}}{\partial x} + \bar{v}\frac{\partial \bar{u}}{\partial y} - \frac{1}{\text{Re}}\left(\frac{\partial^2 \bar{u}}{\partial x^2} + \frac{\partial^2 \bar{u}}{\partial y^2}\right) = 0$$

$$\frac{\partial \bar{v}}{\partial t} + \bar{u}\frac{\partial \bar{v}}{\partial x} + \bar{v}\frac{\partial \bar{v}}{\partial y} - \frac{1}{\text{Re}}\left(\frac{\partial^2 \bar{v}}{\partial x^2} + \frac{\partial^2 \bar{v}}{\partial y^2}\right) = 0 \ . \tag{6.97}$$

The discretised form of (6.97), equivalent to (6.13), is

$$\frac{u_{j,k}^{n+1} - u_{j,k}^{n}}{\Delta t} + Ru_{j,k}^{n+1} = 0 \ ,$$

$$\frac{v_{j,k}^{n+1} - v_{j,k}^{n}}{\Delta t} + Rv_{j,k}^{n+1} = 0 \ , \tag{6.98}$$

where $Ru_{j,k}^{n+1}$ and $Rv_{j,k}^{n+1}$ are given by (6.13).

It may be noted that the steady-state terms have been evaluated at time-level $n + 1$. This is to make the pseudotransient form compatible with Newton's method and to allow the time level $n + 1$ to be directly equivalenced to the iteration level in

Newton's method (Sect. 6.1.1). An examination of (6.13) indicates that $Ru_{j,k}$ and $Rv_{j,k}$ are functions of $u_{j,k}^{n+1}$, $u_{j-1,k}^{n+1}$, $v_{j,k}^{n+1}$, etc. Therefore expanding $Ru_{j,k}^{n+1}$ and $Rv_{j,k}^{n+1}$ about time (iteration) level n, as in Sect. 8.2.2, allows (6.98) to be written as

$$\frac{\Delta u_{j,k}^{n+1}}{\Delta t} + Ru_{j,k}^n + \left(\frac{\partial Ru_{j,k}}{\partial q_{l,m}}\right)^n \Delta q_{l,m} = 0 \ , \tag{6.99}$$

$$\frac{\Delta v_{j,k}^{n+1}}{\Delta t} + Rv_{j,k}^n + \left(\frac{\partial Rv_{j,k}}{\partial q_{l,m}}\right)^n \Delta q_{l,m} = 0 \ ,$$

where $\Delta u^{n+1} = u^{n+1} - u^n$, etc. The term q includes both u and v and the indices l, m denote all possible values of j and k that lead to non-zero partial derivatives. That is $l = j-1, j, j+1; m = k-1, k, k+1$. Equations (6.99) can be combined into a single equation as

$$\left(\frac{\underline{I}}{\Delta t} + \underline{J}\right)\Delta q = -R \ , \tag{6.100}$$

which may be compared with the conventional Newton's method (6.17). The effect of choosing small values of Δt is to make the augmented Jacobian $(\underline{I}/\Delta t + \underline{J})$ more diagonally dominant.

Equation (6.100) is a more effective means of "under-relaxing" Newton's method than (6.19). For the solution of Burgers' equation a typical output produced by the program NEWTBU based on (6.100) is shown in Fig. 6.24. It is

```
NEWTONS METHOD FOR N = 18      ITMX= 50 IRD= 1  IPN= 5
DT=  .10000E-01  EPS=  .100E-04  RE=  .100E+02  OM=  .100E+01
A =   110.13   110.13      .00      .00    1.00


UE=   .9990     .9586     .4888    -.0559    -.0991
UE=   .9990     .9583     .4863    -.0565    -.0991
UE=   .9990     .9572     .4786    -.0584    -.0991
UE=   .9989     .9553     .4655    -.0614    -.0992
UE=   .9988     .9524     .4462    -.0657    -.0992
VE=   .0000     .0000     .0000     .0000     .0000
VE=   .1317     .1277     .0753     .0090     .0012
VE=   .2679     .2598     .1515     .0179     .0023
VE=   .4142     .4010     .2297     .0266     .0034
VE=   .5774     .5577     .3109     .0349     .0045
IT=  0  RMS=  .1496D+00  R= -.2169D+00 -.2850D-01 -.2188D+00
IT=  5  RMS=  .1756D-01  R= -.1071D-01 -.2309D-02 -.1703D-01
IT= 10  RMS=  .2063D-02  R= -.1072D-02 -.2097D-03 -.1748D-02
IT= 15  RMS=  .2465D-03  R= -.1160D-03 -.2103D-04 -.1905D-03
IT= 20  RMS=  .2963D-04  R= -.1308D-04 -.2219D-05 -.2157D-04


AFTER  23 ITERATIONS THE RMS RESIDUAL IS   .83174D-05
U=   .9990     .9586     .4888    -.0559    -.0991
U=   .9990     .9605     .4835    -.0567    -.0991
U=   .9990     .9603     .4748    -.0585    -.0991
U=   .9989     .9577     .4625    -.0615    -.0992
U=   .9988     .9524     .4462    -.0657    -.0992
V=   .0000     .0000     .0000     .0000     .0000
V=   .1317     .1282     .0748     .0090     .0012
V=   .2679     .2605     .1507     .0179     .0023
V=   .4142     .4017     .2290     .0266     .0034
V=   .5774     .5577     .3109     .0349     .0045
```

Fig. 6.24. Typical output from program NEWTBU, incorporating (6.100)

apparent that convergence is much more rapid with (6.100) than with (6.17, 19). The convergence of (6.100) is dependent on the choice of Δt. A large value of Δt implies convergence like the full Newton's method. But for the two-dimensional Burgers' equations this implies a divergence (Sect. 6.1.3). A small value of Δt implies a slower rate of convergence but has the advantage of greatly expanding the radius of convergence. This overcomes a major disadvantage of Newton's method in Table 6.5. As long as the equations being solved, (6.98), have a unique solution, choosing a sufficiently small Δt will find the solution from any reasonable starting solution.

The modifications to NEWTBU to implement (6.100) are given by lines 36 and 39 of subroutine JACBU (Fig. 6.13). As implemented in NEWTBU the major disadvantages of Newton's method are associated with the lack of economy of forming $\underline{J}$ and factorising the augmented Jacobian $\{\underline{I}/\Delta t + \underline{J}\}$. The factorisation of $\underline{J}$ is an $O(N^3)$ process which is prohibitively expensive if N is large. In addition, the retention of all locations of $\underline{J}$ in main memory for reasonable economy places a severe limitation on the largest system of equations that can be solved.

Many entries in $\underline{J}$ are zero, and of those that are non-zero, only a few are large. This implies, in the present problem, that Ru and Rv are strong functions of some nodal unknowns $u_{j,k}$, $v_{j,k}$ but very weak functions of other nodal unknowns. Consequently it is desirable to consider approximations to $\underline{J}$ that capture the strong dependence, but ignore the weak dependence if this leads to smaller Jacobians or structures of $\underline{J}$ that lend themselves to approximate factorisation (Sect. 8.2) in an efficient manner. The price to be paid may be more iterations to reach convergence. But if each iteration is sufficiently economical the price is acceptable.

When the approximate factorisation (splitting) techniques described in Sect. 8.2 are applied to nonlinear problems they can be related to an approximate factorisation of (6.100). This aspect is clarified in Sect. 10.4, where the two-dimensional Burgers' equations (the current problem) are solved by an approximate factorisation of (6.100).

6.5 Strategies for Steady Problems

Most flow problems are governed by systems of coupled partial differential equations (Chap. 11). For flow formulations involving velocity potential (Sect. 14.3), stream function (Sect. 17.3) or incompressible flows involving pressure (Sects. 17.1 and 17.2) one of the governing equations is typically diagonally dominant. In some instances this equation may also be linear. Consequently the iterative techniques (Sect. 6.3) are then applicable, particularly multigrid methods (Sect. 6.3.5), as are the special direct solvers (Sect. 6.2.6) if the grid is uniform.

When velocities or vorticity appear as dependent variables the governing equations usually include nonlinear convective terms (Chap. 10). Discretisation of such equations produce nonlinear systems of equations like (6.1) that are not diagonally dominant, unless the (viscous) diffusion terms are much larger than the convective terms. The choice of solution method is then often between a Newton or quasi-Newton solver and a pseudotransient, approximate factorisation solver. The relative strengths and weaknesses of the two approaches are summarised in Tables 6.5 and 6.6.

Table 6.5. Newton's method

Advantages:	(i) Rapid convergence (few iterations)
	(ii) Can be modified to overcome many explicit disadvantages
	(iii) Approximate solution can be exploited
Disadvantages:	(i) Small radius of convergence if large number of unknowns
	(ii) Factorisation of J at each iteration is computationally expensive
	(iii) Factorisation of J requires large memory for full storage
	(iv) Fails to converge if J becomes ill-conditioned

Table 6.6. Pseudotransient formulation plus approximate factorisation

Advantages:	(i) Each iteration is economical
	(ii) Very large radius of convergence
	(iii) Small memory requirement (splitting uncouples gridlines)
	(iv) Detects if problems not steady
	(v) Approximate solution can be exploited
Disadvantage:	(i) Slow convergence (large number of iterations)

For viscous flow problems at high Reynolds number (Chaps. 17 and 18) J becomes ill-conditioned and this often prevents convergence of Newton's method even when starting from a nearby converged solution.

For some problems it is not known, a priori, if a steady solution exists. By constructing an unsteady formulation an oscillatory pseudosteady solution can be automatically detected.

Historically Newton's method has been used more often with finite element methods and the pseudotransient splitting approach has been used more often with finite difference and finite volume methods. However the pseudotransient approach can be interpreted as providing a diagonally augmented system of equations to be solved iteratively, as in (6.100). Consequently, even if the augmented equations are not formally diagonally dominant (6.52) they are suitable for solution using multigrid techniques (Sect. 6.3.5) and usually produce more rapid convergence.

6.6 Closure

In this chapter we have briefly reviewed techniques for solving the algebraic equations that result from discretisation (Sect. 3.1), particularly for steady problems.

The algebraic equations produced by discretising the fluid dynamic governing equations (Chaps. 14–18) are nonlinear, typically. Consequently some iterative procedure is inevitable. It is conceptually useful to think of an outer iteration to cope with the nonlinear nature of the discretised equations. At each step of the outer iteration a linear system of equations is solved. This system may be solved by direct (Sect. 6.2) or iterative methods (Sect. 6.3).

Newton's method (Sect. 6.1.1) and the pseudotransient formulation (Sect. 6.4) are candidates for the outer iteration. Although Newton's method (Sect. 6.1.1) achieves quadratic convergence when close to the solution, its small radius of convergence for large numbers of nodal unknowns makes it a less effective method, in its basic form, than the pseudotransient formulation (Sect. 6.4).

For multidimensional problems the value of split implicit schemes (Sect. 8.2) is directly proportional to the efficiency of solving systems of equations that are narrowly banded (Sect. 6.2.2–6.2.5) at each stage of the outer iteration.

Iterative techniques at each stage of the outer iteration are usually most effective for systems that are strongly diagonally dominant. This often occurs with discretisations of the transonic inviscid flow governing equations (Sect. 14.3) and this has led to the development of special iterative techniques (Sect. 14.3.5). However a particular feature of the multigrid technique (Sect. 6.3.5) is that it permits solution on the coarsest grid by a direct method in an economical manner. Thus the multigrid technique is available whether the system of equations is diagonally dominant or not.

6.7 Problems

Nonlinear Steady Problems (Sect. 6.1)

6.1 For the solar collector problem considered in program NEWTON, introduce modifications to calculate $\underline{J}$ in (6.10) only every p steps. Compare the number of iterations to convergence and the overall operation count for various values of p. It will be more meaningful to start from a solution for $T^{(0)}$ that requires a larger number of iterations than in Fig. 6.8 to reach convergence.

6.2 Run program NEWTBU for values of Re $= 5, 2, 1$. For each case pick the value of ω that produces the most rapid convergence. What do you notice about the dependence of the number of iterations to convergence on Re? Can you correlate this with the relative magnitude of terms in (6.12) and the consequent diagonal dominance of $\underline{J}$ in (6.17)?

6.3 Modify program NEWTBU so that the solution is chosen from

$$q_m^{(n+1)} = q^{(n)} + \omega_m \Delta q^{(n+1)} ,$$

where ω_m corresponds to the minimum R_{rms} in the search direction defined by $\Delta q^{(n+1)}$. Compare the number of iterations and CPU time to convergence with the results shown in Figs. 6.14 and 6.24.

Direct Method for Linear Systems (Sect. 6.2)

6.4 Write a computer program to solve (6.27) with $R_{cell}=0.5$ using subroutines BANFAC and BANSOL for v_j, $j=2,\ldots,10$. Boundary conditions are $v_1=0$, $v_{11}=1.0$. This problem has the exact solution (9.46).

$$v_j= -0.0060834+0.00365\,(5/3)^j \quad \text{for} \quad j=1,\ldots,11 \ . \tag{6.101}$$

6.5 A five-point scheme corresponding to (6.27) is

$$(1+R_{cell})v_{j-2}-8(2+R_{cell})v_{j-1}+30v_j-8(2-R_{cell})v_{j+1}$$
$$+(1-R_{cell})v_{j+2}=0 \ . \tag{6.102}$$

Modify subroutines BANFAC and BANSOL to solve the pentadiagonal system found from the above equation. Assume sufficient boundary conditions are available to evaluate v_{j-2}, v_{j-1} at the left-hand boundary and v_{j+1}, v_{j+2} at the right-hand boundary. Implement this scheme for $j=3,\ldots,9$ and $R_{cell}=1.0$. Use

$$v_j=(e^{10x_j}-1)(e^{10}-1) \quad \text{with} \quad x_j=0.1(j-1) \tag{6.103}$$

to obtain boundary values for v_1, v_2, v_{10} and v_{11}. Equation (6.103) provides the exact solution of the problem which (6.102) discretises. Thus the solution of (6.102) should be close to (6.103).

6.6 Write a computer program to implement the solution of block tridiagonal systems of equations with 2×2 blocks. Test by discretising.

$$\frac{d^2T}{dx^2}-2\frac{dS}{dx}=0 \quad \text{and} \quad \frac{d^2S}{dx^2}-0.5\frac{dT}{dx}=0 \tag{6.104}$$

for boundary conditions $T=0$ and $S=1$ at $x=0$, $T=2\sinh(1)$, $S=\cosh(1)$ at $x=1$ using three-point centred finite difference formulae. Obtain solutions in the interval $0\le x\le1.0$ with $\Delta x=0.1$. The exact solution to (6.104), and the Dirichlet boundary conditions given, is $T=2\sinh(x)$, $S=\cosh(x)$.

Iterative Methods (Sect. 6.3)

6.7 For $b/a=3$ and 10 obtain solutions to the duct problem (Sect. 6.3.2) corresponding to Table 6.3 for the Jacobi, Gauss-Seidel and SOR methods applied with finite difference discretisation. What is the effect of increasing aspect ratio?

6.8 Modify program DUCT to introduce the ADI scheme (6.66, 67) and confirm the results shown in Table 6.3. Convergence is often faster if a sequence of iteration parameters, $\lambda_y^{(n)}$, is used. Determine the effectiveness of the following choice, for the case $b/a=1$, $\Delta x=\Delta y$:

$$\lambda_y=\left(\frac{0.5}{\sin(0.5\pi l/N)}\right)^2, \quad l=1,\ldots,N-1 \ , \tag{6.105}$$

where N is the number of divisions on each side of the duct. Equation (6.105) gives a cycle, which is repeated if required. The appropriate strategy for choosing the iteration parameter sequence is discussed by Varga (1962) and Wachpress (1966).

6.9 Apply a V-cycle multigrid method to the duct problem (Sect. 6.3.2) for conditions corresponding to Table 6.4 and compare the number of iterations to convergence and the overall operation count with the results shown in Table 6.4.

Pseudotransient Methods (Sect. 6.4)

6.10 Apply the pseudotransient Newton's method (program NEWTBU) with $\underline{J}$ only evaluated every p iterations. Determine the influence of p on the number of iterations to convergence and the overall operation count.

6.11 Modify program NEWTBU so that individual elements of $\underline{J}$ are set to zero if they are less than TOL. For TOL $= 10^{-5}, 10^{-3}, 10^{-1}$, determine how many elements of $\underline{J}$ are set to zero and the influence on the number of iterations to convergence. Would the resulting pattern of non-zero elements in $\underline{J}$ permit a more efficient solution of (6.100) than using subroutines FACT and SOLVE; for example using an iterative technique based on (6.51).

6.12 Can program NEWTBU be made to converge faster if a sequence of Δt's is used? Introduce a geometric cycle of time steps

$$\Delta t = \Delta t^{(0)} r^l, \qquad l = 1, \dots, p,$$

where the geometric ratio, $r \approx 1.2$ and $p \approx 5$ to 40. Experiment with different combinations of $\Delta t^{(0)}$, r and p.

7. One-Dimensional Diffusion Equation

From a computational perspective the diffusion equation contains the same dissipative mechanism as is found in flow problems with significant viscous or heat conduction effects. Consequently computational techniques that are effective for the diffusion equation will provide guidance in choosing appropriate algorithms for viscous fluid flow (Chaps. 15–18).

In this chapter the one-dimensional diffusion equation will be used as a vehicle for developing explicit and implicit schemes. Attention will be given to the stability and accuracy of the various schemes. The problem of accurately implementing boundary and initial conditions will also be considered.

The one-dimensional diffusion or heat conduction equation

$$\frac{\partial \bar{T}}{\partial t} - \alpha \frac{\partial^2 \bar{T}}{\partial x^2} = 0 \tag{7.1}$$

has already been introduced as a model parabolic partial differential equation (Sect. 2.3) and used to illustrate the discretisation process (Sect. 3.1) and the implementation of the finite difference method (Sect. 3.5).

In (7.1) $\bar{T}$ may be interpreted as the velocity, vorticity, temperature or concentration depending on whether the diffusion of momentum, vorticity, heat or mass is being considered. If $\bar{T}$ is the temperature, (7.1) governs the flow of heat in a rod which is insulated along its edges but can transfer heat to the surroundings via the ends of the rod (A and B in Fig. 7.1).

Two types of boundary condition are common. First the dependent variable is a known function of time. In the notation of (7.1) this would be (for the end A)

$$\bar{T}(0, t) = b(t) \ . \tag{7.2}$$

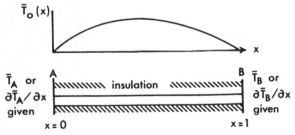

Fig. 7.1. One-dimensional, unsteady heat conduction

This is a Dirichlet boundary condition (Sect. 2.1.2). In practice b is often a constant. Second, the spatial derivative of the dependent variable may be specified. To suit (7.1) this could be written (for the end A),

$$\frac{\partial \bar{T}}{\partial x}(0, t) = c(t) \ . \tag{7.3}$$

This is a Neumann boundary condition (Sect. 2.1.2). As with Dirichlet boundary conditions, c is often a constant.

To obtain unique solutions of (7.1) it is also necessary to specify initial conditions. These are given by

$$\bar{T}(x, 0) = T_0(x) \ . \tag{7.4}$$

The exact solution $\bar{T}(x, t)$ satisfies (7.1) in conjunction with (7.4) and (7.2) or (7.3) applied at $x = 0$ and $x = 1.0$.

To obtain the approximate solution, (7.1) is discretised (Sect. 3.1) and the resulting algebraic equation is manipulated to generate an algorithm. The algorithm gives the solution at the $(n+1)$-th time level (Fig. 3.2) in terms of the known solution at the nth and earlier time levels. The overall procedure is described in Sect. 3.5.

7.1 Explicit Methods

For explicit methods a single unknown, e.g. T_j^{n+1}, appears on the left hand side of the algebraic formula resulting from discretisation.

7.1.1 FTCS Scheme

If a two-point forward difference approximation is introduced for the time derivative and a three-point centred difference approximation is introduced for the spatial derivative in (7.1), the result is

$$\frac{T_j^{n+1} - T_j^n}{\Delta t} - \frac{\alpha(T_{j-1}^n - 2T_j^n + T_{j+1}^n)}{\Delta x^2} = 0 \ . \tag{7.5}$$

Equation (7.5) will be referred to as the FTCS (forward time centred space) scheme. It can be seen that the spatial derivative has been evaluated at the nth time level, i.e. at a known time level. Rearranging (7.5) gives the algorithm

$$T_j^{n+1} = sT_{j-1}^n + (1 - 2s) T_j^n + sT_{j+1}^n \ , \tag{7.6}$$

where the discretisation parameter $s = \alpha \Delta t / \Delta x^2$.

The grid points connected together by (7.6) are shown in Fig. 7.2. Substitution of $\bar{T}$, the exact solution of (7.1), into (7.5) and expansion of the various terms as a

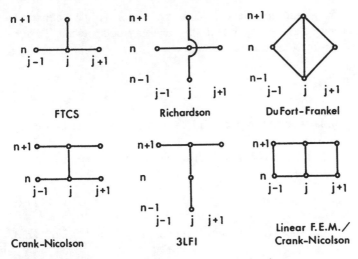

Fig. 7.2. Active nodes for diffusion equation algebraic schemes

Taylor series about the (j, n)-th node (as in Sect. 3.2.1) gives

$$\left[\frac{\partial \bar{T}}{\partial t} - \alpha \frac{\partial^2 \bar{T}}{\partial x^2}\right]_j^n + E_j^n = 0 \; ,$$

where the truncation error E_j^n is given by

$$E_j^n = \left[\frac{\Delta t}{2} \frac{\partial^2 \bar{T}}{\partial t^2} - \alpha \frac{\Delta x^2}{12} \frac{\partial^4 \bar{T}}{\partial x^4}\right]_j^n + O(\Delta t^2, \Delta x^4) \; . \tag{7.7}$$

It can be seen that (7.5) is consistent (Sect. 4.2) with (7.1).

From the leading term in E_j^n the FTCS scheme will be referred to as being first-order accurate in time and second-order accurate in space. However, it should be remembered that this statement is strictly only correct in the limit $\Delta t, \Delta x \rightarrow 0$. In practice, solutions are obtained on a grid of finite spacings and the magnitude of terms like $\partial^2 \bar{T}/\partial t^2$ is not known, a priori.

Application of the von Neumann stability analysis (Sect. 4.3) indicates that the amplification factor G is given by

$$G = 1 - 4s \sin^2\left(\frac{\theta}{2}\right) \; ,$$

where $s = \alpha \Delta t/\Delta x^2$ and $\theta = m\pi\Delta x$. For any value of θ, $|G| \leq 1$ if $s \leq 0.5$. Thus (7.6) will produce stable solutions if $s \leq 0.5$. It has already been seen (Sect. 4.2.1) that the particular choice $s = 1/6$ introduces a cancellation of terms in the truncation error and (7.6) then has a truncation error of $O(\Delta t^2, \Delta x^4)$.

The above properties of the FTCS scheme applied to the diffusion equation (7.1) are summarised in Table 7.1. It may be noted that the form of the truncation

Table 7.1. Algebraic (discretised) schemes for the diffusion equation $\partial\bar{T}/\partial t - \alpha\partial^2\bar{T}/\partial x^2 = 0$

Scheme	Algebraic form	Truncation error[a] (E) (leading term)	Amplification factor $G(\theta = m\pi\Delta x)$	Stability restrictions	Remarks
FTCS	$\dfrac{\Delta T_j^{n+1}}{\Delta t} - \alpha L_{xx} T_j^n = 0$	$\alpha(\Delta x^2/2)\left(s - \dfrac{1}{6}\right)\dfrac{\partial^4 T}{\partial x^4}$	$1 - 4s\sin^2\left(\dfrac{\theta}{2}\right)$	$s \leq 0.5$	$s = \alpha\dfrac{\Delta t}{\Delta x^2}$ $L_{xx} = \dfrac{1}{\Delta x^2}[1, -2, 1]$
DuFort-Frankel	$\dfrac{T_j^{n+1} - T_j^{n-1}}{2\Delta t} - \dfrac{\alpha}{\Delta x^2}[T_{j-1}^n$ $- (T_j^{n-1} + T_j^{n+1}) + T_{j+1}^n] = 0$	$\alpha\Delta x^2\left(s^2 - \dfrac{1}{12}\right)\dfrac{\partial^4 T}{\partial x^4}$	$\dfrac{2s\cos\theta + (1 - 4s^2\sin^2\theta)^{1/2}}{(1+2s)}$	None	$\Delta T_j^{n+1} = T_j^{n+1} - T_j^n$
Crank-Nicolson	$\dfrac{\Delta T_j^{n+1}}{\Delta t} - \alpha L_{xx}\left(\dfrac{T_j^n + T_j^{n+1}}{2}\right) = 0$	$-\alpha\left(\dfrac{\Delta x^2}{12}\right)\dfrac{\partial^4 T}{\partial x^4}$	$\dfrac{1 - 2s\sin^2(\theta/2)}{1 + 2s\sin^2(\theta/2)}$	None	
Three-level fully implicit	$\dfrac{3}{2}\dfrac{\Delta T_j^{n+1}}{\Delta t} - \dfrac{1}{2}\dfrac{\Delta T_j^n}{\Delta t} - \alpha L_{xx}T_j^{n+1} = 0$	$-\alpha\left(\dfrac{\Delta x^2}{12}\right)\dfrac{\partial^4 T}{\partial x^4}$	$\dfrac{1 + \frac{4}{3}i\left[\frac{3}{16} + s(1-\cos\theta)\right]^{1/2}}{2\left[1 + \frac{4}{3}s(1-\cos\theta)\right]}$	None	$\Delta T_j^n = T_j^n - T_j^{n-1}$
Linear F.E.M. /Crank-Nicolson	$M_x\dfrac{\Delta T_j^{n+1}}{\Delta t} - \alpha L_{xx}\left(\dfrac{T_j^n + T_j^{n+1}}{2}\right) = 0$	$\alpha\left(\dfrac{\Delta x^2}{12}\right)\dfrac{\partial^4 T}{\partial x^4}$	$\dfrac{(2 - 3s) + \cos\theta(1 + 3s)}{(2 + 3s) + \cos\theta(1 - 3s)}$	None	$M_x = \{\frac{1}{6}, \frac{2}{3}, \frac{1}{6}\}$

[a] The truncation error has been expressed solely in terms of Δx and x-derivatives as in the modified equation method (Section 9.2.2). Thus the algebraic scheme is equivalent to $\partial T/\partial t - \alpha\partial^2 T/\partial x^2 + E(T) = 0$

error shown in Table 7.1 is equivalent to (7.7). The accuracy of numerical solutions using the FTCS scheme is shown in Table 7.3.

7.1.2 Richardson and DuFort-Frankel Schemes

In (7.6) the use of a two-point one-sided difference formula produces a first-order contribution to the truncation error and the use of a three-point centred difference formula produces a second-order contribution to the truncation error. Therefore a logical improvement to (7.6) would be the scheme

$$\frac{T_j^{n+1} - T_j^{n-1}}{2\Delta t} - \frac{\alpha(T_{j-1}^n - 2T_j^n + T_{j+1}^n)}{\Delta x^2} = 0 \ , \tag{7.8}$$

due to Richardson (Fig. 7.2). However, although the scheme is of $O(\Delta t^2, \Delta x^2)$, a von Neumann stability analysis (Noye 1983, p. 138) indicates that the scheme is unconditionally unstable for $s > 0$. Thus it is of no practical use. It may be noted that the unstable behaviour refers to the equation as a whole. When the centred difference approximation for the time derivative is introduced into the convection equation (9.2) a stable algorithm can be obtained, (9.15).

The Richardson scheme (7.8) can be modified to produce a stable algorithm. This is achieved by replacing T_j^n in (7.8) with $0.5(T_j^{n-1} + T_j^{n+1})$. The resulting equation is

$$\frac{T_j^{n+1} - T_j^{n-1}}{2\Delta t} - \frac{\alpha[T_{j-1}^n - (T_j^{n-1} + T_j^{n+1}) + T_{j+1}^n]}{\Delta x^2} = 0 \tag{7.9}$$

Equation (7.9), which is known as the DuFort-Frankel scheme (Fig. 7.2), can be manipulated to give the explicit algorithm

$$T_j^{n+1} = \left(\frac{2s}{1+2s}\right)(T_{j-1}^n + T_{j+1}^n) + \left(\frac{1-2s}{1+2s}\right)T_j^{n-1} \ . \tag{7.10}$$

The DuFort-Frankel scheme is three-level in time unless $s = 0.5$, for which it coincides with the FTCS scheme. For a three-level scheme, two time-levels of the solution must be stored and an alternative two-level scheme is required for the first time step.

Application of the von Neumann stability analysis (Sect. 4.3) to (7.10) produces the amplification factor G given in Table 7.1. Since $|G| \leq 1$ for any value of θ with $s > 0$, it follows that the DuFort-Frankel scheme is stable for any value of Δt. There is a price to pay for this very favorable stability result. A Taylor expansion of the exact solution substituted into (7.9) about the (j, n)-th node produces the result

$$\left[\frac{\partial \bar{T}}{\partial t} - \alpha \frac{\partial^2 \bar{T}}{\partial x^2} + \alpha \left(\frac{\Delta t}{\Delta x}\right)^2 \frac{\partial^2 \bar{T}}{\partial t^2}\right]_j^n + O(\Delta t^2, \Delta x^2) = 0 \ . \tag{7.11}$$

Thus for consistency $\Delta t/\Delta x$ must $\rightarrow 0$ as $\Delta t, \Delta x \rightarrow 0$, i.e. it is required that $\Delta t \ll \Delta x$

for consistency. However $\alpha(\Delta t/\Delta x)^2 = s\Delta t$ and we expect s to be of $O(1)$ for diffusion problems. Therefore the DuFort-Frankel scheme is consistent with (7.1) but will be inaccurate if $s\Delta t$ is large. The alternative form of the truncation error (Table 7.1) indicates that if $s = (1/12)^{1/2}$, the DuFort-Frankel scheme has a truncation error of $O(\Delta x^4)$. A corresponding solution accuracy is indicated in Table 7.3.

From a practical point of view there is still an effective restriction on the size of Δt when using the DuFort-Frankel scheme, even though it arises from an accuracy restriction rather than a stability restriction, as with the FTCS scheme.

7.1.3 Three-Level Scheme

A general explicit three-level discretisation of (7.1) can be written as

$$aT_j^{n+1} + bT_j^n + cT_j^{n-1} - (dL_{xx}T_j^n + eL_{xx}T_j^{n-1}) = 0 \ , \tag{7.12}$$

where

$$L_{xx}T_j = (T_{j-1} - 2T_j + T_{j+1})/\Delta x^2 \ .$$

The parameters a, b, c, d and e may be determined by expanding each term in (7.12) as a Taylor series about node (j, n) and requiring that (7.12) is consistent with (7.1). Examples of this procedure are provided in Sects. 3.2.2 and 3.2.3. This procedure permits (7.12) to be rewritten with only two disposable constants, γ and β, instead of five. Thus (7.12) is replaced by

$$\frac{(1+\gamma)(T_j^{n+1} - T_j^n)}{\Delta t} - \frac{\gamma(T_j^n - T_j^{n-1})}{\Delta t} - \alpha[(1-\beta)L_{xx}T_j^n + \beta L_{xx}T_j^{n-1}] = 0 \ . \tag{7.13}$$

A Taylor series expansion of (7.13) about node (j, n) indicates consistency with (7.1) and a truncation error given by

$$E_j^n = \alpha s \Delta x^2 \frac{\partial^4 \bar{T}}{\partial x^4} \left(0.5 + \gamma + \beta - \frac{1}{12s} \right) + O(\Delta x^4) \ , \tag{7.14}$$

where $s = \alpha \Delta t/\Delta x^2$. In (7.14) all time derivatives have been replaced with spatial derivatives, using the governing equation, as in Table 7.1.

Clearly (7.13) has a truncation error of fourth order if β is given by

$$\beta = -0.5 - \gamma + \frac{1}{12s} \ . \tag{7.15}$$

Equation (7.13) produces the algorithm

$$T_j^{n+1} = \left(\frac{1+2\gamma}{1+\gamma} \right) T_j^n - \left(\frac{\gamma}{1+\gamma} \right) T_j^{n-1} + \left(\frac{s}{1+\gamma} \right) L'_{xx}((1-\beta)T_j^n + \beta T_j^{n-1}) \ , \tag{7.16}$$

where $L'_{xx}T_j = T_{j-1} - 2T_j + T_{j+1}$.

Equation (7.16) turns out to be only conditionally stable, with the maximum value of s for stability being a function of γ. This can be established by applying the von Neumann stability analysis (Sect. 4.3) to (7.13). This requires solution of the following quadratic equation for G:

$$G^2(1+\gamma)-G[1+2\gamma+2s(1-\beta)(\cos\theta-1)]+[\gamma-2\beta s(\cos\theta-1)]=0 \ . \qquad (7.17)$$

For stability it is necessary that $|G|\leq 1$ for all values of θ. This generates the stability map as a function of γ and s shown in Fig. 7.3. At $\gamma=0$, (7.17) has a more restrictive stability limitation on s ($s\leq 0.34$) than the FTCS scheme. For very large values of γ the stability limitation becomes $s\leq 0.5$, the same as for the FTCS scheme. The accuracy of the three-level scheme (7.13) is examined in Sect. 7.1.4.

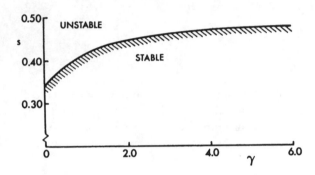

Fig. 7.3. Stability map for (7.13 and 15)

7.1.4 DIFEX: Numerical Results for Explicit Schemes

In this section the FTCS scheme (Sect. 7.1.1), the DuFort-Frankel scheme (Sect. 7.1.2) are compared with the three-level scheme (Sect. 7.1.3). All three methods are incorporated into program DIFEX (Fig. 7.4), which is an extension of DIFF (Fig. 3.13).

Solutions are sought in the computational domain $0\leq x\leq 1.0$ and $2.00\leq t\leq 9.00$, with initial conditions given at $t=2.00$ by the exact solution (3.42) divided by 100. Boundary conditions are

$$T(0,t)=T(1,t)=1.0 \ . \qquad (7.18)$$

The accuracy of the various schemes is assessed by evaluating the rms difference between the computed and exact solutions at $T=9.00$. The exact solution is computed in subroutine EXTRA (Fig. 7.5). For the DuFort-Frankel scheme and the three-level scheme two levels of initial data are required at $t=2.00$ and $t=2.00-\Delta t$.

The major parameters used by program DIFEX (Fig. 7.4) are defined in Table 7.2. A typical solution, produced by the three-level scheme on a relatively coarse grid, is shown in Fig. 7.6.

```
 1   C     DIFEX SOLVES THE DIFFUSION (1D TRANSIENT HEAT CONDUCTION)
 2   C     EQUATION USING VARIOUS EXPLICIT SCHEMES
 3   C
 4         DIMENSION TN(41),DUM(41),TD(41),TE(41),TOL(41),EL(3)
 5         DOUBLE PRECISION SUM,AVS,RMS,DSQRT,DMP
 6   C
 7         OPEN(1,FILE='DIFEX.DAT')
 8         OPEN(6,FILE='DIFEX.OUT')
 9         READ(1,1)ME,IPR,JMAX,MAXEX,NMAX,ALPH,S,TMAX,TST,GAM
10       1 FORMAT(5I5,E10.3,F5.3,3F5.2)
11   C
12         PI = 3.1415927
13         JMAP = JMAX - 1
14         AJM = JMAP
15         DELX = 1./AJM
16         DELT = DELX*DELX*S/ALPH
17   C
18         IF(ME .EQ. 1)WRITE(6,2)
19         IF(ME .EQ. 2)WRITE(6,3)
20         IF(ME .EQ. 3)WRITE(6,4)
21       2 FORMAT(' FTCS SCHEME',//)
22       3 FORMAT(' DUFORT-FRANKEL SCHEME',//)
23       4 FORMAT(' 3-LEVEL, 4TH-ORDER SCHEME',//)
24         WRITE(6,5)JMAX,MAXEX,NMAX,TMAX,TST,GAM
25       5 FORMAT(' JMAX=',I5,'  MAXEX=',I5,'  NMAX=',I5,
26      1' TMAX=',F5.2,'  TST=',F5.2,'  GAM=',F5.2)
27         WRITE(6,6)S,ALPH,DELT,DELX
28       6 FORMAT(' S=',F5.3,'  ALPH = ',E10.3,'  DELT = ',E10.3,
29      1'  DELX = ',E10.3,//)
30   C
31         IF(ME .EQ. 2)AS = 2.*S/(1.+2.*S)
32         IF(ME .EQ. 2)BS = (1. - 2.*S)/(1. + 2.*S)
33         IF(ME .NE. 3)GOTO 7
34         AS = (1.+2.*GAM)/(1.+GAM)
35         BS = -GAM/(1.+GAM)
36         CS = S/(1.+GAM)
37         DS = 1.5 + GAM - 1./12./S
38         ES = -0.5 - GAM + 1./12./S
39         EL(1) = 1.
40         EL(2) = -2.
41         EL(3) = 1.
42       7 CONTINUE
43   C
44   C     OBTAIN INITIAL CONDITIONS FROM EXACT SOLUTION
45   C
46         T = TST - 2.*DELT
47         DO 9 I = 1,2
48         T = T + DELT
49   C
50         CALL EXTRA(JMAX,MAXEX,DELX,PI,ALPH,T,TE)
51   C
52         DO 8 J = 2,JMAP
53         IF(I .EQ. 1)TOL(J) = TE(J)/100.
54         IF(I .EQ. 2)TN(J) = TE(J)/100.
55       8 CONTINUE
56       9 CONTINUE
57   C
58   C     SET BOUNDARY CONDITIONS
59   C
```

Fig. 7.4. Listing of program DIFEX

```
60          TOL(1) = 1.
61          TOL(JMAX) = 1.
62          TN(1) = 1.
63          TN(JMAX) = 1.
64          TD(1) = 100.*TN(1)
65          TD(JMAX) = 100.*TN(JMAX)
66          N = 0
67    C
68    C     EACH TIME STEP STARTS AT STATEMENT 10
69    C
70       10 N = N + 1
71          IF(ME .EQ. 1)GOTO 15
72          IF(ME .EQ. 2)GOTO 13
73    C
74    C     ME = 3, 3-LEV., 4TH ORDER
75    C
76          DO 12 J =2,JMAP
77          DUM(J) = AS*TN(J) + BS*TOL(J)
78          DO 11 K = 1,3
79          KJ = J - 2 + K
80          DUM(J) = DUM(J) + CS*EL(K)*(DS*TN(KJ) + ES*TOL(KJ))
81       11 CONTINUE
82       12 CONTINUE
83          GOTO 17
84    C
85    C     ME = 2, DUFORT-FRANKEL
86    C
87       13 DO 14 J = 2,JMAP
88       14 DUM(J) = AS*(TN(J-1) + TN(J+1)) + BS*TOL(J)
89          GOTO 17
90    C
91    C     ME = 1, FTCS
92    C
93       15 DO 16 J = 2,JMAP
94          DUM(J) = (1.-2.*S)*TN(J) + S*(TN(J-1) + TN(J+1))
95       16 CONTINUE
96       17 DO 18 J = 2,JMAP
97          IF(ME .GT. 1)TOL(J) = TN(J)
98       18 TN(J) = DUM(J)
99    C
100         DO 19 J = 2,JMAP
101      19 TD(J) = 100.*TN(J)
102         T = T + DELT
103         IF(IPR .EQ. 1)WRITE(6,20)T,(TD(J),J=1,JMAX)
104      20 FORMAT(' T= ',F5.2,' TD=',11F6.2)
105   C
106   C     IF MAXIMUM TIME OR MAXIMUM NUMBER OF TIME-STEPS
107   C     EXCEEDED EXIT FROM LOOP
108   C
109         IF(N .GE. NMAX)GOTO 21
110         IF(T .LT. TMAX)GOTO 10
111   C
112   C     OBTAIN EXACT SOLUTION AND COMPARE
113   C
114      21 SUM = 0.
115   C
116         CALL EXTRA(JMAX,MAXEX,DELX,PI,ALPH,T,TE)
117   C
118         DO 22 J = 1,JMAX
119         DMP = TE(J) - TD(J)
120         SUM = SUM + DMP*DMP
121      22 CONTINUE
122         IF(IPR .NE. 1)WRITE(6,20)T,(TD(J),J=1,JMAX)
123         WRITE(6,23)T,(TE(J),J=1,JMAX)
124      23 FORMAT(/,' T= ',F5.2,' TE=',11F6.2,//)
125   C
126   C     RMS IS THE RMS ERROR
127   C
128         AVS = SUM/(1. + AJM)
129         RMS = DSQRT(AVS)
130         WRITE(6,24)RMS
131      24 FORMAT(' RMS DIF = ',D11.4,//)
132         STOP
133         END
```

Fig. 7.4. (cont.) Listing of program DIFEX

```
1
2          SUBROUTINE EXTRA(JMAX,MAXEX,DELX,PI,ALPH,T,TE)
3    C
4    C     EXACT SOLUTION OF THE TRANSIENT HEAT CONDUCTION PROBLEM
5    C
6          DIMENSION X(41),TE(41)
7    C
8          DO 2 J = 1,JMAX
9          AJ = J - 1
10         X(J) = DELX*AJ
11         TE(J) = 100.
12         DO 1 M = 1,MAXEX
13         AM = M
14         DAM = (2.*AM - 1.)
15         DXM = DAM*PI*X(J)
16         DTM = -ALPH*DAM*DAM*PI*PI*T
17   C
18   C     LIMIT THE ARGUMENT SIZE OF EXP(DTM)
19   C
20         IF(DTM .LT. - 25.)DTM = -25.0
21         DTM = EXP(DTM)
22         IF(DTM .LT. 1.0E-10)GOTO 2
23       1 TE(J) = TE(J) - 400./DAM/PI*SIN(DXM)*DTM
24       2 CONTINUE
25         RETURN
26         END
```

Fig. 7.5. Listing of subroutine EXTRA

Table 7.2. Parameters used by program DIFEX

Parameter	Description
ME	=1, FTCS scheme
	=2, DUFORT-FRANKEL scheme
	=3, 3-level, fourth-order scheme
IPR	controls output; IPR =1 for TD solution
JMAX	number of points in the x direction
MAXEX	number of terms in the exact solution
NMAX	maximum number of time-steps
ALPH	thermal diffusivity, α
S	$= \alpha \Delta t / \Delta x^2$
TMAX	maximum time
TST	initial time
GAM	parameter γ in (7.16)
TD	dimensional temperature array
TN	nondimensional temperature array, T^n in (7.16)
TOL	nondimensional temperature array, T^{n-1} in (7.16)
TE	exact (dimensional) temperature array
DELX	Δx
DELT	Δt
EL	coefficients in the difference operator L'_{xx}
T	time
X	location $0 \leq x \leq 1.0$
RMS	rms error in the solution

3-LEVEL, 4TH-ORDER SCHEME

JMAX=   11 MAXEX=   10 NMAX= 500 TMAX= 9.00 TST= 2.00 GAM= 0.00
S= .300 ALPH =   .100E-01 DELT =   .300E+00 DELX =   .100E+00

T= 9.20 TD=100.00 84.12 69.80 58.45 51.16 48.65 51.16 58.45 69.80 84.12100.00

T= 9.20 TE=100.00 84.12 69.80 58.45 51.17 48.66 51.17 58.45 69.80 84.12100.00

RMS DIF =   .4169D-02   **Fig. 7.6.** Typical output produced by program DIFEX

The accuracy of various solutions, using program DIFEX, is compared in Table 7.3. Two values of $s=0.3$ and 0.41 have been used to allow the three-level fourth-order scheme (7.13) to be compared with the FTCS and DuFort-Frankel schemes. In addition the two special cases, FTCS with $s=1/6$ and DuFort-Frankel with $s=(1/12)^{1/2}$, are included. Both of these cases correspond to truncation error cancellation.

Table 7.3. Computational solutions of (7.1) using explicit schemes

Case	s	γ	RMS ($\Delta x = 0.2$)	RMS ($\Delta x = 0.1$)	RMS ($\Delta x = 0.05$)	approx. conv. rate, r
FTCS	1/6		0.7266×10^{-2}	0.4921×10^{-3}	0.3279×10^{-4}	3.9
FTCS	0.30		0.6439	0.1634	0.0413	2.9
FTCS	0.41		1.2440	0.3023	0.0755	2.0
D.-F.	0.289		0.0498	0.2328×10^{-2}	0.1152×10^{-3}	4.3
D.-F.	0.30		0.0244	0.0136	0.00395	1.8
D.-F.	0.41		0.8481	0.2085	0.05250	2.0
3L-4TH	0.30	0.0	0.0711	0.00416	0.00022	4.2
3L-4TH	0.30	0.5	0.1372	0.00665	0.00029	4.5
3L-4TH	0.30	1.0	0.2332	0.00916	0.00054	4.1
3L-4TH	0.41	1.0	0.7347	0.0229	0.00140	4.0

At $s=0.3$ the FTCS scheme demonstrates a second-order convergence rate. The approximate convergence rate has been obtained from the ratio of the rms error on the $\Delta x = 0.1$ and 0.05 grids, i.e.

$$r = \log\{\text{RMS}_{\Delta x = 0.1}/\text{RMS}_{\Delta x = 0.05}\}/\log(2)$$

The DuFort-Frankel scheme is considerably more accurate than the FTCS scheme. This is related to the proximity of the special value, $s = 1/(12)^{1/2} = 0.289$, for which a fourth-order rate of convergence is achieved. The three-level fourth-order scheme (with $\gamma = 0$) is less accurate than the DuFort-Frankel scheme on a coarse grid ($\Delta x = 0.20$). The tendency for higher-order schemes to require a finer grid to demonstrate superior accuracy is common. As γ is increased the accuracy of the three-level fourth-order scheme on a given grid reduces although a fourth-order rate of convergence is maintained, approximately. For $s=0.41$ and $\gamma=1.0$ the three-level fourth-order scheme demonstrates comparable accuracy to the DuFort-Frankel scheme on a coarse grid but is substantially more accurate on a fine grid.

The results shown in Table 7.3 indicate that it is straightforward to construct higher-order schemes by truncation error cancellation. Such schemes, typically, produce higher accuracy on a fine grid, but often with more severe stability restrictions. The degree of improvement in accuracy may not be as great for the nonlinear equations governing fluid motion as for the diffusion equation, which has a very smooth solution in the present situation.

7.2 Implicit Methods

For implicit schemes the spatial term $\partial^2 \bar{T}/\partial x^2$ in (7.1) is evaluated, at least partially, at the unknown time level $(n+1)$. In practice this leads to a coupling of the equations for each node $(j, n+1)$ at the $(n+1)$-th time level, and the need to solve a system of algebraic equations to advance the solution.

7.2.1 Fully Implicit Scheme

The simplest implicit finite difference representation for (7.1) is

$$\frac{(T_j^{n+1} - T_j^n)}{\Delta t} - \frac{\alpha(T_{j-1}^{n+1} - 2T_j^{n+1} + T_{j+1}^{n+1})}{\Delta x^2} = 0 \ . \tag{7.19}$$

To generate a useful algorithm (7.19) is rewritten as

$$-sT_{j-1}^{n+1} + (1+2s)\,T_j^{n+1} - sT_{j+1}^{n+1} = T_j^n \ . \tag{7.20}$$

A Taylor expansion about the (j,n)-th node indicates that this scheme has a truncation error

$$E_j^n = -\frac{\Delta t}{2}\left(1 + \frac{1}{6s}\right)\left[\frac{\partial^2 \bar{T}}{\partial t^2}\right]_j^n + O(\Delta t^2, \Delta x^4) \ .$$

This is the same order as for the explicit (FTCS) scheme, (7.5) with $s \neq 1/6$, although the multiplying constant is larger.

Application of the von Neumann stability analysis (Sect. 4.3) produces the following expression for the amplification factor:

$$G = [1 + 2s(1 - \cos\theta)]^{-1} \ .$$

For any choice of θ, $|G| \leq 1$ if $s > 0$. That is, (7.20) is unconditionally stable. This is clearly an improvement over the conditionally stable explicit scheme (7.5).

However, to solve (7.20) it is necessary to consider all the nodes j and the corresponding equations. Thus a matrix of equations can be written for the unknown values T_j^{n+1}:

$$
\begin{bmatrix}
(1+2s) & -s & & & & \\
-s & (1+2s) & -s & & & \\
\cdots & \cdots & \cdots & & & \\
& \cdots & \cdots & \cdots & & \\
& & -s & (1+2s) & -s & \\
& & & \cdots & \cdots & \cdots \\
& & & & \cdots & \cdots \\
& & & & -s & (1+2s)
\end{bmatrix}
\begin{bmatrix}
T_2^{n+1} \\
T_3^{n+1} \\
\cdots \\
\cdots \\
T_j^{n+1} \\
\cdots \\
\cdots \\
T_{J-1}^{n+1}
\end{bmatrix}
=
\begin{bmatrix}
d_2 \\
d_3 \\
\cdots \\
\cdots \\
d_j \\
\cdots \\
\cdots \\
d_{J-1}
\end{bmatrix}
\tag{7.21}
$$

In (7.21)

$$d_2 = T_2^n + sT_1^{n+1} \ ,$$

$$d_j = T_j^n , \quad d_{J-1} = T_{J-1}^n + sT_J^{n+1} \ ,$$

where T_1^{n+1} and T_J^{n+1} are known from the Dirichlet boundary conditions. It is apparent that the system of equations is tridiagonal. Consequently the Thomas algorithm (Sect. 6.2.2) can be used to solve (7.21) in $5(J-2)-4$ operations (only multiplications and divisions are counted).

In practice, allowing for the setting up of the equations, the solution of the implicit system of equations (7.21) via the Thomas algorithm, requires twice as much computer time, typically, as solving the same number of explicit equations (7.6). The time-step can be made considerably larger than the limiting explicit time-step, $\Delta t_{\mathrm{exp}} = 0.5\Delta x^2/\alpha$; however, then the accuracy of the solution will be less.

7.2.2 Crank-Nicolson Scheme

An alternative implicit algorithm for (7.1) is provided by the Crank-Nicolson scheme (Fig. 7.2) which is

$$\frac{(T_j^{n+1} - T_j^n)}{\Delta t} - \alpha(0.5L_{xx}T_j^n + 0.5L_{xx}T_j^{n+1}) = 0 \ , \tag{7.22}$$

where

$$L_{xx}T = \frac{T_{j-1} - 2T_j + T_{j+1}}{\Delta x^2} \ .$$

Effectively this scheme evaluates the spatial derivative at the average of the nth and $(n+1)$-th time levels, i.e. at the $(n+1/2)$-th time level. If a Taylor expansion is made about $(j, n+1/2)$, (7.22) is found to be consistent with (7.1) with a truncation error of

$O(\Delta t^2, \Delta x^2)$. This is a considerable improvement over the fully implicit and FTCS schemes that are only first-order accurate in time.

A von Neumann stability analysis indicates that the Crank-Nicolson scheme is unconditionally stable, Table 7.1. A rearrangement of (7.22) gives the algorithm

$$-0.5s\,T_{j-1}^{n+1} + (1+s)\,T_j^{n+1} - 0.5s\,T_{j+1}^{n+1}$$
$$= 0.5s\,T_{j-1}^n + (1-s)\,T_j^n + 0.5s\,T_{j+1}^n \ , \tag{7.23}$$

which may be compared with (7.20). By considering all spatial nodes (7.23) produces a tridiagonal system of equations which can be solved efficiently using the Thomas algorithm.

Because of the second-order temporal accuracy, the Crank-Nicolson scheme is a very popular method for solving parabolic partial differential equations. The properties of the Crank-Nicolson scheme are summarised in Table 7.1.

A generalisation of (7.22) can be obtained by writing

$$\frac{\Delta T_j^{n+1}}{\Delta t} - \alpha\left[(1-\beta)\,L_{xx}\,T_j^n + \beta L_{xx}\,T_j^{n+1}\right] = 0 \ , \tag{7.24}$$

where $\Delta T_j^{n+1} = T_j^{n+1} - T_j^n$ and $0 \leq \beta \leq 1$. If $\beta = 0$ the FTCS scheme is obtained. If $\beta = 0.5$ the Crank-Nicolson scheme is obtained and if $\beta = 1.0$ the fully implicit scheme is obtained.

A von Neumann stability analysis of (7.24) indicates that a stable solution is possible for

$$\Delta t \leq \frac{0.5\,\Delta x^2}{\alpha(1-2\beta)} \qquad \text{if } 0 \leq \beta < 1/2$$

$$\text{no restriction} \qquad \text{if } 1/2 \leq \beta \leq 1 \ .$$

It may be noted that the Crank-Nicolson scheme is on the boundary of the unconditionally stable regime. For many steady flow problems it is efficient to solve an equivalent transient problem until the solution no longer changes (Sect. 6.4). However, often the solution in different parts of the computational domain approaches the steady-state solution at significantly different rates; the equations are then said to be stiff (Sect. 7.4). Unfortunately the Crank-Nicolson scheme often produces an oscillatory solution in this situation which, although stable, does not approach the steady state rapidly. Certain three-level (in time) schemes are more effective than the Crank-Nicolson scheme in this regard.

7.2.3 Generalised Three-Level Scheme

For the diffusion equation, a generalised three-level scheme that includes (7.24) can be written

$$\frac{(1+\gamma)\Delta T_j^{n+1}}{\Delta t} - \frac{\gamma \Delta T_j^n}{\Delta t} - \alpha\left[(1-\beta)\,L_{xx}\,T_j^n + \beta L_{xx}\,T_j^{n+1}\right] = 0 \ , \tag{7.25}$$

where $\Delta T^n = T_j^n - T_j^{n-1}$. The inclusion of the extra time-level implies a larger memory requirement to store the solution. However, the modern trend is for computer memories to become cheaper and larger. A second effect is that additional execution time is required, typically 10%–15%, to manipulate the additional terms.

A particularly effective three-level scheme is given by the choice: $\gamma = 0.5$, $\beta = 1.0$. This scheme has a truncation error of $O(\Delta t^2, \Delta x^2)$, is unconditionally stable, can be solved using the Thomas algorithm and damps out the spurious oscillations, discussed above in relation to stiff problems. We will refer to this scheme as the three-level fully implicit (3LFI) scheme and will make further use of it when discussing approximate factorisation (Sects. 8.2 and 8.3). The 3LFI scheme has the useful feature of being A-stable (Sect. 7.4).

The properties of some of the various numerical schemes for representing the diffusion equation (7.1) are shown in Table 7.1. Many more schemes are given by Richtmyer and Morton (1967, p. 189).

7.2.4 Higher-Order Schemes

The starting point for this section will be the discretised equation, (7.25), modified to embrace both finite difference and finite element three-level schemes. Thus (7.25) is replaced by

$$(1+\gamma) M_x \left(\frac{\Delta T_j^{n+1}}{\Delta t} \right) - \frac{\gamma M_x \Delta T_j^n}{\Delta t} - \alpha \left[\beta L_{xx} T_j^{n+1} + (1-\beta) L_{xx} T_j^n \right] = 0 \ , \qquad (7.26)$$

where

$$\Delta T_j^{n+1} = T_j^{n+1} - T_j^n \ , \qquad \Delta T_j^n = T_j^n - T_j^{n-1} \ , \quad \text{and}$$

$$L_{xx} T_j = \frac{T_{j-1} - 2T_j + T_{j+1}}{\Delta x^2} \ .$$

The similarity in the structure of (7.26) and (7.13) is noteworthy. For the finite element method, $M_x = \{\frac{1}{6}, \frac{2}{3}, \frac{1}{6}\}$, so that

$$M_x \Delta T_j = \frac{1}{6} \Delta T_{j-1} + \frac{2}{3} \Delta T_j + \frac{1}{6} \Delta T_{j+1} \ . \qquad (7.27)$$

For the finite difference method, $M_x = \{0, 1, 0\}$. The parameters γ and β may be chosen to provide particular levels of accuracy and/or stability. In Sect. 7.2.2 the various schemes correspond to $\gamma = 0$. The particular choice $\gamma = 0$, $\beta = 0.5$ gives the Crank-Nicolson method. Results for both finite element and finite difference forms of the Crank-Nicolson method are provided in Table 7.3. The choice $\gamma = 0.5$, $\beta = 1.0$ is discussed, briefly, in Sect. 7.2.3. In this section γ will be treated as a free parameter but β will be treated as a function of γ.

A Taylor series expansion of (7.26) about node (j, n) produces the following expression for the truncation error leading term:

$$E_j^n = \alpha s \Delta x^2 \frac{\partial^4 T}{\partial x^4}\left(0.5 + \gamma + \frac{\delta - \frac{1}{12}}{s} - \beta\right) + O(\Delta x^4) , \tag{7.28}$$

where the mass operator is written as

$$M_x = \{\delta, 1 - 2\delta, \delta\} . \tag{7.29}$$

This includes both the finite difference $(\delta = 0)$ and finite element $(\delta = \frac{1}{6})$ formulations. The form of the truncation error (7.28) has eliminated all time derivatives, as in Table 7.1.

The schemes considered previously (Sect. 7.2.3, etc.) have corresponded to the choice $\beta = 0.5 + \gamma$. However, it is clear that fourth-order accuracy should be possible for the choice

$$\beta = 0.5 + \gamma + \frac{\delta - \frac{1}{12}}{s} . \tag{7.30}$$

In turn this motivates the choice $M_x = (\frac{1}{12}, \frac{5}{6}, \frac{1}{12})$, since this will produce a fourth-order truncation error with $\beta = 0.5 + \gamma$.

Equation (7.26) is applied at every node producing a tridiagonal system of equations of the form

$$A_j T_{j-1}^{n+1} + B_j T_j^{n+1} + C_j T_{j+1}^{n+1} = (1 + 2\gamma) M_x T_j^n - \gamma M_x T_j^{n-1} + (1 - \beta) s L'_{xx} T_j^n , \tag{7.31}$$

where

$$A_j = C_j = (1 + \gamma)\delta - s\beta , \quad B_j = (1 + \gamma)(1 - 2\delta) + 2s\beta \quad \text{and}$$

$$L'_{xx} T_j^n = T_{j-1}^n - 2T_j^n + T_{j+1}^n .$$

Although the solution of (7.31) requires a higher operation count than the fully implicit or Crank-Nicolson schemes it is considerably more accurate (Sect. 7.2.5).

7.2.5 DIFIM: Numerical Results for Implicit Schemes

In this section the accuracy of the higher-order scheme (7.31) is compared with the accuracy of the lower-order implicit schemes and the low and high-order explicit schemes of Sect. 7.1.

Program DIFIM is an extension of program DIFF to obtain the computational solution of (7.1) subject to the boundary conditions (7.18) by repeatedly solving (7.31) to advance the solution in time for all interior nodes. The solution of (7.31) is undertaken by subroutines BANFAC and BANSOL. Since A_j, B_j and C_j are independent of time it is only necessary to call BANFAC once, at the first time-step. A listing of program DIFIM is provided in Fig. 7.7. Program DIFIM can invoke five options as shown in Table 7.4, which correspond to different choices of

```
1
2    C
3    C      DIFIM.FOR SOLVES THE DIFFUSION EQUATION USING
4    C      VARIOUS IMPLICIT SCHEMES
5    C
6           DOUBLE PRECISION SUM,AVS,RMS,DSQRT,DMP
7           DIMENSION DUM(65),TE(41),A(5,65),D(65)
8           DIMENSION ELX(3),EMX(3),TOL(41),TN(41),TD(41)
9    C
10          OPEN(1,FILE='DIFIM.DAT')
11          OPEN(6,FILE='DIFIM.OUT')
12          READ(1,1)ME,IPR,JMAX,MAXEX,NMAX,ALPH,S,TMAX,TST,GAM
13        1 FORMAT(5I5,E10.3,F5.3,3F5.2)
14   C
15          PI = 3.1415927
16          JMAP = JMAX - 1
17          AJM = JMAP
18          DELX = 1./AJM
19          DELT = DELX*DELX*S/ALPH
20   C
21          IF(ME .EQ. 1)WRITE(6,4)
22          IF(ME .EQ. 2)WRITE(6,5)
23          IF(ME .EQ. 3)WRITE(6,6)
24          IF(ME .EQ. 4)WRITE(6,7)
25          IF(ME .EQ. 5)WRITE(6,8)
26          WRITE(6,2)JMAX,MAXEX,NMAX,TMAX,TST,GAM
27        2 FORMAT(' JMAX=',I5,'  MAXEX=',I5,'  NMAX=',I5,
28          1' TMAX=',F8.2,'  TST=',F5.2,'  GAM=',F5.2)
29          WRITE(6,3)S,ALPH,DELT,DELX
30        3 FORMAT(' S= ',F5.3,'  ALPH = ',E10.3,'  DELT = 'E10.3,
31          1' DELX = ',E10.3,//)
32        4 FORMAT(' DIFFUSION EQUATION:  FDM-2ND',//)
33        5 FORMAT(' DIFFUSION EQUATION:  FEM-2ND',//)
34        6 FORMAT(' DIFFUSION EQUATION:  FDM-4TH',//)
35        7 FORMAT(' DIFFUSION EQUATION:  FEM-4TH',//)
36        8 FORMAT(' DIFFUSION EQUATION:  COMP',//)
37   C
38   C      IMPLICIT PARAMETERS
39   C
40          BET = 0.5 + GAM
41          IF(ME .EQ. 3)BET = BET - 1./12./S
42          IF(ME .EQ. 4)BET = BET + 1./12./S
43          IF(ME .EQ. 1 .OR. ME .EQ. 3)EMX(1) = 0.
44          IF(ME .EQ. 2 .OR. ME .EQ. 4)EMX(1) = 1./6.
45          IF(ME .EQ. 5)EMX(1) = 1./12.
46          EMX(2) = 1. - 2.*EMX(1)
47          EMX(3) = EMX(1)
48          AD = EMX(1)*(1. + GAM) - BET*S
49          BD = EMX(2)*(1. + GAM) + 2.*BET*S
50          CD = AD
51          ELX(1) = 1.
52          ELX(2) = -2.
53          ELX(3) = 1.
54          JMAF = JMAX - 2
55   C
56   C      OBTAIN INITIAL CONDITIONS FROM THE EXACT SOLUTION
57   C
```

Fig. 7.7. Listing of program DIFIM

β as a function of γ and different choices for M_x. Case 1 with $\gamma = 0$ corresponds to the Crank-Nicolson scheme (7.26). Many of the major parameters used by program DIFIM are the same as used in program DIFEX (Table 7.1). Additional parameters are shown in Table 7.5. A typical solution corresponding to ME = 3 and $\gamma = 0$ is shown in Fig. 7.8.

```
58            T = TST - 2.*DELT
59            DO 10 I = 1,2
60            T = T + DELT
61   C
62            CALL EXTRA(JMAX,MAXEX,DELX,PI,ALPH,T,TE)
63   C
64            DO 9 J = 2,JMAP
65            IF(I .EQ. 1)TOL(J) = TE(J)/100.
66            IF(I .EQ. 2)TN(J) = TE(J)/100.
67          9 CONTINUE
68         10 CONTINUE
69   C
70   C      SET BOUNDARY CONDITIONS
71   C
72            TOL(1) = 1.
73            TOL(JMAX) = 1.
74            TN(1) = 1.
75            TN(JMAX) = 1.
76            TD(1) = 100.*TN(1)
77            TD(JMAX) = 100.*TN(JMAX)
78            N = 0
79   C
80   C      EACH TIME STEP STARTS AT STATEMENT 11
81   C
82         11 N = N + 1
83   C
84   C      SET UP THE TRIDIAGONAL SYSTEM OF EQUATIONS
85   C
86            DO 14 J = 2,JMAP
87            JM = J - 1
88            IF(N .GT. 1)GOTO 12
89            A(1,JM) = 0.
90            A(2,JM) = AD
91            A(3,JM) = BD
92            A(4,JM) = CD
93            A(5,JM) = 0.
94         12 D(JM) = 0.
95            DO 13 K = 1,3
96            KJ = J - 2 + K
97            D(JM) = D(JM) + EMX(K)*((1. + 2.*GAM)*TN(KJ) - GAM*TOL(KJ))
98            D(JM) = D(JM) + S*ELX(K)*(1.-BET)*TN(KJ)
99         13 CONTINUE
100        14 CONTINUE
101           D(1) = D(1) - A(2,1)*TN(1)
102           D(JMAF) = D(JMAF) - A( ,JMAF)*TN(JMAX)
103  C
104  C      SOLVE BANDED SYSTEM OF EQUATIONS
105  C
106           IF(N .EQ. 1)CALL BANFAC(A,JMAF,1)
107  C
108           CALL BANSOL(D,DUM,A,JMAF,1)
109  C
110           DO 15 J = 2,JMAP
111           TOL(J) = TN(J)
112        15 TN(J) = DUM(J-1)
113  C
114           DO 16 J = 2,JMAP
115        16 TD(J) = 100.*TN(J)
116           T = T + DELT
117           IF(IPR .EQ. 1)WRITE(6,17)T,(TD(J),J=1,JMAX)
118        17 FORMAT(' T= ',F5.2,' TD=',11F6.2)
```

Fig. 7.7. (cont.) Listing of program DIFIM

```
119 C
120 C     IF MAXIMUM TIME OR MAXIMUM NUMBER OF TIME-STEPS EXCEEDED
121 C     EXIT FROM LOOP
122 C
123       IF(N .GE. NMAX)GOTO 18
124       IF(T .LT. TMAX)GOTO 11
125 C
126 C     OBTAIN EXACT SOLUTION AND COMPARE
127 C
128    18 CALL EXTRA(JMAX,MAXEX,DELX,PI,ALPH,T,TE)
129 C
130       SUM = 0.
131       DO 19 J = 1,JMAX
132       DMP = TE(J) - TD(J)
133       SUM = SUM + DMP*DMP
134    19 CONTINUE
135       IF(IPR .NE. 1)WRITE(6,17)T,(TD(J),J=1,JMAX)
136       WRITE(6,20)T,(TE(J),J=1,JMAX)
137    20 FORMAT(/,' T= ',F5.2,' TE=',11F6.2,//)
138 C
139 C     RMS IS THE RMS ERROR
140 C
141       AVS = SUM/(1. + AJM)
142       RMS = DSQRT(AVS)
143       WRITE(6,21)RMS
144    21 FORMAT(' RMS DIF = ',D11.4,//)
145       STOP
146       END
```

Fig. 7.7. (cont.) Listing of program DIFIM

Table 7.4. Different options implemented in program DIFIM

Case (ME)	Descriptor	M_x	β
1	FDM-2ND order	$(0, 1, 0)$	$0.5 + \gamma$
2	FEM-2ND order	$(1/6, 2/3, 1/6)$	$0.5 + \gamma$
3	FDM-4TH order	$(0, 1, 0)$	$0.5 + \gamma - 1/(12s)$
4	FEM-4TH order	$(1/6, 2/3, 1/6)$	$0.5 + \gamma + 1/(12s)$
5	COMPosite	$(1/12, 5/6, 1/12)$	$0.5 + \gamma$

Table 7.5. Additional parameters used in program DIFIM

Parameter	Description
ME	parameter controlling choice of method (Table 7.4)
GAM	parameter γ used in (7.26)
BET	parameter β used in (7.26) and Table 7.4
EMX	mass operator, M_x
ELX	difference operator, L'_{xx}
AD, BD, CD	A_j, B_j and C_j used in (7.31)

7.2 Implicit Methods 235

```
DIFFUSION EQUATION: FEM-4TH

JMAX=   11  MAXEX=   10 NMAX=  500  TMAX=   12.00  TST= 4.50  GAM=  .00
S= 1.000  ALPH =   .100E-01  DELT =   .100E+01  DELX =   .100E+00

T= 12.50 TD=100.00 88.55 78.22 70.02 64.76 62.94 64.76 70.02 78.22 88.55100.00

T= 12.50 TE=100.00 88.54 78.21 70.00 64.74 62.92 64.74 70.00 78.21 88.54100.00

RMS DIF =   .1522D-01
```
Fig. 7.8. Typical output produced by program DIFIM

Table 7.6. Accuracy of the implicit schemes (Table 7.4) when solving (7.1)

Case (ME)	γ	RMS ($\Delta x = 0.2$)	RMS ($\Delta x = 0.1$)	RMS ($\Delta x = 0.05$)	approx conv. rate, r
1	0.	0.3895	0.1466	0.03993	1.9
2	0.	0.8938	0.1787	0.04185	2.1
3	0.	0.2393	0.01526	0.001053	3.9
4	0.	0.2393	0.01522	0.000897	4.1
5	0.	0.2393	0.01525	0.001034	3.9
1	1.0	2.090	0.03003	0.03245	3.0[a]
2	1.0	1.760	0.2475	0.04668	2.4
3	1.0	2.367	0.1246	0.008129	3.9
4	1.0	1.395	0.09269	0.005912	4.0
5	1.0	1.867	0.1087	0.007097	3.9

[a] Based on $\mathrm{RMS}_{\Delta x = 0.2} / \mathrm{RMS}_{\Delta x = 0.05}$

Program DIFIM has been used to compare the various methods shown in Table 7.4. Comparative results are presented in Table 7.6 for the solution accuracy on various grids ($\Delta x = 0.2, 0.1$ and 0.05) at $t = 12.0$. These results were obtained with $\alpha = 0.01$ and the initial condition defined by the exact solution (3.42) at $t = 4.5$. Dirichlet boundary conditions have been imposed on T at $x = 0$ and 1.0. The approximate convergence rate has been obtained as in Table 7.3.

All the results presented in Table 7.6 have been obtained at $s = 1.0$. The nominally second-order methods (cases 1 and 2) are showing approximately second-order convergence and the nominally fourth-order methods (cases 3, 4 and 5) are showing fourth-order convergence. This is occurring at both $\gamma = 0$ and $\gamma = 1$.

Generally, increasing γ reduces the level of accuracy while maintaining approximately the same convergence rate. For coarse grids the accuracy demonstrated by the higher-order schemes (cases 3–5) is not significantly higher than that demonstrated by the lower-order schemes (cases 1 and 2), particularly at large γ. However on a fine grid the difference in accuracy is substantial. Cases 3, 4 and 5 show comparable accuracy on coarse and fine grids at $\gamma = 0$. However at $\gamma = 1$, case 4 is more accurate than cases 3 or 5, particularly on a coarse grid.

Table 7.7. Comparison of implicit and explicit schemes, $s = 0.41$

Case		γ	RMS $(\Delta x = 0.2)$	RMS $(\Delta x = 0.1)$	RMS $(\Delta x = 0.05)$	Approx conv. rate, r
FDM-2ND,	imp	0.	0.7502	0.2004	0.05127	2.0
FDM-4TH,	imp	0.	0.03718	0.002407	0.000206	3.5
FTCS,	exp	0.	1.2440	0.30230	0.07550	2.0
FDM-2ND,	imp	1.	0.7625	0.1681	0.04938	1.8
FDM-4TH,	imp	1.	0.6482	0.03155	0.00192	4.0
FDM-4TH,	exp	1.	0.7347	0.02290	0.00140	4.0

Some of the implicit and explicit schemes are compared in Table 7.7, for both $\gamma = 0$ and $\gamma = 1.0$ with $s = 0.41$. These results have been obtained at $t = 9.00$ with the initial specification of T at $t = 2.00$. On a coarse grid the various schemes demonstrate comparable accuracy. However, on a fine grid the implicit schemes are considerably more accurate.

The generally high levels of accuracy achieved with fine grids are partly a reflection of the smoothness of the exact solution and the relative simplicity of the governing equation. As noted in Sect. 7.1, such high levels of accuracy should not be expected when dealing with fluid dynamic problems. However, the selection of the coefficients in the discretised equation, as in (7.26), to reduce the truncation error is a valid technique, as long as the stability of the algorithm is sufficient. Generally the improved stability behaviour of implicit schemes, compared with explicit schemes, permits more flexibility in the choice of the free parameters, i.e. γ and β in (7.26).

7.3 Boundary and Initial Conditions

In Sects. 7.1 and 7.2 Dirichlet boundary conditions are used and where two levels of data are necessary as initial conditions, these are provided by the exact solution. Consequently the only source of error in the solutions arises from the discretisation of the derivatives in the governing equations (7.1). In this section boundary and initial conditions will be considered in situations where additional errors are introduced in implementing the boundary and initial conditions.

7.3.1 Neumann Boundary Conditions

The algorithms developed so far, for example (7.6), are appropriate to internal nodes. To use these formulae for boundary values, i.e. T_1^{n+1} in Fig. 7.9, would require knowledge of the solution outside the computational domain. Therefore special formulae must be developed at the boundaries. For Dirichlet boundary conditions (7.2) there is no difficulty. Where a value for T_1^n is required the substitution $T_1^n = b^n$ is made, from (7.2).

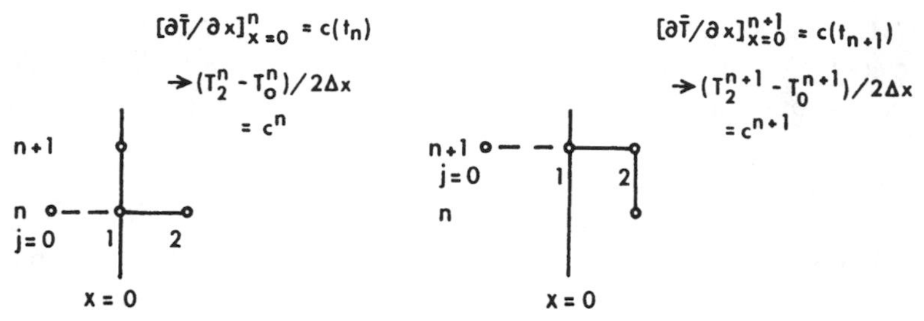

Explicit treatment   Implicit treatment

Fig. 7.9. Treatment of Neumann boundary conditions

Neumann boundary conditions (7.3) pose a greater problem. A one-sided finite difference expression, using only information inside the domain, can be introduced for $[\partial T/\partial x]_1$ in (7.3). The result is

$$\frac{T_2^{n+1}-T_1^{n+1}}{\Delta x}=c^{n+1} \ . \tag{7.32}$$

In a typical application the FTCS scheme (7.6) would be used to obtain the value of T_j^{n+1} at all interior nodes, $j=2, \ldots, J-1$. At the boundary, $x=0$, (7.32) gives

$$T_1^{n+1}=T_2^{n+1}-c^{n+1}\Delta x \ . \tag{7.33}$$

The major problem here is that (7.33) has a truncation error of $O(\Delta x)$ whereas the FTCS scheme has a truncation error of $O(\Delta x^2)$. Since the diffusion equation is a parabolic equation (Sect. 2.3) the lower accuracy of the solution at the boundary will affect the accuracy of the solution in the interior for all later time.

Therefore it is desirable to represent the Neumann condition (7.3) with an algebraic expression that has the same truncation error as the expression used for the interior nodes. This can be achieved in the following way. Equation (7.3) is represented by

$$\frac{T_2^n-T_0^n}{2\Delta x}=c^n \ . \tag{7.34}$$

To achieve a truncation error of $O(\Delta x^2)$ (7.34) includes the fictitious node (0, n) which lies outside the computational domain (Fig. 7.9). However, if the computational domain is extended, notionally, to include this point, then (7.34) can be combined with an interior equation, for example (7.5) centred at $(1, n)$, to eliminate T_0^n. The result is

$$T_1^{n+1}=-2s\Delta x c^n+(1-2s)\,T_1^n+2s\,T_2^n \ , \tag{7.35}$$

and the truncation error is of $O(\Delta t, \Delta x^2)$ everywhere in the domain. If an implicit scheme is being used in the interior, for example (7.19), (7.34) is evaluated at time-level $n+1$ and combined with (7.20) to give (Fig. 7.9)

$$(1+2s)\,T_1^{n+1}-2s\,T_2^{n+1}=T_1^n-2s\Delta x\,c^{n+1} \ , \tag{7.36}$$

which becomes the first equation in the tridiagonal system equivalent to (7.21). This system can be solved by the Thomas algorithm. The construction used to create (7.35 and 36) is also applicable if Neumann boundary conditions are required at $x = 1$ (Fig. 7.1).

It is expected that the use of different formulae at the boundaries may alter the stability properties. Strictly the relatively straightforward von Neumann stability analysis (Sect. 4.3) is only applicable to internal equations; although Trapp and Ramshaw (1976) suggest that the von Neumann analysis can be applied, heuristically, at boundaries. Alternatively the matrix method may be used to determine the stability of the complete system of equations, including those formed at the boundaries.

For the diffusion equation Mitchell and Griffiths (1980, pp. 47–53) discuss the application of the matrix method to a Crank-Nicolson formulation with various boundary conditions. They show that with Dirichlet boundary conditions the Crank-Nicolson scheme is unconditionally stable. With Neumann boundary conditions the Crank-Nicolson scheme is stable but one eigenvalue of the amplification matrix is unity which produces an oscillatory solution. For the general mixed boundary condition an additional restriction is necessary on the parameters in the boundary condition to ensure stability.

7.3.2 Accuracy of Neumann Boundary Condition Implementation

The implementation of Neumann boundary conditions (Sect. 7.3.1) with the diffusion equation will be combined with some of the schemes described in Sects. 7.1 and 7.2 and the impact on the overall accuracy will be indicated. Here solutions to the diffusion equation

$$\frac{\partial \bar{T}}{\partial t} - \alpha \frac{\partial^2 \bar{T}}{\partial x^2} = 0 \tag{7.37}$$

will be obtained in the spatial interval $0.1 \le x \le 1.0$, with boundary conditions

$$\frac{\partial \bar{T}}{\partial x} = c = 2 - 2\pi \sin(0.05\pi) \exp\left[-\alpha \left(\frac{\pi}{2}\right)^2 t \right] \qquad \text{at } x = 0.1 \tag{7.38}$$

and $\bar{T} = 2$ at $x = 1.0$.

Initial conditions are chosen to be

$$\bar{T} = 2x + 4\cos(0.5\pi x) \quad \text{at} \quad t = 0 . \tag{7.39}$$

This problem has the exact solution

$$\bar{T} = 2x + 4\cos(0.5\pi x) \exp\left[-\alpha \left(\frac{\pi}{2}\right)^2 t \right] , \tag{7.40}$$

which will be used to assess the accuracy of the computational solution.

Numerical solutions of the above problem have been obtained using a modified form of program DIFEX (Fig. 7.4). The rms errors for different size grids are shown in Table 7.8. The rms errors were computed at $t=9.00$. To avoid errors in implementing the initial conditions the exact solution (7.40) has supplied the initial conditions at $t=0.80$. All solutions shown in Table 7.8 have been obtained with $s=0.30$.

Table 7.8. Explicit interior schemes with Neumann boundary condition at $x=0.1$

Interior method	Boundary condition formula	rms error			Approximate convergence rate, r
		$\Delta x=0.225$	$\Delta x=0.1125$	$\Delta x=0.05625$	
FTCS	(7.32)	0.1958	0.07978	0.03539	1.2
FTCS	(7.34)	0.1753×10^{-2}	0.4235×10^{-3}	0.1064×10^{-3}	2.0
3L-4TH, $\gamma=0$	(7.34)	0.4244×10^{-2}	0.9142×10^{-3}	0.2144×10^{-3}	2.1
3L-4TH, $\gamma=1$	(7.34)	0.8684×10^{-2}	0.2034×10^{-2}	0.4867×10^{-3}	2.1

The combination of boundary condition implementation (7.32) and the FTCS scheme is seen to produce a relatively inaccurate solution which increases in accuracy roughly linearly with grid refinement. Thus the first-order spatial accuracy associated with (7.32) is over-riding the second-order spatial accuracy associated with the FTCS scheme. This is confirmed by the error distribution for the $\Delta x=0.225$ grid shown in Table 7.9. The solution is least accurate at $x=0.1$ where the Neumann boundary condition is applied and becomes progressively more accurate as $x=1.00$ is approached.

Table 7.9. Error distribution for $\Delta x=0.225$ in Table 7.8

$x=$	0.1	0.325	0.550	0.775	1.000
FTCS + (7.32)	-0.3799	-0.1989	-0.08366	-0.02610	0.000
FTCS + (7.34)	0.332×10^{-3}	-0.221×10^{-2}	-0.272×10^{-2}	-0.173×10^{-2}	0.000
3L-4TH, $\gamma=0+$(7.34)	0.832×10^{-2}	0.421×10^{-2}	0.166×10^{-2}	0.572×10^{-3}	0.000

The second-order first derivative specification (7.34) can be combined with the FTCS scheme to produce the algorithm (7.35) at $x=0.1$. The rms errors for this combination are shown in Table 7.8. Clearly more accurate solutions are being obtained than with the first-order Neumann boundary condition specification (7.32). Equation (7.34) has a second-order truncation error as has the FTCS scheme. The combination is demonstrating, Table 7.8, second-order convergence. An examination of the error distribution, Table 7.9, for this combination indicates that the error at $x=0.1$, where the Neumann boundary condition is applied, is smaller than the error elsewhere.

The second-order boundary condition specification (7.34) can be combined with the fourth-order internal scheme (7.16). This is done most easily by using (7.34) to obtain T_0^n explicitly, and then applying (7.16) centred at node 1. The rms errors for various degrees of grid refinement are shown in Table 7.8 for two values of γ. Generally, increasing γ reduces the accuracy. The second-order truncation error of the boundary condition specification is sufficient to reduce the global convergence rate to second order. In fact the accuracy for both values of γ is less than the accuracy of the FTCS scheme. A consideration of the error distribution, Table 7.9, indicates that the error produced by the 3L-4TH scheme is significantly larger close to $x = 0.1$, where the derivative boundary condition is applied, but is smaller close to $x = 1.00$, than the error produced by the FTCS scheme.

In principle a higher-order representation for $\partial T/\partial x$ than (7.34) could be introduced that involved additional nodal values, T^n, etc. However, besides producing a more complicated algorithm it would be necessary to establish the stability restriction on s. Strictly this should be undertaken with the matrix method (Sect. 4.3). Often introducing higher-order boundary condition formulae reduces the working range of s associated with the interior algorithm.

The first- and second-order boundary condition specifications (7.32, 34) have also been combined with some of the implicit schemes described in Sect. 7.2, to produce the solution errors shown in Table 7.10. The rms errors are computed at $t = 15.00$. The exact solution (7.40) has been used to provide the initial solution at $t = 5.20$. The solutions shown in Table 7.10 have been obtained at $s = 1.0$. The two interior methods, FDM-2ND and FDM-4TH, correspond to ME$= 1$ and 3, respectively, in Table 7.4.

Table 7.10. Implicit interior schemes with Neumann boundary condition at $x = 0.1$

Interior method	Boundary condition formulae	rms error			Approx. convergence rate, r
		$\Delta x = 0.225$	$\Delta x = 0.1125$	$\Delta x = 0.05625$	
FDM-2ND (ME$=1$)	(7.32)	0.1448	0.07138	0.03344	1.1
FDM-2ND (ME$=1$)	(7.34)	0.01478	0.00363	0.00087	2.1
FDM-4TH, $\gamma = 0$	(7.34)	0.00912	0.00223	0.00053	2.1
FDM-4TH, $\gamma = 1.0$	(7.34)	0.01478	0.00539	0.00141 ·	1.9

The combination of the first-order boundary condition specification (7.32) and the second-order FDM scheme produces an overall first-order convergence rate and a solution accuracy that is comparable to that produced by the corresponding explicit scheme, Table 7.8. This comparable accuracy is in spite of the larger s used to produce the implicit results. This provides indirect confirmation that the first-order boundary condition specification is dominating the rms error evaluation.

A significant reduction in error is achieved by utilising the second-order boundary condition specification (7.34) particularly on the finest grid. The accuracy is not as great as produced in combination with the FTCS scheme, although the convergence rate is second order. Switching to a fourth-order interior scheme, FDM-4TH, produces a slightly more accurate solution but with only a second-order convergence rate. For $\gamma = 1.0$ the accuracy is less than for $\gamma = 0$, which is consistent with the behaviour of the explicit scheme.

To achieve a higher-order convergence rate it would be necessary to introduce a more accurate finite difference representation for the Neumann boundary condition at $x = 0.10$. Since this would couple, typically, four or five grid points it would necessitate a modification to the Thomas algorithm for solving the matrix of discretised equations.

7.3.3 Initial Conditions

Initial conditions, of the form (7.4) do not cause any difficulty for a two-level scheme, like (7.6 or 20), except at the boundary if a Dirichlet boundary condition is enforced. It can happen that, at a boundary $x = 0$ say, the value of $\bar{T}(0, 0)$ specified by the initial condition is different from the value of $\bar{T}(0, 0)$ specified by the boundary condition. The preferred strategy is to average the values of $\bar{T}(0, 0)$ for the first time-level, $n = 0$, but to revert to the proper boundary condition for subsequent time. In Table 3.7 this strategy is compared with the alternative of specifying $\bar{T}(0, 0)$ by the initial conditions for the first time level, $n = 0$. For this particular example the averaging strategy is seen to produce a more accurate solution.

Three-level schemes, like (7.25), require two levels of known data; therefore they must be supplemented at the start of the integration. This should be done with a two-level scheme that achieves the same or better accuracy. Thus a second-order two-level scheme like the Crank-Nicolson method is appropriate. Alternatively a first-order temporal scheme like the FTCS scheme can be used as long as the second level of initial data is computed on a fine grid to compensate for the first-order accuracy. Richardson extrapolation (Sect. 4.4.1) is useful for this purpose.

7.4 Method of Lines

Algorithms developed in Sects. 7.1 and 7.2 introduce discretisation formulae for both the time derivative and the spatial derivative terms in (7.1) simultaneously. This often permits cancellation of terms in the respective truncation error expressions so that higher-order accuracy is achieved.

However, an alternative philosophy is to discretise the spatial term first reducing the partial differential equation to a system of ordinary differential equations for the nodal values. Thus (7.1) becomes, noting the change in dependent variable from T to u,

$$\frac{du_j}{dt} - \frac{\alpha(u_{j-1} - 2u_j + u_{j+1})}{\Delta x^2} = 0 \ , \tag{7.41}$$

where a second-order three-point finite difference expression has been used to discretise $\partial^2 \bar{u}/\partial x^2$ in (7.1). The process of reducing (7.1) to (7.41) is an example of the method of lines (Holt 1984) or semi-discretisation.

An attractive feature of the method of lines is that the various techniques for solving systems of ordinary differential equations (Gear 1971; Lambert 1973; Seward et al. 1984) are available to solve the semi-discrete form of the original partial differential equation. However, it should be emphasised that the construction of the semi-discrete form introduces an error associated with the spatial discretisation. Consequently the "best" choice for solving the resulting system will usually be an algorithm of lower order than if the system of ordinary differential equations contains no approximation. The more effective methods for solving initial value problems governed by ordinary differential equations can usually be categorised as either linear multistep methods or Runge-Kutta methods.

The repeated application of (7.41) at all internal nodes produces a system that can be written

$$\frac{d\boldsymbol{u}}{dt} = \underline{A}\boldsymbol{u} \ . \tag{7.42}$$

If the spatial term in (7.1) had been nonlinear (7.42) would be written more generally as

$$\frac{d\boldsymbol{u}}{dt} = \boldsymbol{F} \ . \tag{7.43}$$

The scalar equivalents of (7.42 and 43) are

$$\frac{du}{dt} = \lambda u \quad \text{and} \tag{7.44}$$

$$\frac{du}{dt} = f \ . \tag{7.45}$$

A general linear multistep method applied to (7.45) can be written

$$\sum_{l=0}^{m} \alpha_l u^{n+l} = \Delta t \sum_{l=0}^{m} \beta_l f^{n+l} \ , \tag{7.46}$$

where the solution u^{n+m} is being sought. If $\beta_m = 0$, this solution is obtained explicitly; if $\beta_m \neq 0$, (7.46) is an implicit algorithm for u^{n+m}.

The simplest case, $m = 1$, gives

$$\alpha_1 u^{n+1} + \alpha_0 u^n = \Delta t(\beta_0 f^n + \beta_1 f^{n+1}) \ , \tag{7.47}$$

which includes the Euler scheme ($\alpha_1 = -\alpha_0 = 1$, $\beta_0 = 1$, $\beta_1 = 0$),

$$u^{n+1} - u^n = \Delta t f^n \ . \tag{7.48}$$

Applied to the linear system (7.42), in the form of (7.41), this coincides with the FTCS scheme (7.5). Equation (7.47) also includes the trapezoidal scheme

$$u^{n+1} - u^n = \Delta t(0.5f^n + 0.5f^{n+1}) \ . \tag{7.49}$$

Applied to (7.41) the trapezoidal scheme coincides with the Crank-Nicolson scheme (7.22).

The two-step interpretation of (7.46) is

$$\alpha_2 u^{n+2} + \alpha_1 u^{n+1} + \alpha_0 u^n = \Delta t(\beta_0 f^n + \beta_1 f^{n+1} + \beta_2 f^{n+2}) \ . \tag{7.50}$$

When applied to (7.41) with $\beta_2 = 0$, (7.50) provides the solution u_j^{n+2}, explicitly, and is identical with (7.12). If $\beta_2 \neq 0$, (7.50) applied to (7.41) includes both (7.25) and (7.26). For example, the 3LFI scheme (Sect. 7.2.3) corresponds to the choice $\alpha_2 = 1.5$, $\alpha_1 = -2.0$, $\alpha_2 = 0.5$, $\beta_0 = \beta_1 = 0$, $\beta_2 = 1.0$. In fact all of the schemes developed in Sects. 7.1 and 7.2 can be interpreted as linear multistep methods.

For the solution of ordinary differential equations it is customary to use linear multistep methods of considerably higher order (involving more steps) and in a predictor-corrector mode. An explicit scheme provides the predictor and an implicit scheme the corresponding corrector (Gear 1971, p. 136). However, if a high-order strategy is applied to the system of equations arising from the semi-discretisation of a partial differential equation, e.g. (7.41), a large storage area is required to store the u^{n+l} and F^{n+l} vectors. In addition the use of a high-order scheme to march in time would be inefficient since the solution errors would probably be dominated by the low-order discretisation of the spatial terms, unless the exact solution is very smooth.

The need to avoid the storage of a large number of solutions at previous time steps motivates the use of one-step methods, such as Runge-Kutta schemes to solve semi-discrete systems. The general R-stage Runge-Kutta scheme applied to (7.45) can be written

$$u^{n+1} = u^n + \Delta t \sum_{r=1}^{R} c_r f^r \ , \tag{7.51a}$$

with

$$f^r = f(t + \Delta t a_r, y + \Delta t \sum_{s=1}^{R} b_{rs} f^s) \ , \qquad r = 1, 2, \ldots, R \ , \tag{7.51b}$$

and

$$a_r = \sum_{s=1}^{R} b_{rs} \ , \qquad r = 1, 2, \ldots, R \ . \tag{7.51c}$$

Equation (7.51) includes implicit Runge-Kutta schemes which are computationally expensive since (7.51b) is a nonlinear equation which must be solved iteratively for each of the f^r at each time-step. Consequently there is more interest in explicit Runge-Kutta schemes for which the upper summation limits in (7.51b, c) are replaced by $r-1$. Thus the Euler scheme (7.48) can be interpreted as a single-stage ($R=1$) explicit Runge-Kutta scheme with $c_1=1$, $b_{11}=0$, $a_1=0$.

There is an infinite number of second-order two-stage explicit Runge-Kutta schemes. A typical one is the improved Euler scheme

$$u^* = u^n + \Delta t f^n \quad \text{and} \tag{7.52}$$

$$u^{n+1} = u^n + 0.5\,\Delta t(f^n + f^*) \ .$$

Many third-order three-stage explicit Runge-Kutta schemes are available (Lambert 1973). The following fourth-order, four-stage explicit Runge-Kutta scheme has been widely used for solving initial value problems governed by ordinary differential equations:

$$u^{n+1} = u^n + \frac{\Delta t}{6}(f^n + 2f^* + 2f^{**} + f^{***}) \ , \quad \text{where} \tag{7.53}$$

$$u^* \quad = u^n + 0.5\Delta t f^n \ , \qquad f^* \quad = f(t^{n+1/2}, u^*)$$

$$u^{**} \quad = u^n + 0.5\Delta t f^* \ , \qquad f^{**} \quad = f(t^{n+1/2}, u^{**})$$

$$u^{***} = u^n + \Delta t f^{**} \ , \qquad f^{***} = f(t^n, u^{***}) \ .$$

Clearly higher-order Runge-Kutta schemes require more evaluations of f. When solving semi-discrete systems, e.g. (7.43), the repeated evaluation of all the terms associated with the spatial discretisation often dominates the overall execution time.

If the scalar equation (7.44) is considered along with the initial condition, $u(0)=u_0$, the corresponding exact solution has the form

$$u(t) = u_0\,e^{\lambda t} + u_{ss} \ . \tag{7.54}$$

For the physical solution to correspond to a stable process for the interval $0 \leq t \leq \infty$, it is necessary that $\lambda \leq 0$. Absolute stability of a linear multistep method (7.46) is analysed by applying it to (7.44) and constructing the stability polynomial

$$\sum_{l=0}^{m} (\alpha_l - \lambda\Delta t\beta_l)r^l = 0 \ . \tag{7.55}$$

Absolute stability is obtained for a given $\lambda\Delta t$ (Lambert 1973, p. 66) if all the roots, r_s, of (7.55) satisfy

$$|r_s| < 1 \ . \tag{7.56}$$

This is essentially equivalent to the restriction required for stability of a discretised partial differential equation using the von Neumann stability analysis (Sect. 4.3).

For the Euler scheme, absolute stability is achieved if $-2 \leq \lambda\Delta t \leq 0$. Since $\lambda < 0$, this corresponds to $\Delta t < 2/|\lambda|$. For the trapezoidal scheme absolute stability implies $\lambda\Delta t < 0$; thus the trapezoidal scheme is stable for any choice of Δt.

An equivalent criterion of absolute stability to (7.56) is provided by Lambert (1973, p. 136) and leads to the limit $-2 \leq \lambda\Delta t \leq 0$ for both the one-stage (Euler scheme) and two-stage explicit Runge-Kutta schemes (7.52). However, the four-stage explicit Runge-Kutta scheme (7.53) has the slightly enlarged stability interval $-2.78 \leq \lambda \Delta t \leq 0$.

In the method of lines, systems of ordinary differential equations occur. Thus it is appropriate to test the absolute stability of equations of the form (7.42). The stability polynomial (7.55) can still be derived if λ is replaced by λ_k, the eigenvalues of $\underline{A}$. In general λ_k will be complex. However, for a physically stable problem it will be necessary that the real part of λ_k, $Rl\lambda_k$, is non-positive.

As an example, the application of the Euler scheme (7.48) to (7.41) produces the following criterion for absolute stability:

$$-2 \leq Rl\lambda_k \Delta t \leq 0 \ . \tag{7.57}$$

For a system of M equations based on (7.41), $\underline{A}$ is tridiagonal with eigenvalues given by (9.48), i.e.

$$\lambda_k = \frac{\alpha\{-2 + 2\cos[(k\pi)/(M+1)]\}}{\Delta x} \tag{7.58}$$

Consequently the severest restriction on Δt coming from (7.56) is

$$-2 \leq -\frac{4\alpha\Delta t}{\Delta x^2} \quad \text{or} \quad \Delta t \leq \frac{0.5\Delta x^2}{\alpha} \tag{7.59}$$

Thus the same stability restriction on Δt is obtained as for the FTCS scheme (7.5). This is to be expected since the Euler time-stepping method applied to (7.41) coincides with the FTCS scheme. Application of the trapezoidal rule (7.49) to (7.41) leads to the restriction $Rl\lambda_k \Delta t \leq 0$. But since $Rl\lambda_k \leq 0$ from (7.58) there is no restriction on Δt. The trapezoidal rule applied to (7.41) coincides with the Crank-Nicolson scheme (7.22). As these examples suggest, the absolute stability of the algorithm solving a system of ordinary differential equations is equivalent to the stability of the algorithm solving the related discretised partial differential equation (Sect. 4.3).

If the system obtained from applying the method of lines is nonlinear, e.g. (7.43), the determination of the stability of the algorithm depends on the eigenvalues of the Jacobian $\partial F/\partial U$. However, now the eigenvalues depend on the solution and may need to be monitored while the solution is being obtained.

The exact solution of the system of equations (7.42) formed from (7.41) can be written as

$$U = \sum_{k=1}^{m} \alpha_k e^{\lambda_k t} e_k + U_{ss} \ , \tag{7.60}$$

where the eigenvalues λ_k are given by (7.58) and e^k are the corresponding eigenvectors. The coefficients α_k are determined by the initial conditions $U_0(x_j)$. From (7.58) it is clear that the eigenvalues cover a considerable range, particularly on a refined grid. This implies that the contributions to the solution of different eigenvectors in (7.60) will decay at different rates.

If an explicit scheme is used to solve (7.42) the restriction on the time-step will be determined by $|Rl\lambda_k|_{max}$. However, the time over which the integration will be required will be determined by $|Rl\lambda_k|_{min}$. The ratio $|Rl\lambda_k|_{max}/|Rl\lambda_k|_{min}$ is called the stiffness ratio. If this ratio is very large, perhaps 10^4–10^6, the system of ordinary differential equations is said to be stiff. Clearly algorithms that have no restriction on the time-step are desirable for solving stiff systems.

Such unconditional stability is guaranteed if the numerical method is A-stable (Dahlquist 1963), i.e. its region of absolute stability contains the whole of the left-hand half plane $Rl\lambda\Delta t < 0$. But A-stability is rather demanding. No explicit linear multistep or explicit Runge-Kutta scheme is A-stable. A second-order A-stable implicit linear multistep method is the highest order possible.

A related, but less severe, stability concept is $A(\alpha)$ stability (Widlund 1967), which is illustrated in Fig. 7.10. A method is $A(\alpha)$ stable if an infinite wedge, symmetric about the negative real axis and of half-angle α, can be constructed within which the method is absolutely stable. If all the eigenvalues of $\underline{A}$ are real, as is the case for (7.41), a numerical method that is $A(0)$ stable will be effective even if the system is stiff.

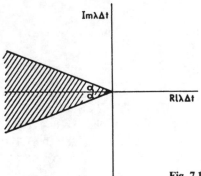

Fig. 7.10. $A(\alpha)$ stability

To obtain the solution of steady flow problems the numerical algorithm is often based on an equivalent pseudo-transient construction (Sect. 6.4). To achieve the steady-state solution as efficiently as possible, implicit schemes are constructed to exploit the known stability behaviour of equivalent algorithms for ordinary differential equations (Beam and Warming 1979). Such implicit algorithms (Chaps 8–10, 14, 17 and 18) are usually equivalent to linear multistep methods of low order.

The matching of computational algorithms to the characteristics of vector and parallel processors (Ortega and Voigt 1985; Gentzsch and Neves 1988) has renewed interest in explicit methods, even for obtaining steady-state solutions via a pseudo-transient formulation. These are often based on Runge-Kutta time-marching schemes in conjunction with a method-of-lines construction (Chaps. 14 and 18).

7.5 Closure

Explicit schemes, up to fourth-order, have been constructed for the one-dimensional diffusion equation. Higher-order accuracy has been achieved by balancing the truncation errors associated with discretising the spatial term $\partial^2 \bar{T}/\partial x^2$ at different time levels n and the time derivative $\partial \bar{T}/\partial t$ at different spatial locations j. In practice this requires choosing certain free parameters, as in (7.15), so that low-order terms in the truncation error are identically zero.

Explicit schemes usually impose a maximum value on $\varDelta t$ (through the s parameter) for stable solutions. Higher-order explicit schemes typically have more severe stability restrictions on $\varDelta t$. For the DuFort-Frankel scheme it is necessary to limit $\varDelta t$ to obtain accurate solutions even though stable solutions can be obtained with very large values of $\varDelta t$.

Implicit schemes often achieve unconditional stability at the price of additional computational effort. As long as the implicit scheme can be restricted to tridiagonal form the total computational effort is about double that of solving an equivalent explicit scheme. The inclusion of additional terms at the implicit $(n+1)$ time level allows more flexibility in constructing higher-order schemes. However, the results (Table 7.6) suggest that a fine grid may be necessary to obtain a significant improvement in accuracy.

If the problem governed by the diffusion equation requires a derivative boundary condition specification, additional errors arise in discretising the boundary condition. For problems governed by a parabolic partial differential equation it is important that derivative boundary conditions be discretised with formulae of the same order of accuracy as those used in the interior.

By discretising the spatial term only, the governing partial differential equation (PDE) is converted to the semi-discrete form, consisting of a system of ordinary differential equations (ODEs) in time. Consequently various algorithms developed specifically for ODEs become available for solving the PDE. However, it is emphasised that the equivalent system of ODEs contains errors arising from the spatial discretisation. This mitigates against the use of very high-order ODE solvers. The use of the semi-discrete form is convenient when a spectral method is employed to discretise the spatial terms (Sects. 5.6, 15.3.3 and 17.1.6).

7.6 Problems

Explicit Methods (Sect. 7.1)

7.1 Derive a) (7.14), b) (7.17).

7.2 Carry out an approximate operation count for the FTCS, DuFort-Frankel and 3L-4TH ($\gamma = 0$) schemes at $s = 0.30$ and compare the computational efficiency.

7.3 Construct a two-level five-point scheme that is fourth-order accurate. Asymmetric five-point formulae will be necessary adjacent to the boundaries. Test the stability of the scheme using the von Neumann analysis. Obtain numerical results to confirm the stability restrictions and convergence rate if possible.

7.4 Consider (7.13) with negative values of γ to see if stable and accurate solutions can be obtained. For $\gamma = -0.5$ can (7.13) be modified to look more like a DuFort-Frankel scheme, i.e. by changing the β structure of the spatial derivative discretisation? If so, determine whether a stable, higher-accuracy "DuFort-Frankel" scheme, can be constructed.

Implicit Methods (Sect. 7.2)

7.5 Derive (7.30).

7.6 Apply a von Neumann stability analysis to (7.31), with β given by (7.30), to determine the γ, s envelope for stable operation for the cases: a) $\delta = 0$, b) $\delta = 1/12$, c) $\delta = 1/6$.

7.7 For $\delta = 1/6$, $\gamma = 0$, $\beta = 1$ and $s = 1/6$, (7.31) becomes explicit. Compare solutions for this case with those produced by the FTCS scheme. Explore other parameter combinations that make (7.31) explicit.

7.8 Use program DIFIM to obtain solutions for cases ME$=4$, 5 with $\gamma = 0.5$. Compare the accuracy and convergence rate with the data shown in Table 7.6.

7.9 Obtain an approximate operation count for the cases ME$=1,4$ and $\gamma = 0$ in Table 7.6 and compare the computational efficiency with $s = 1.0$ with that of the FTCS scheme at $s = 0.5$ and the DuFort-Frankel scheme at $s = 0.3$.

Boundary and Initial Conditions (Sect. 7.3)

7.10 Modify program DIFEX to implement:
a) first-order accurate Neumann boundary condition (7.32),
b) second-order accurate Neumann boundary condition (7.34), and reproduce the results given in Table 7.8.
Programs DIFEX and EXTRA will need to be modified to suit the problem defined by Eqs. (7.37) to (7.40)

7.11 Modify program DIFIM to implement:
a) first-order accurate Neumann boundary condition (7.32),
b) second-order accurate Neumann boundary condition (7.34), and obtain the results indicated in Table 7.10. For JMAX$=6$ compare the error distribution with that shown in Table 7.9.
Programs DIFIM and EXTRA will need to be modified to suit the problem defined by Eqs. (7.37) to (7.40).

7.12 For the DuFort-Frankel scheme starting from the exact solution (subroutine EXTRA) at $t = 2.00$, introduce the FTCS scheme on a finer x and t grid to obtain the second level of starting data. For $s = 0.3$ and $s = 0.5$ compare the accuracy of the FTCS-started DuFort-Frankel solution at $t = 9.00$ with that obtained by starting from two levels of data given by the exact solution at $t = 2.00$ and $2.00 - \Delta t$.

Method of Lines (Sect. 7.4)

7.13 Modify program DIFEX to apply the four-stage Runge-Kutta method and obtain results to compare with Table 7.3, for $s = 0.41$.

7.14 For the scheme developed in Problem 7.13 carry out:
i) a Taylor series expansion to determine the truncation error leading term
ii) a von Neumann stability analysis to determine any restriction on s
iii) an operation count to compare the computational efficiency with that of the FTCS scheme.

8. Multidimensional Diffusion Equation

A broad conclusion from Chap. 7 is that implicit schemes are more effective than explicit schemes for problems with significant dissipation, as exemplified by the one-dimensional diffusion equation.

In extending implicit schemes to multidimensions, special procedures are necessary if economical algorithms are to be obtained. The special procedures are often built around some means of splitting the equation on a convenient coordinate basis (Sects. 8.2, 8.3 and 8.5). The use of splitting constructions also requires careful attention being given to the implementation of derivative (Neumann) boundary conditions (Sect. 8.4). The splitting techniques developed in this chapter are applicable to finite difference, finite element and finite volume methods.

8.1 Two-Dimensional Diffusion Equation

In two dimensions, the diffusion equation is written

$$\frac{\partial \bar{T}}{\partial t} - \alpha_x \frac{\partial^2 \bar{T}}{\partial x^2} - \alpha_y \frac{\partial^2 \bar{T}}{\partial y^2} = 0 \ . \tag{8.1}$$

For the region shown in Fig. 8.1, Dirichlet boundary conditions are

$$\bar{T}(0, y, t) = a(y, t) \ ,$$
$$\bar{T}(1, y, t) = b(y, t) \ , \tag{8.2}$$
$$\bar{T}(x, 0, t) = c(x, t) \ ,$$
$$\bar{T}(x, 1, t) = d(x, t) \ ,$$

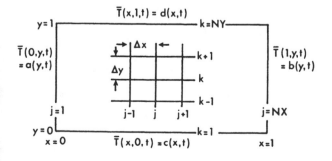

Fig. 8.1. Two-dimensional domain and Dirichlet boundary conditions

and initial conditions

$$\bar{T}(x, y, 0) = T_0(x, y) \ . \tag{8.3}$$

In this section, typical explicit and implicit schemes for the one-dimensional diffusion equation are extended to two dimensions to see if they are directly applicable.

8.1.1 Explicit Methods

The FTCS scheme, in two dimensions, is

$$\frac{\Delta T_{j,k}^{n+1}}{\Delta t} - \alpha_x L_{xx} T_{j,k}^n - \alpha_y L_{yy} T_{j,k}^n = 0 \quad \text{where} \tag{8.4}$$

$$L_{xx} T_{j,k}^n = \frac{T_{j-1,k}^n - 2T_{j,k}^n + T_{j+1,k}^n}{\Delta x^2} \ , \text{ etc.}$$

As an algorithm this becomes

$$T_{j,k}^{n+1} = s_x T_{j-1,k}^n + (1 - 2s_x - 2s_y) T_{j,k}^n + s_x T_{j+1,k}^n$$

$$+ s_y T_{j,k-1}^n + s_y T_{j,k+1}^n \ , \tag{8.5}$$

where $s_x = \alpha_x \Delta t / \Delta x^2$ and $s_y = \alpha_y \Delta t / \Delta y^2$. A Taylor series expansion about the (j, k, n) node indicates that (8.5) is consistent (Sect. 4.2) with (8.1) and has a truncation error of $O(\Delta t, \Delta x^2, \Delta y^2)$.

A von Neumann stability analysis demonstrates that (8.5) will be stable if

$$s_x + s_y \leq 0.5 \ . \tag{8.6}$$

It may be noted that if $s_x = s_y = s$, (8.6) gives $s \leq 0.25$, which is more restrictive than the corresponding expression in one dimension (Sect. 7.1.1). However, if small time-steps must be used to obtain a sufficiently accurate solution, the restrictive stability condition may not be critical.

For the case $\alpha = \alpha_x = \alpha_y$ and $\Delta x = \Delta y$, Mitchell and Griffiths (1980, p. 55) provide an extension to (8.4), i.e.

$$\frac{\Delta T_{j,k}^{n+1}}{\Delta t} - \alpha L_{xx} T_{j,k}^n - \alpha L_{yy} T_{j,k}^n - \alpha^2 \Delta t L_{xx} L_{yy} T_{j,k}^n = 0 \ , \tag{8.7}$$

which is stable in the range $0 < s \leq 0.5$. Although (8.7) is a nine-point scheme it can be implemented economically in two stages

$$T_{j,k}^* = (1 + \alpha \Delta t L_{yy}) T_{j,k}^n \quad \text{and} \tag{8.8}$$

$$T_{j,k}^{n+1} = (1 + \alpha \Delta t L_{xx}) T_{j,k}^* \ . \tag{8.9}$$

This scheme extends to a three-stage algorithm in three dimensions while maintaining the "one-dimensional" stability range $0 < s \leq 0.5$. By contrast, the three-dimensional FTCS scheme with $\Delta x = \Delta y = \Delta z$ is stable in the range $0 < s \leq 1/6$.

One of the more interesting explicit algorithms is the hopscotch method (Gourlay 1970). This method can be interpreted, in its simplest form, as a two-stage FTCS scheme. In the first stage, (8.5) is applied at all grid points for which $j + k + n$ is even; that is, on a grid pattern corresponding to the black squares on a chess-board. In the second-stage the following equation is solved at all grid points for which $j + k + n$ is odd, i.e. on the red squares of an "equivalent" chess-board.

$$(1 + 2s_x + 2s_y) T_{j,k}^{n+1} = T_{j,k}^n + s_x (T_{j-1,k}^{n+1} + T_{j+1,k}^{n+1}) + s_y (T_{j,k-1}^{n+1} + T_{j,k+1}^{n+1}) . \quad (8.10)$$

The terms evaluated at t_{n+1} on the right-hand side of (8.10) are known from the first stage. The simple hopscotch method has a truncation error of $O(\Delta t, \Delta x^2, \Delta y^2)$ but, in contrast to the FTCS scheme, is unconditionally stable. Mitchell and Griffiths (1980, p. 77) discuss the hopscotch family of methods in more detail.

8.1.2 Implicit Method

Following the same procedure as for one-dimensional problems, it is possible to obtain an implicit scheme by evaluating the spatial derivatives in (8.1) at the time-level $(n + 1)$. The resulting algorithm is

$$-s_x T_{j-1,k}^{n+1} + (1 + 2s_x + 2s_y) T_{j,k}^{n+1} - s_x T_{j+1,k}^{n+1}$$

$$-s_y T_{j,k-1}^{n+1} - s_y T_{j,k+1}^{n+1} = T_{j,k}^n . \quad (8.11)$$

This scheme has a truncation error of $O(\Delta t, \Delta x^2, \Delta y^2)$ and is unconditionally stable. However, the difficulty here is in obtaining, economically, the solution of the equations that result from applying (8.11) at every node.

For the present equation, (8.11), it is possible to number the nodes so that three terms are on or adjacent to the main diagonal but the other two terms are displaced, effectively, by the number of internal nodes across the grid (say, $NX - 2$ in Fig. 8.1). Consequently, the Thomas algorithm cannot be used. Using conventional Gauss elimination (Sect. 6.2.1) would be prohibitively expensive; using sparse Gauss elimination would still be unacceptably uneconomical for refined grids.

8.2 Multidimensional Splitting Methods

The problem with the two-dimensional implicit scheme can be overcome by splitting the solution algorithm (system of algebraic equations) into two half-steps to advance one time-step. At each half-step, only terms associated with a particular coordinate direction are treated implicitly. Consequently, only three implicit terms appear and these can be grouped adjacent to the main diagonal. As a result, the very efficient Thomas algorithm can be used to obtain the solution. The overall process of treating each time-step as a sequence of simpler substeps is referred to as (time-)splitting.

8.2.1 ADI Method

The best known example of a splitting technique is the alternating direction implicit (ADI) method (Peaceman and Rachford 1955). We will examine the ADI method in detail and then introduce a generalisation of the splitting approach.

The ADI interpretation of (8.1) is written in two half time-steps as follows. During the first half-step the following discretisation is used:

$$\frac{T_{j,k}^* - T_{j,k}^n}{\Delta t/2} - \alpha_x L_{xx} T_{j,k}^* - \alpha_y L_{yy} T_{j,k}^n = 0 \ , \tag{8.12}$$

and during the second

$$\frac{T_{j,k}^{n+1} - T_{j,k}^*}{\Delta t/2} - \alpha_x L_{xx} T_{j,k}^* - \alpha_y L_{yy} T_{j,k}^{n+1} = 0 \ . \tag{8.13}$$

During the first half-step the solution T is known at time-level n but is unknown at the $(n+1/2)$ time-level, denoted by $*$. However, unknown nodal values T^* are associated with the x-direction only (i.e. constant value of k in Fig. 8.2). Equation (8.12) can be rewritten as one member of a system of equations as

$$-0.5s_x T_{j-1,k}^* + (1+s_x) T_{j,k}^* - 0.5s_x T_{j+1,k}^*$$
$$= 0.5s_y T_{j,k-1}^n + (1-s_y) T_{j,k}^n + 0.5s_y T_{j,k+1}^n \ . \tag{8.14}$$

Other members of the system are formed around the other nodes in the same row k. Thus, the solution of the system of equations gives the intermediate solution $T_{j,k}^*$, $j = 2, \ldots, NX-1$, for one value of k only. Sequentially, systems of equations are solved for $T_{j,k}^*, j = 2, \ldots, NX-1$, for each row, $k = 2, \ldots, NY-1$, using the Thomas algorithm.

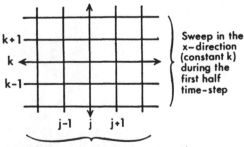

Sweep in the x-direction (constant k) during the first half time-step

Sweep in the y-direction (constant j) during the second half time-step

Fig. 8.2. ADI implementation

During the second half-step (8.13) is used, but in the form

$$-0.5s_y\, T^{n+1}_{j,k-1} + (1+s_y)\, T^{n+1}_{j,k} - 0.5s_y\, T^{n+1}_{j,k+1}$$

$$= 0.5s_x\, T^{*}_{j-1,k} + (1-s_x)\, T^{*}_{j,k} + 0.5s_x\, T^{*}_{j+1,k} \ . \tag{8.15}$$

During the second half-step the solution at time-level $n+1$ is unknown but the solution at the intermediate time-level * is known. A system of equations associated with all the nodes along one grid line in the y-direction (j fixed) is solved for $T^{n+1}_{j,k}$, $k=2, \ldots,$ NY -1. The process is repeated for each grid line, $j=2, \ldots,$ NX -1, in turn.

The stability of the ADI scheme can be ascertained by applying the von Neumann stability analysis to obtain the amplification factor for each half-step. The stability of the complete time-step is determined by the product of the two half-step amplification factors, i.e.

$$G = G'G'' = \left[\frac{1-2s_y\sin^2(\theta_y/2)}{1+2s_x\sin^2(\theta_x/2)}\right]\left[\frac{1-2s_x\sin^2(\theta_x/2)}{1+2s_y\sin^2(\theta_y/2)}\right]. \tag{8.16}$$

An examination of (8.16) indicates that $|G| \leq 1$ for any value of s_x, s_y, θ_x, θ_y. However, a consideration of $|G'|$ and $|G''|$ indicates that whereas the full step is unconditionally stable, each half-step is only conditionally stable. But only the full step is of interest.

The composite scheme (8.12 and 13) is consistent with (8.1) and has a truncation error of $O(\Delta t^2, \Delta x^2, \Delta y^2)$. The second-order time accuracy follows from the symmetry of the scheme, just as the second-order accuracy of the Crank-Nicolson scheme followed from the symmetry about the time-level $(n+1/2)$.

However, to achieve a global truncation error of $O(\Delta t^2)$, it is necessary to introduce boundary values for the intermediate solution $T^{*}_{j,k}$ that are compatible with the interior algorithms (8.14 and 15). For example, if Dirichlet boundary conditions are imposed, the evaluation of $T^{*}_{NX,k} = b^{n+1/2}_k$ at $x=1$ produces an algorithm that has a truncation error of $O(\Delta t)$. To produce a truncation error of $O(\Delta t^2)$ it is necessary to evaluate $T^{*}_{NX,k}$ from (8.12 and 13) as

$$T^{*}_{NX,k} = 0.5(b^{n}_k + b^{n+1}_k) - 0.25\,\Delta t\,\alpha_y L_{yy}(b^{n+1}_k - b^{n}_k). \tag{8.17}$$

A similar problem arises with approximate factorisation schemes; appropriate boundary condition specifications are discussed in Sects. 8.3.2 and 8.4.

Thus, it may be concluded that the ADI scheme, in two dimensions, has the desirable attributes of being unconditionally stable, second-order accurate and economical to solve. The ADI scheme extends to the three-dimensional diffusion equation where three steps, each occupying $\Delta t/3$, replace (8.14 and 15). In three dimensions, the ADI scheme is economical, spatially second-order accurate but is only conditionally stable. It is necessary that s_x, s_y, $s_z \leq 1.5$ for stability where $s_z = \alpha_z \Delta t / \Delta z^2$.

8.2.2 Generalised Two-Level Scheme

We now seek to generalise the splitting concept. A general two-level implicit finite difference scheme for (8.1) can be written

$$\frac{\Delta T_{j,k}^{n+1}}{\Delta t} - (1-\beta)(\alpha_x L_{xx} T_{j,k}^n + \alpha_y L_{yy} T_{j,k}^n)$$

$$-\beta(\alpha_x L_{xx} T_{j,k}^{n+1} + \alpha_y L_{yy} T_{j,k}^{n+1}) = 0 \ , \quad \text{where} \tag{8.18}$$

$$\Delta T_{j,k}^{n+1} = T_{j,k}^{n+1} - T_{j,k}^n \ .$$

Here $\Delta T_{j,k}^{n+1}$ can be thought of as the correction to the solution at time-level n in order to advance to time-level $(n+1)$. In order to minimize the build-up of round-off error it is useful to have $\Delta T_{j,k}^{n+1}$ appear explicitly in the computer program. The role of β in (8.18) is to weight the time-levels n and $(n+1)$; this same idea was used in (7.24).

Equation (8.18) will be manipulated to give an implicit algorithm in $\Delta T_{j,k}^{n+1}$. First the term $T_{j,k}^{n+1}$ is expanded as a Taylor series about the nth time-level, as in (3.16), viz.

$$T_{j,k}^{n+1} = T_{j,k}^n + \Delta t \left[\frac{\partial T}{\partial t}\right]_{j,k}^n + 0.5 \Delta t^2 \left[\frac{\partial T^2}{\partial t^2}\right]_{j,k}^n + \cdots \ ,$$

which can be approximated by

$$T_{j,k}^{n+1} = T_{j,k}^n + \Delta t \left(\frac{\Delta T_{j,k}^n}{\Delta t}\right) + O(\Delta t^2) \ . \tag{8.19}$$

Substituting (8.19) into (8.18) gives

$$\frac{\Delta T_{j,k}^{n+1}}{\Delta t} - (\alpha_x L_{xx} T_{j,k}^n + \alpha_y L_{yy} T_{j,k}^n)$$

$$-\beta(\alpha_x L_{xx} \Delta T_{j,k}^{n+1} + \alpha_y L_{yy} \Delta T_{j,k}^{n+1}) = 0 \ , \tag{8.20}$$

or, after rearrangement,

$$[1 - \beta \Delta t(\alpha_x L_{xx} + \alpha_y L_{yy})] \Delta T_{j,k}^{n+1} = \Delta t(\alpha_x L_{xx} + \alpha_y L_{yy}) T_{j,k}^n \ . \tag{8.21}$$

Algebraic operators appropriate to both directions appear on the left hand side of (8.21). In order to make use of the Thomas algorithm, (8.21) is replaced by the *approximate factorisation*

$$(1 - \beta \Delta t \alpha_x L_{xx})(1 - \beta \Delta t \alpha_y L_{yy}) \Delta T_{j,k}^{n+1} = \Delta t(\alpha_x L_{xx} + \alpha_y L_{yy}) T_{j,k}^n \ . \tag{8.22}$$

Comparing (8.21) and (8.22) indicates that (8.22) contains an additional term, on the left hand side,

$$\beta^2 \Delta t^2 \alpha_x \alpha_y L_{xx} L_{yy} \Delta T_{j,k}^{n+1} \ .$$

That is, (8.22) approximates (8.21) to $O(\Delta t^2)$.

Equation (8.22) can be implemented as a two-stage algorithm at each time-step. As the first stage the following system of equations is solved on every grid-line in the x-direction (constant k in Fig. 8.2):

$$(1 - \beta \Delta t \alpha_x L_{xx}) \Delta T_{j,k}^* = \Delta t (\alpha_x L_{xx} + \alpha_y L_{yy}) T_{j,k}^n \tag{8.23}$$

for $\Delta T_{j,k}^*$, which can be thought of as an intermediate approximation to $\Delta T_{j,k}^{n+1}$. When L_{xx} is a three-point centred difference operator, (8.23) is a tridiagonal system that can be solved efficiently using the Thomas algorithm (Sect. 6.2.2). During the second stage, the following system of equations is solved:

$$(1 - \beta \Delta t \alpha_y L_{yy}) \Delta T_{j,k}^{n+1} = \Delta T_{j,k}^* \tag{8.24}$$

on every grid-line in the y-direction (constant j in Fig. 8.2). The structure of (8.23, 24) is similar to that of the ADI scheme (8.14, 15). For more complicated equations the evaluation of all the spatial terms on the right hand side of (8.23) is a major contribution to the execution time. The present implementation only requires one evaluation of the spatial terms per time-step. In contrast, the ADI scheme requires two evaluations. The algorithm (8.23, 24) is due to Douglas and Gunn (1964).

The two-stage algorithm (8.23, 24) is unconditionally stable for $\beta \geq 0.5$ and has a truncation error of $O(\Delta t^2, \Delta x^2, \Delta y^2)$ if $\beta = 0.5$. The approximate factorisation construction extends to three dimensions and, in contrast to the ADI scheme, is unconditionally stable for $\beta \geq 0.5$.

If the present formulation is used to obtain solutions of steady problems as the steady-state limit of the transient solution (Sect. 6.4), it is useful to define

$$\text{RHS} = (\alpha_x L_{xx} + \alpha_y L_{yy}) T_{j,k}^n \tag{8.25}$$

in (8.23). As the steady-state solution is approached RHS tends to zero; thus monitoring the value of RHS indicates the proximity to the steady-state solution.

8.2.3 Generalised Three-Level Scheme

For the one-dimensional diffusion equation a generalised three-level implicit scheme was given by (7.25). A corresponding three-level scheme is written in two dimensions as

$$\frac{(1 + \gamma) \Delta T_{j,k}^{n+1}}{\Delta t} - \frac{\gamma \Delta T_{j,k}^n}{\Delta t} = (1 - \beta)(\alpha_x L_{xx} + \alpha_y L_{yy}) T_{j,k}^n$$

$$+ \beta(\alpha_x L_{xx} + \alpha_y L_{yy}) T_{j,k}^{n+1} \ , \quad \text{where} \tag{8.26}$$

$$\Delta T_{j,k}^n = T_{j,k}^n - T_{j,k}^{n-1} \ .$$

With the same construction as was used to develop (8.23, 24) from (8.18), the following two-stage algorithm is obtained from (8.26). During the first stage

$$\left(1 - \frac{\beta}{(1+\gamma)} \Delta t \alpha_x L_{xx}\right) \Delta T^*_{j,k} = \frac{\Delta t}{(1+\gamma)} (\alpha_x L_{xx} + \alpha_y L_{yy}) T^n_{j,k} + \frac{\gamma}{(1+\gamma)} \Delta T^n_{j,k} \qquad (8.27)$$

generates a tridiagonal system of equations associated with each grid line in the x-direction.

During the second stage of every time-step, the following equation is used:

$$\left(1 - \frac{\beta}{(1+\gamma)} \Delta t \alpha_y L_{yy}\right) \Delta T^{n+1}_{j,k} = \Delta T^*_{j,k} \; . \qquad (8.28)$$

For the particular choice $\beta = 1$, $\gamma = 0.5$ the two-stage algorithm, given by (8.27, 28), is consistent with (8.1) with a truncation error of $O(\Delta t^2, \Delta x^2, \Delta y^2)$, and is unconditionally stable. For the first time-step ($n = 0$), a two-level scheme, such as given by (8.23, 24), is required.

The splitting schemes discussed in this section extend naturally to three dimensions (Mitchell and Griffiths 1980, p. 85). The modern approach to splitting whereby a term, typically of $O(\Delta t^2)$, is added to the implicit equation to construct the factorisation is discussed at length by Gourlay (1977). Higher-order split schemes are possible for the diffusion equation in two and three dimensions. Some of these are discussed by Mitchell and Griffiths (1980, pp. 61, 87).

8.3 Splitting Schemes and the Finite Element Method

Here we apply the Galerkin finite element method (Sect. 5.3) to the two-dimensional diffusion equation (8.1) with boundary and initial conditions given by (8.2, 3) and determine whether the splitting schemes developed in Sect. 8.2 must be modified to include the finite element form of the discretised equations. Rectangular elements are used with bilinear interpolating functions (5.59) in each element. If the Galerkin finite element method is applied on a grid that is uniform in the x and y directions, the result can be written (after dividing all terms by $\Delta x \Delta y$) as

$$M_x \otimes M_y \left[\frac{\partial T}{\partial t}\right]_{j,k} = \alpha_x M_y \otimes L_{xx} T_{j,k} + \alpha_y M_x \otimes L_{yy} T_{j,k} \; , \qquad (8.29)$$

where $\otimes$ denotes the tensor (or outer) product (Mase 1970, p. 15); M_x and M_y are directional mass operators and L_{xx} and L_{yy} are directional difference operators (Appendix A.2). The directional mass operators are

$$M_x = (\tfrac{1}{6}, \tfrac{2}{3}, \tfrac{1}{6}) \quad \text{and} \quad M_y = (\tfrac{1}{6}, \tfrac{2}{3}, \tfrac{1}{6})^T \; , \qquad (8.30)$$

and the directional difference operators are

$$L_{xx} = \left(\frac{1}{\Delta x^2}, -\frac{2}{\Delta x^2}, \frac{1}{\Delta x^2}\right) \quad \text{and} \quad L_{yy} = \left(\frac{1}{\Delta y^2}, -\frac{2}{\Delta y^2}, \frac{1}{\Delta y^2}\right)^T. \tag{8.31}$$

Thus, the term $M_y \otimes L_{xx} T_{j,k}$ gives a nine-point representation of $\partial^2 T/\partial x^2$. With reference to Fig. 8.3,

$$M_y \otimes L_{xx} T_{j,k} = \frac{1}{6}\left\{\frac{T_{j-1,k+1} - 2T_{j,k+1} + T_{j+1,k+1}}{\Delta x^2}\right\} + \frac{2}{3}\left\{\frac{T_{j-1,k} - 2T_{j,k} + T_{j+1,k}}{\Delta x^2}\right\}$$

$$+ \frac{1}{6}\left\{\frac{T_{j-1,k-1} - 2T_{j,k-1} + T_{j+1,k-1}}{\Delta x^2}\right\}. \tag{8.32}$$

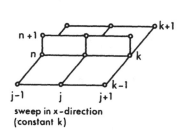
sweep in x-direction
(constant k)

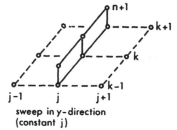
sweep in y-direction
(constant j)

(a)   Finite element formulation

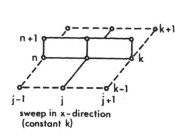
sweep in x-direction
(constant k)

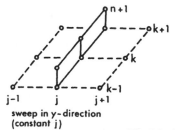
sweep in y-direction
(constant j)

(b)   Finite difference formulation

Fig. 8.3a, b. Active nodes for split schemes (8.23, 24); (8.39, 40)

The finite difference representation introduced earlier, for example (8.4) and following, can be incorporated into the framework of (8.29) by defining the finite difference directional mass operators as

$$M_x^{fd} = (M_y^{fd})^T = (0, 1, 0) ,$$

so that the finite difference expressions will contain only three terms.

The major difference between finite element and finite difference discretisation in multidimensions is that the finite element method evaluates spatial derivatives over a region whereas the finite difference method evaluates spatial derivatives along a particular coordinate direction. Generally, the finite element formulation is more accurate but also less economical, since more terms must be evaluated.

If quadratic Lagrange elements are used instead of linear Lagrange elements each of the mass and difference operators contains five terms so that a typical tensor product, $M_x \otimes L_{yy} T_{j,k}$, contains 25 terms. However, such an expression is consistent with the derivative $\partial^2 \bar{T}/\partial y^2$ being represented to $O(\Delta y^3)$ typically.

8.3.1 Finite Element Splitting Constructions

In this section, two means of splitting the finite element equations will be developed. The first produces an algorithm similar to the ADI scheme (8.14, 15). The second algorithm obtains solutions in terms of the corrections $\Delta T_{j,k}$ as in (8.23, 24).

As with the treatment of the one-dimensional diffusion equation a finite difference representation will be introduced for $\partial T/\partial t$ in (8.29) and the spatial terms will be evaluated as a weighted average of the nth and $(n+1)$-th time-levels. Thus, (8.29) is replaced by

$$M_x \otimes M_y \frac{\Delta T_{j,k}^{n+1}}{\Delta t} = (\alpha_x M_y \otimes L_{xx} + \alpha_y M_x \otimes L_{yy})[(1-\beta) T_{j,k}^n + \beta T_{j,k}^{n+1}] \ . \quad (8.33)$$

Equation (8.33) is the finite element equivalent of (8.18). The ADI finite element splitting is obtained by adding $\Delta t \alpha_x \alpha_y L_{xx} \otimes L_{yy}[\beta^2 T^{n+1} - (1-\beta)^2 T^n]$ to the left hand side of (8.33) so that it becomes

$$(M_x - \alpha_x \beta \Delta t L_{xx}) \otimes (M_y - \alpha_y \beta \Delta t L_{yy}) T_{j,k}^{n+1}$$
$$= [M_x + \alpha_x(1-\beta)\Delta t L_{xx}] \otimes [M_y + \alpha_y(1-\beta)\Delta t L_{yy}] T_{j,k}^n \quad (8.34)$$

Equation (8.34) can be implemented very efficiently in two stages. In the first stage,

$$(M_x - \alpha_x \beta \Delta t L_{xx}) T_{j,k}^* = [M_y + \alpha_y(1-\beta)\Delta t L_{yy}] T_{j,k}^n \quad (8.35)$$

produces a tridiagonal system on each x gridline. In the second stage,

$$(M_y - \alpha_y \beta \Delta t L_{yy}) T_{j,k}^{n+1} = [M_x + \alpha_x(1-\beta)\Delta t L_{xx}] T_{j,k}^* \quad (8.36)$$

produces a tridiagonal system on each y gridline.

If $\beta = 0.5$ the only difference between the finite element ADI construction (8.35, 36) and the finite difference ADI construction (8.14, 15), is the appearance of the mass operators in (8.35, 36). The ADI finite element algorithm is economical, second-order accurate in time and space and is unconditionally stable if $\beta \geq 0.5 + (\delta - 0.25)/s$ where δ is defined by (8.44) and s is s_x or s_y.

An alternative splitting can be developed in the following way. The term $T_{j,k}^{n+1}$, on the right hand side of (8.33) is expanded about the nth time-level as a Taylor series and terms of $O(\Delta t^2)$ are discarded. After rearrangement, the result can be written

$$[M_x \otimes M_y - \beta \Delta t(\alpha_x M_y \otimes L_{xx} + \alpha_y M_x \otimes L_{yy})] \Delta T_{j,k}^{n+1}$$
$$= \Delta t(\alpha_x M_y \otimes L_{xx} + \alpha_y M_x \otimes L_{yy}) T_{j,k}^n \ . \quad (8.37)$$

If the additional term $\beta^2 \Delta t^2 \alpha_x \alpha_y L_{xx} \otimes L_{yy} \Delta T_{j,k}^{n+1}$ is added to the left hand side of (8.37) the following splitting is obtained:

$$
\begin{aligned}
(M_x - \beta \Delta t \alpha_x L_{xx}) &\otimes (M_y - \beta \Delta t \alpha_y L_{yy}) \Delta T_{j,k}^{n+1} \\
&= \Delta t (\alpha_x M_y \otimes L_{xx} + \alpha_y M_x \otimes L_{yy}) T_{j,k}^n ,
\end{aligned}
\tag{8.38}
$$

which is implemented as

$$
(M_x - \beta \Delta t \alpha_x L_{xx}) \Delta T_{j,k}^* = \Delta t (\alpha_x M_y \otimes L_{xx} + \alpha_y M_x \otimes L_{yy}) T_{j,k}^n
\tag{8.39}
$$

for all grid lines in the x-direction and as

$$
(M_y - \beta \Delta t \alpha_y L_{yy}) \Delta T_{j,k}^{n+1} = \Delta T_{j,k}^*
\tag{8.40}
$$

for all grid lines in the y-direction.

Equations (8.39 and 40) may be compared with (8.23 and 24). The right hand side of (8.39) takes longer to evaluate than the right hand side of (8.23). The right hand sides are evaluated at time-level n. As can be seen from Fig. 8.3, nine nodes are active at time-level n for the finite element method compared with five for the finite difference method. However, the evaluation of (8.39) and the solutions of the tridiagonal systems of equations is negligibly different for the two formulations. Computational experiments indicate that the finite element method is slower, overall, than the finite difference method but usually provides more accurate solutions.

The computational algorithm (8.39, 40) is consistent with (8.1) with a truncation error of $O(\Delta t^2, \Delta x^2, \Delta y^2)$ if $\beta = 0.5$. The same splitting (8.39, 40) would be obtained if quadratic Lagrange elements (Sect. 5.3.2) were used instead of linear Lagrange elements, but the operators M_x, L_{yy}, etc., would have a maximum of five terms instead of three (Fletcher 1984, p. 122).

8.3.2 TWDIF: Generalised Finite Difference/Finite Element Implementation

The various splitting schemes considered in Sects. 8.2.2, 8.2.3 and 8.3.1 can be combined into a single generalised three-level scheme that embraces both finite difference and finite element formulations. This will be demonstrated here for the region shown in Fig. 8.1.

The temperature distribution in the domain $0 \le x \le 1$, $0 \le y \le 1$ is governed by (8.1) with the particular boundary conditions

$$
\begin{aligned}
\bar{T}(0, y, t) &= 20 + 80y , \\
\bar{T}(1, y, t) &= 20 + 80 [y - e^{-0.5 a \pi^2 t} \sin(0.5 \pi y)] , \\
\bar{T}(x, 0, t) &= 20 , \\
\bar{T}(x, 1, t) &= 20 + 80 [y - e^{-0.5 a \pi^2 t} \sin(0.5 \pi x)] .
\end{aligned}
\tag{8.41}
$$

For this example, the following analytic solution is available:

$$\bar{T} = 20 + 80[y - e^{-0.5\alpha\pi^2 t} \sin(0.5\pi x) \sin(0.5\pi y)] \ . \tag{8.42}$$

The generalised scheme can be introduced by applying a three-level finite difference discretisation to (8.29) with $\alpha = \alpha_x = \alpha_y$ to give

$$M_x \otimes M_y \frac{(1+\gamma)\Delta T_{j,k}^{n+1} - \gamma \Delta T_{j,k}^n}{\Delta t}$$

$$= \alpha(M_y \otimes L_{xx} + M_x \otimes L_{yy})[(1-\beta) T_{j,k}^n + \beta T_{j,k}^{n+1}] \ , \tag{8.43}$$

which is equivalent to (8.26). However, here the directional mass operators (8.30) will be generalised to

$$M_x = M_y^T = \{\delta, \ 1 - 2\delta, \ \delta\} \ . \tag{8.44}$$

Clearly, the finite difference scheme, Sects. 8.2.2 and 8.2.3, corresponds to $\delta = 0$ and the finite element scheme, Sect. 8.3.1, corresponds to $\delta = 1/6$.

The approximate factorisation of (8.43) produces the following two-stage algorithm in place of (8.39, 40).

First stage:

$$\left[M_x - \left(\frac{\alpha\beta\Delta t}{(1+\gamma)}\right) L_{xx}\right] \Delta T_{j,k}^* = \text{RHS}^n \ , \quad \text{where} \tag{8.45}$$

$$\text{RHS}^n = [\alpha\Delta t(M_y \otimes L_{xx} + M_x \otimes L_{yy}) T_{j,k}^n + \gamma M_x \otimes M_y \Delta T_{j,k}^n]/[1+\gamma] \ . \tag{8.46}$$

Second stage:

$$\left[M_y - \left(\frac{\alpha\beta\Delta t}{(1+\gamma)}\right) L_{yy}\right] \Delta T_{j,k}^{n+1} = \Delta T_{j,k}^* \ . \tag{8.47}$$

For equations formed at points adjacent to boundaries, boundary values of ΔT^* in (8.45) and ΔT^{n+1} in (8.47) must be provided. The appropriate form for these boundary values can be determined most easily by forming (8.45, 47) into a single composite scheme by eliminating ΔT^*. It is then clear that boundary values for ΔT^{n+1} required to solve (8.47) can be obtained directly from (8.41), whereas boundary values for ΔT^* must be computed from (8.41) via (8.47) reversed, e.g. on $j = NX(x=1)$,

$$\Delta T_{NX,k}^* = \left[M_y - \left(\frac{\alpha\beta\Delta t}{(1+\gamma)}\right) L_{yy}\right] \Delta T_{NX,k}^{n+1} \ . \tag{8.48}$$

If the boundary conditions (8.41) are not time dependent then zero boundary values for ΔT^* and ΔT^{n+1} are generated. Known boundary values of dependent variables, ΔT^* and ΔT^{n+1}, are transferred to the right hand side of (8.45) and (8.47), respectively, before solution.

The above algorithm is implemented in program TWDIF (Fig. 8.4). In the first stage, (8.45) generates tridiagonal systems of equations associated with each x-gridline. These are solved using subroutines BANFAC and BANSOL (Figs. 6.18 and 6.19). Equation (8.46) is evaluated in subroutine REDIF (Fig. 8.5). In the second stage, (8.47) produces tridiagonal systems of equations associated with each y gridline which are solved using subroutines BANFAC and BANSOL. The major parameters used in program TWDIF are defined in Table 8.1. Typical output for the finite difference Crank-Nicolson scheme on a coarse grid is shown in Fig. 8.6.

Program TWDIF is used to compare the Crank-Nicolson, two-level fully implicit (2LFI) and three-level fully implicit (3LFI) finite difference schemes with the Crank-Nicolson finite element scheme. The results are summarised in Table 8.2. For the three-level fully implicit scheme (3LFI) the first step is evaluated using the Crank-Nicolson scheme. Also included in Table 8.2 are rms errors obtained with a modified Crank-Nicolson scheme with $\delta = 1/12$ in (8.44).

The Crank-Nicolson and 3LFI finite difference schemes are of comparable accuracy particularly on a fine grid. This is to be expected since both schemes have second-order truncation errors. The very accurate 3LFI solution on a coarse grid is considered to be an anomaly. The two-level fully implicit scheme (2LFI) is considerably less accurate, and this is consistent with its truncation error of $O(\Delta t, \Delta x^2, \Delta y^2)$. The Crank-Nicolson finite element scheme has a second-order truncation error and demonstrates comparable accuracy to the Crank-Nicolson finite difference scheme. However, a Crank-Nicolson scheme with $\delta = 1/12$ in (8.44) produces a significantly more accurate solution on both coarse and fine grids. This is to be expected since this scheme has a truncation error of $O(\Delta t^2, \Delta x^4, \Delta y^4)$.

The schemes shown in Table 8.2 are coded as nine-point schemes. However, subroutine REDIF can be evaluated more efficiently for finite difference schemes by recognising that M_x and M_y only contain one non-zero entry each. Consequently, the non-zero contributions to DMX and DMY in subroutine REDIF can be grouped into three-component vectors.

Although the finite element scheme is formally a nine-point scheme, the symmetry of the mass operators (8.44) and the requirement of evaluating (8.46) only on sweeps in the x-direction permit the following economical procedure to be employed. First (8.46) is rewritten as

$$RHS = \sum_{l'=1}^{3} \{ L_{xxl'} \, MYT_{l'} + M_{xl'} \, LYT_{l'} \} , \quad \text{where} \tag{8.49}$$

$$MYT_{l'} = \left(\frac{\alpha \Delta t}{1+\gamma} \right) \sum_{m=k-1}^{k+1} M_{ym'} \, T_{l,m}^n , \tag{8.50}$$

$$LYT_{l'} = \left(\frac{\alpha \Delta t}{1+\gamma} \right) \sum_{m=k-1}^{k+1} L_{yym'} \, T_{l,m}^n + \left(\frac{\gamma}{1+\gamma} \right) \sum_{m=k-1}^{k+1} M_{ym'} \, \Delta T_{l,m}^n \tag{8.51}$$

```
 1 C
 2 C      TWDIF APPLIES APPROXIMATION FACTORISATION TO SOLVE
 3 C      THE UNSTEADY HEAT CONDUCTION PROBLEM  FOR T(X,Y) .
 4 C      REDIF EVALUATES THE RIGHT-HAND SIDE.
 5 C
 6        DOUBLE PRECISION SUMT,RMST,DSQRT,AN
 7        DIMENSION T(41,41),DT(41,41),R(41,41),EMX(3),EMY(3),
 8       1B(5,41),RRT(41),DDT(41),SIX(41),SIY(41),Y(41),ERR(41,41)
 9        COMMON DX,DY,EMX,EMY,NX,NY,R,T,DT
10        OPEN(1,FILE='TWDIF.DAT')
11        OPEN(6,FILE='TWDIF.OUT')
12        READ(1,1)IBC,NX,NY,ME,GAM,BET
13        READ(1,2)ALF,DTIM,DTMCH,TMAX
14      1 FORMAT(4I5,2F5.2)
15      2 FORMAT(4E10.3)
16 C
17        PI = 3.141592654
18        PIH = 0.5*PI
19        NXS = NX + 1
20        NYS = NY + 1
21        NXP = NX - 1
22        NYP = NY - 1
23        IF(IBC .EQ. 1)NXN = NXP
24        IF(IBC .EQ. 1)NYN = NYP
25        IF(IBC .EQ. 2)NXN = NX
26        IF(IBC .EQ. 2)NYN = NY
27        NXPP = NXN - 1
28        NYPP = NYN - 1
29        AN = NXPP*NYPP
30        ANX = NXP
31        DX = 1./ANX
32        ANY = NYP
33        DY = 1./ANY
34        SX = ALF*DTIM/DX/DX
35        SY = ALF*DTIM/DY/DY
36        EMX(1) = 0.
37        IF(ME .EQ. 2)EMX(1) = 1./6.
38        IF(ME .EQ. 3)EMX(1) = 1./12.
39        EMX(2) = 1. - 2.*EMX(1)
40        EMX(3) = EMX(1)
41        DO 3 J = 1,3
42      3 EMY(J) = EMX(J)
43        WRITE(6,4)NX,NY,ME,GAM,BET,SX,SY
44      4 FORMAT(' UNSTEADY HEAT CONDUCTION WITH NX,NY =',2I3,/,
45       1' ME =',I3,' GAM =',F5.2,' BETA =',F5.2,' SX,SY =',2F6.3)
46        IF(ME .EQ. 1)WRITE(6,5)EMX
47      5 FORMAT(' APPROX. FACT.,   3PT FDM,  EMX=',3E10.3)
48        IF(ME .EQ. 2)WRITE(6,6)EMX
49      6 FORMAT(' APPROX. FACT.,  LINEAR FEM,  EMX=',3E10.3)
50        IF(IBC .EQ. 1)WRITE(6,7)
51        IF(IBC .EQ. 2)WRITE(6,8)
52      7 FORMAT(' DIRICHLET B.C.')
53      8 FORMAT(' DIRICHLET AND NEUMANN B.C.')
54        WRITE(6,9)DTIM,DTMCH,TMAX
55      9 FORMAT(' DTIM =',E10.3,' PRINT INT. =',E10.3,' TMAX =',E10.3,//)
56 C
57 C      GENERATE INITIAL SOLUTION
58 C
59        DO 11 J = 1,NX
60        AJ = J - 1
61        X = AJ*DX
62        SIX(J) = SIN(PIH*X)
```

Fig. 8.4. Listing of program TWDIF

```
63           DO 10 K = 1,NY
64           AK = K - 1
65           Y(K) = AK*DY
66           SIY(K) = SIN(PIH*Y(K))
67        10 T(K,J) = 80.*(Y(K) - SIX(J)*SIY(K)) + 20.
68        11 CONTINUE
69           IF(IBC .EQ. 1)GOTO 14
70           DO 12 J = 1,NX
71        12 T(NYS,J) = T(NYP,J) + 160.*DY
72           DO 13 K = 1,NYS
73        13 T(K,NXS) = T(K,NXP)
74        14 CONTINUE
75           GAMH = GAM
76           GAM = 0.
77           BETH = BET
78           BET = 0.5
79           TIM = 0.
80           TMCH = -0.00001 + DTMCH
81           DIM = - 0.5*ALF*PI*PI
82           DO 17 J = 1,NX
83           DO 15 K = 1,NY
84           DT(K,J) = 0.
85        15 R(K,J) = 0.
86           DO 16 K = 1,5
87        16 B(K,J) = 0.
88        17 CONTINUE
89        18 CCXA = BET*DTIM*ALF/DX/DX/(1.+GAM)
90           CCYA = BET*DTIM*ALF/DY/DY/(1.+GAM)
91           TFAC = EXP(DIM*(TIM+DTIM)) - EXP(DIM*TIM)
92 C
93           CALL REDIF(ALF,GAM,DTIM,NXN,NYN)
94 C
95 C         TRIDIAGONAL SYSTEMS IN THE X-DIRECTION
96 C
97           DO 21 K = 2,NYN
98           DO 19 J = 2,NXN
99           JM = J - 1
100          B(2,JM) = EMX(1) - CCXA
101          B(3,JM) = EMX(2) + 2.*CCXA
102          B(4,JM) = EMX(3) - CCXA
103       19 RRT(JM) = R(K,J)
104          B(2,1) = 0.
105          IF(IBC .EQ. 1)DT(K,NX) = - 80.*SIY(K)*TFAC
106          IF(IBC .EQ. 1)DUM=(EMY(1)-CCYA)*SIY(K-1)+(EMY(2)+2.*CCYA)*SIY(K)
107        1 + (EMY(3)-CCYA)*SIY(K+1)
108          IF(IBC .EQ. 1)RRT(JM) = RRT(JM) + B(4,JM)*80.*TFAC*DUM
109          IF(IBC .EQ. 2)DT(K,NXS) = DT(K,NXP)
110          IF(IBC .EQ. 2)B(2,JM) = B(2,JM) + B(4,JM)
111          B(4,JM) = 0.
112 C
113          CALL BANFAC(B,NXPP,1)
114          CALL BANSOL(RRT,DDT,B,NXPP,1)
115 C
116          DO 20 J = 2,NXN
117          JM = J - 1
118       20 R(K,J) = DDT(JM)
119       21 CONTINUE
120 C
121 C        TRIDIAGONAL SYSTEMS IN THE Y-DIRECTION
122 C
123          DO 24 J = 2,NXN
124          DO 22 K = 2,NYN
125          KM = K - 1
```

Fig. 8.4. (cont.) Listing of program TWDIF

```
126          B(2,KM) = EMY(1) - CCYA
127          B(3,KM) = EMY(2) + 2.*CCYA
128          B(4,KM) = EMY(3) - CCYA
129       22 RRT(KM) = R(K,J)
130          B(2,1) = 0.
131          IF(IBC .EQ. 1)DT(NY,J) = - 80.*SIX(J)*TFAC
132          IF(IBC .EQ. 1)RRT(KM) = RRT(KM) - B(4,KM)*DT(NY,J)
133          IF(IBC .EQ. 2)DT(NYS,J) = DT(NYP,J)
134          IF(IBC .EQ. 2)B(2,KM) = B(2,KM) + B(4,KM)
135          B(4,KM) = 0.
136  C
137          CALL BANFAC(B,NYPP,1)
138          CALL BANSOL(RRT,DDT,B,NYPP,1)
139  C
140  C       INCREMENT T
141  C
142          DO 23 K = 2,NYN
143          KM = K - 1
144          DT(K,J) = DDT(KM)
145       23 T(K,J) = T(K,J) + DDT(KM)
146       24 CONTINUE
147          IF(IBC .EQ. 1)DT(NY,NX) = - 80.*TFAC
148          IF(IBC .EQ. 2)DT(NYS,NXS) = DT(NYP,NXP)
149          TIM = TIM + DTIM
150          TFAC = EXP(DIM*TIM)
151          IF(IBC .EQ. 2)GOTO 26
152          DO 25 K = 1,NY
153          T(K,NX) = 20. + 80.*(Y(K) - TFAC*SIY(K))
154       25 T(NY,K) = 20. + 80.*(1. - TFAC*SIY(K))
155          GOTO 29
156       26 DO 27 J = 1,NX
157       27 T(NYS,J) = T(NYP,J) + 160.*DY
158          DO 28 K = 1,NYS
159       28 T(K,NXS) = T(K,NXP)
160       29 GAM = GAMH
161          BET = BETH
162          IF(TIM .GE. TMCH)GOTO 30
163          IF(TIM .LE. TMAX)GOTO 18
164          GOTO 37
165  C
166  C       COMPARE WITH EXACT SOLUTION
167  C
168       30 TMCH = TMCH + DTMCH
169          SUMT = 0.
170          DO 33 KA = 1,NY
171          K = 1 + NY - KA
172          DO 31 J = 1,NX
173          TE = 20. + 80.*(Y(K) - TFAC*SIX(J)*SIY(K))
174          ERR(K,J) = T(K,J) - TE
175       31 SUMT = SUMT + (T(K,J) - TE)**2
176          WRITE(6,32)(T(K,J),J=1,NX)
177       32 FORMAT(' T=',11F7.3)
178       33 CONTINUE
179          DO 34 KA = 1,NY
180          K = 1 + NY - KA
181       34 WRITE(6,35)(ERR(K,J),J=1,NX)
182       35 FORMAT(' E=',11F7.3)
183          RMST = DSQRT(SUMT/AN)
184          WRITE(6,36)TIM,RMST
185       36 FORMAT(' TIME =',E11.4,'     RMST =',D11.4)
186          IF(TIM .LE. TMAX)GOTO 18
187       37 CONTINUE
188          STOP
189          END
```

Fig. 8.4. (cont.) Listing of program TWDIF

```
1          SUBROUTINE REDIF(ALF,GAM,DTIM,NXN,NYN)
2  C
3  C       EVALUATES RIGHT-HAND SIDE OF THE
4  C       UNSTEADY HEAT CONDUCTION EQUATION
5  C
6          DIMENSION DT(41,41),T(41,41),R(41,41),
7         1DMX(3,3),DMY(3,3),EMX(3),EMY(3)
8          COMMON DX,DY,EMX,EMY,NX,NY,R,T,DT
9          CCX = ALF/DX/DX
10         CCY = ALF/DY/DY
11         DO 1 I = 1,3
12         DMX(I,1) = CCX*EMY(I)
13         DMX(I,2) = -2.*DMX(I,1)
14         DMX(I,3) = DMX(I,1)
15         DMY(1,I) = CCY*EMX(I)
16         DMY(2,I) = -2.*DMY(1,I)
17       1 DMY(3,I) = DMY(1,I)
18  C
19         DO 5 J = 2,NXN
20         DO 4 K = 2,NYN
21         RTD = 0.
22         DO 3 M = 1,3
23         MJ = J - 2 + M
24         DO 2 N = 1,3
25         NK = K - 2 + N
26       2 RTD = RTD + (DMX(N,M)+DMY(N,M))*T(NK,MJ)*DTIM
27       1 + GAM*EMX(M)*EMY(N)*DT(NK,MJ)
28       3 CONTINUE
29       4 R(K,J) = RTD/(1. + GAM)
30       5 CONTINUE
31         RETURN
32         END
```

Fig. 8.5. Listing of subroutine REDIF

Table 8.1. Parameters used by program TWDIF

Parameter	Description
ALF	$\alpha = \alpha_x = \alpha_y$ in (8.1)
GAM, BET	γ, β in (8.43)
EMX, EMY	M_x, M_y in (8.44)
IBC	= 1, Dirichlet boundary conditions everywhere
	= 2, Neumann boundary conditions at $x = 1$ and $y = 1$
ME	= 1, finite difference method;
	= 2, linear finite element method
NX, NY	maximum number of grid points in the x and y directions
DX, DY	Δx, Δy
SX, SY	s_x, s_y after (8.5)
DTIM	Δt
DTMCH	increment of time for printing output
TIM	time
TMCH	comparative time for determining printing
TMAX	maximum time
T	nodal temperature, $T^n_{j,k}$ in (8.43)
TE	$\bar{T}$ in (8.42)
DT	$\Delta T^n_{j,k}$ in (8.46)
DDT	$\Delta T^*_{j,k}$ and $\Delta T^{n+1}_{j,k}$ on return from BANSOL
B	tridiagonal matrix; left-hand sides of (8.45) and (8.47)
R	RHS^n in (8.46); also used to store $\Delta T^*_{j,k}$ in (8.47)
RMST	$\lVert T - \bar{T} \rVert_{\mathrm{rms}}$

```
UNSTEADY HEAT CONDUCTION WITH NX,NY = 6   6
ME =   2 GAM =      .00 BETA =   .50 SX,SY = 1.000 1.000
APPROX. FACT., LINEAR FEM,    EMX=  .167E+00  .667E+00    .167E+00
DIRICHLET B.C.
DTIM =  .400E+04 PRINT INT. =           .160E+05 TMAX =  .161E+05


T=100.000 88.775 78.650 70.614 65.455 63.677
T= 84.000 73.351 63.742 56.109 51.193 49.455
T= 68.000 58.954 50.790 44.300 40.109 38.614
T= 52.000 45.433 39.505 34.790 31.742 30.650
T= 36.000 32.549 29.433 26.954 25.351 24.775
T= 20.000 20.000 20.000 20.000 20.000 20.000
E=    .000    .000    .000    .000    .000  .000
E=    .000    .026    .047    .057    .047  .000
E=    .000    .035    .063    .074    .057  .000
E=    .000    .031    .054    .063    .047  .000
E=    .000    .017    .031    .035    .026  .000
E=    .000    .000    .000    .000    .000  .000
TIME =  .1600E+05 RMST =      .471OD-01
```

Fig. 8.6. Typical output produced by program TWDIF

Table 8.2. rms errors for Dirichlet boundary conditions at $t = 24,000$ with $\alpha = 1 \times 10^{-5}$ and $s_x = s_y = 1.00$

$\Delta x = \Delta y$	Crank-Nicolson, FDM(ME=1) $\gamma = 0,\ \beta = 0.50$	2LFI FDM(ME=1) $\gamma = 0,\ \beta = 1.00$	3LFI FDM(ME=1) $\gamma = 0.5,\ \beta = 1.00$	Crank-Nicolson FEM(ME=2) $\gamma = 0,\ \beta = 0.50$	Crank-Nicolson $\delta = 1/12$ $\gamma = 0,\ \beta = 0.50$
0.20	0.02682	0.40880	0.00306	0.03364	0.002820
0.10	0.00648	0.08874	0.00528	0.00686	0.000134
0.05	0.00157	0.02065	0.00150	0.00159	0.000082

and $m' = m - k + 2$, $l' = l - j + 2$. MYT and LYT are three-component arrays associated with grid points $j - 1, j$ and $j + 1$ when RHS is centred at node (j, k). The structure of (8.49) follows from evaluating all y operators in (8.46) first.

However, the evaluation of MYT and LYT from (8.50, 51) is the same whether the Galerkin node is located at $j - 1, j$ or $j + 1$. Consequently, it is more efficient to evaluate only MYT_3 using (8.50) and only LYT_3 using (8.51). Values of MYT_2 and MYT_1 required in (8.49) are obtained by passing along the values of MYT_3 and MYT_2 from the evaluation at the previous Galerkin node, i.e. $j - 1$ instead of j. LYT_2 and LYT_1 are obtained in an equivalent manner.

The above technique has been used by Fletcher and Srinivas (1984) in obtaining the incompressible viscous flow past a backward-facing step using the stream function vorticity formulation described in Sect. 17.3.3.

8.4 Neumann Boundary Conditions

In Sects. 8.2 and 8.3, the splitting schemes are described and implemented with Dirichlet boundary conditions. For the one-dimensional diffusion equation first- and second-order implementations of Neumann boundary conditions are considered in Sect. 7.3.1.

Here, procedures will be described for combining Neumann boundary conditions with the splitting schemes of Sects. 8.2 and 8.3. A typical multidiffusion problem has mixed Dirichlet/Neumann boundary conditions. To provide a concrete example, the first and third boundary conditions in (8.2) are retained but the second and fourth boundary conditions are replaced with

$$\frac{\partial \bar{T}}{\partial x}(1, y, t) = g(y, t) \quad \text{and} \quad \frac{\partial \bar{T}}{\partial y}(x, 1, t) = h(x, t) , \tag{8.52}$$

where $g(y, t)$ and $h(x, t)$ are known functions.

8.4.1 Finite Difference Implementation

The various schemes considered in Sects. 8.2 and 8.3 are spatially second-order accurate. A second-order accurate implementation of the Neumann boundary conditions (8.52) is

$$\frac{T_{j+1,k} - T_{j-1,k}}{2\Delta x} = g_k(t) \quad \text{and} \quad \frac{T_{j,k+1} - T_{j,k-1}}{2\Delta y} = h_j(t) . \tag{8.53}$$

The boundary conditions (8.53) will be implemented in conjunction with the generalised two-level scheme, Sect. 8.2.2.

For points lying on the boundaries $x = 1$ and $y = 1$, the right hand side of (8.23) can be evaluated after (8.53) has been used to generate additional points $T_{j+1,k}$ and $T_{j,k+1}$ that lie just outside the computational domain proper. However, the operators on the left hand sides of (8.23 and 24) require the construction of corrections $\Delta T^*_{j+1,k}$ and $\Delta T^{n+1}_{j,k+1}$. These can be obtained from (8.53) by evaluation at successive time intervals, i.e.

$$\Delta T^{n+1}_{j+1,k} = \Delta T^{n+1}_{j-1,k} + 2\Delta x \, \Delta g^{n+1}_k , \quad \text{where} \tag{8.54a}$$

$$\Delta g^{n+1}_k = g^{n+1}_k - g^n_k \quad \text{and} \tag{8.54b}$$

$$\Delta T^{n+1}_{j,k+1} = \Delta T^{n+1}_{j,k-1} + 2\Delta y \, \Delta h^{n+1}_j . \tag{8.54c}$$

The splitting into (implicit) one-dimensional operators then requires modification of the relevant components of L_{xx} and L_{yy}, using (8.54). Thus, at $j = NX$ the introduction of (8.54a) leads to the following equation replacing (8.23) during the first stage:

$$(1 - \alpha_x \beta \Delta t L^m_{xx}) \Delta T^*_{j,k} = \Delta t (\alpha_x L_{xx} + \alpha_y L_{yy}) T^n_{j,k}$$

$$- 2\Delta x (1 - \alpha_x \beta \Delta t L_{xx3})(1 - \alpha_y \beta \Delta t \, L_{yy}) \Delta g^{n+1}_k , \tag{8.55}$$

where $L^m_{xx} = \{2, -2, 0\}/\Delta x^2$. However, during the second stage, (8.24) is used without modification. At $k = NY$, (8.23) is used without modification during the first stage. However, during the second stage, (8.24) is replaced by

$$(1 - \alpha_y \beta \Delta t L_{yy}^m) \Delta T_{j,k}^{n+1} = \Delta T_{j,k}^* - (1 - \alpha_y \beta \Delta t L_{yy3}) 2 \Delta y \Delta h_j^{n+1} \tag{8.56}$$

where $L_{yy}^m \equiv \{0, -2, 2\}^T / \Delta y^2$.

The above procedures have been applied to the example considered in Sect. 8.3.2 with $\partial \bar{T}/\partial x(1, y, t) = 0$ and $\partial \bar{T}/\partial y(x, 1, t) = 80$, replacing the relevant Dirichlet boundary conditions. However, for this particular choice of boundary conditions g and h in (8.52) are constants so that the right hand sides of (8.55 and 56) coincide with the right hand sides of (8.23 and 24). For the present case, the exact solution is given by (8.42). Solutions have been obtained with the schemes indicated in Sect. 8.3.2; the rms errors are summarised in Table 8.3.

Table 8.3. rms errors for Dirichlet/Neumann boundary conditions at $t = 24,000$ with $\alpha = 1 \times 10^{-5}$ and $s_x = s_y = 1.00$

$\Delta x (= \Delta y)$	Crank-Nicolson FDM(ME=1) $\gamma = 0, \beta = 0.5$	2LFI FDM(ME=1) $\gamma = 0, \beta = 1.0$	3LFI FDM(ME=1) $\gamma = 0.5, \beta = 1.0$	Crank-Nicolson FEM(ME=2) $\gamma = 0, \beta = 0.5$	Crank-Nicolson $\delta = 1/12$ $\gamma = 0, \beta = 0.5$
0.20	0.12940	1.58100	0.03985	0.15720	0.01341
0.10	0.03199	0.41120	0.02640	0.03358	0.00070
0.05	0.00779	0.10080	0.00744	0.00788	0.00021

Trends are similar to those associated with Dirichlet boundary conditions (Table 8.2) except that the rms error levels are larger for all methods when Neumann boundary conditions must be satisfied. A typical error distribution is shown in Fig. 8.7. It is clear that the largest errors are occurring on and adjacent to boundaries where Neumann boundary conditions are applied. A similar effect was apparent (Table 7.9) with the one-dimensional diffusion equation and with the Sturm-Liouville problem (Sect. 5.4). The largest errors when all boundary

```
UNSTEADY HEAT CONDUCTION WITH NX,NY =  6  6
ME =  1 GAM =  .00 BETA =  .50 SX,SY = 1.000 1.000
APPROX. FACT.,  3PT FDM, EMX=  .000E+00  .100E+01  .000E+00
DIRICHLET AND NEUMANN B.C.
DTIM =  .400E+04 PRINT INT. =  .240E+05 TMAX =  .250E+05


T=100.000 92.370 85.487 80.024 76.517 75.309
T= 84.000 76.743 70.197 65.002 61.667 60.517
T= 68.000 61.827 56.259 51.839 49.002 48.024
T= 52.000 47.515 43.469 40.259 38.197 37.487
T= 36.000 33.642 31.515 29.827 28.743 28.370
T= 20.000 20.000 20.000 20.000 20.000 20.000
E=   .000  -.067  -.127  -.174  -.205  -.216
E=   .000  -.063  -.121  -.166  -.195  -.205
E=   .000  -.054  -.103  -.141  -.166  -.174
E=   .000  -.039  -.074  -.103  -.121  -.127
E=   .000  -.021  -.039  -.054  -.063  -.067
E=   .000   .000   .000   .000   .000   .000
TIME =  .2400E+05      RMST =  .1294D+00
```

Fig. 8.7. Error distribution with Dirichlet/Neumann boundary conditions

conditions are of Dirichlet type occur in the interior at the points furthest from the boundaries.

An interesting feature of the results shown in Table 8.3 is the high accuracy associated with the "fourth-order" interior scheme, $\delta = 1/12$ in (8.44). It may be recalled that, for the one-dimensional diffusion equation (Sect. 7.3.2), the introduction of a second-order accurate boundary condition limited the accuracy even if a higher-order interior scheme was used.

8.4.2 Finite Element Implementation

For governing equations, such as (8.1), in which second derivatives are present the implementation of Neumann boundary conditions enters into the finite element discretisation process. This was illustrated for the one-dimensional Sturm-Liouville problem in Sect. 5.4.1.

Application of the Galerkin finite element method to (8.1) produces the following weighted integral statement (Sect. 5.1):

$$\iint \phi_m \frac{\partial \bar{T}}{\partial t}\, dx\, dy = \alpha \iint \phi_m \left(\alpha_x \frac{\partial^2 \bar{T}}{\partial x^2} + \alpha_y \frac{\partial^2 \bar{T}}{\partial y^2} \right) dx\, dy \ . \tag{8.57}$$

In contrast to the development given in Sect. 5.1, the approximate solution for T has not yet been introduced in (8.57). Here ϕ_m is the weight function associated with the Galerkin node. Two cases are of interest in relation to implementing Neumann boundary conditions. First, m is a node on the boundary $x = 1$ (Fig. 8.1); second, m is a node on the boundary $y = 1$. It will be assumed that ϕ_m is a bilinear interpolating function (Sect. 5.3) although the approach is applicable to any order of interpolation. Because of the local nature of $\phi_m(x, y)$ only two elements make a non-zero contribution to the eventual evaluation of (8.57). The two cases of interest are shown in Fig. 8.8.

The terms on the right hand side of (8.57) are integrated by parts. That is,

$$\iint \phi_m \frac{\partial^2 T}{\partial x^2}\, dx\, dy = \int \left[\phi_m \frac{\partial T}{\partial x} \right]_L^R dy - \iint \frac{\partial \phi_m}{\partial x} \frac{\partial T}{\partial x}\, dx\, dy \ , \tag{8.58}$$

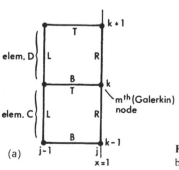

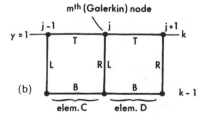

Fig. 8.8. Implementation of Galerkin finite element method at boundaries. (a) boundary $x = 1$; (b) boundary $y = 1$

$$\int\int \phi_m \frac{\partial^2 T}{\partial y^2}\,dx\,dy = \int\left[\phi_m \frac{\partial T}{\partial y}\right]_B^T dx - \int\int \frac{\partial \phi_m}{\partial y}\frac{\partial T}{\partial y}\,dx\,dy \ . \tag{8.59}$$

For the boundary $x=1$ the evaluation over the line integrals in (8.58 and 59) only produce a net non-zero contribution on boundary R in each element. The weighting function ϕ_m is zero on boundaries B and L of element C and L and T of element D. The line integral contribution from boundary B of element D identically cancels the contribution from boundary T of element C.

A conventional approximate solution is introduced for T appearing in the area integrals

$$T = \sum_l \phi_l(x,y)T_l \ . \tag{8.60}$$

From (8.52), $\partial T/\partial x|_{x=1} = g(t)$. A one-dimensional interpolation of $g(t)$ is introduced by

$$g(t) = \sum_q \phi_q^y(y)\,g_q(t) \ . \tag{8.61}$$

For Lagrange interpolation, $\phi_l(x,y) = \phi_p^x(x)\,\phi_q^y(y)$. Evaluation of the various integrals in (8.57–59) on a uniform grid and division by $\Delta x \Delta y$ gives

$$\left[M_x \otimes M_y \frac{\partial T}{\partial t}\right]_{j,k} = \left[\alpha_x\left(\frac{M_y g_k}{\Delta x} + M_y \otimes L_{xx}\,T_{j,k}\right) + \alpha_y\,M_x \otimes L_{yy}\,T_{j,k}\right] . \tag{8.62}$$

The structure of (8.62) is the same as the interior form (8.29) except for the additional term $\alpha_x M_y g_k/\Delta x$ associated with the Neumann boundary condition. But since only two elements contribute, the operators M_x and L_{xx} have the form

$$M_x = \left\{\frac{1}{6},\frac{1}{3},0\right\} \quad \text{and} \quad L_{xx} = \left\{\frac{1}{\Delta x^2}, -\frac{1}{\Delta x^2}, 0\right\} . \tag{8.63}$$

Operators M_y and L_{yy} are the same as in (8.30 and 31).

Implementation of the Neumann boundary condition on $y=1$ follows an equivalent path except that the line integrals in (8.58, 59) only produce a net non-zero contribution on boundaries T in elements C and D in Fig. 8.8b. The result equivalent to (8.62) is

$$\left[M_x \otimes M_y \frac{\partial T}{\partial t}\right]_{j,k} = \left[\alpha_x\,M_y \otimes L_{xx}\,T_{j,k} + \alpha_y\left(\frac{M_x h_j}{\Delta y} + M_x \otimes L_{yy}\,T_{j,k}\right)\right] . \tag{8.64}$$

In (8.64) operators M_x and L_{xx} are given by (8.30 and 31) but M_y and L_{yy} have the components

$$M_y = \left\{0,\frac{1}{3},\frac{1}{6}\right\}^T \quad \text{and} \quad L_{yy} = \left\{0, -\frac{1}{\Delta y^2}, \frac{1}{\Delta y^2}\right\}^T . \tag{8.65}$$

It is of interest to compare the treatment of Neumann boundary conditions by the finite difference and finite element methods. The equivalent finite difference form to (8.29) is

$$
\left[\frac{\partial T}{\partial t}\right]_{j,k} = \alpha_x\, L_{xx}\, T_{j,k} + \alpha_y\, L_{yy}\, T_{j,k}\ , \tag{8.66}
$$

where L_{xx} and L_{yy} are given by (8.31). Introduction of (8.53) on $x = 1$ produces the following local form for (8.66):

$$
\left[\frac{\partial T}{\partial t}\right]_{j,k} = 2\alpha_x\left(\frac{g_k}{\varDelta x} + L_{xx}\, T_{j,k}\right) + \alpha_y\, L_{yy}\, T_{j,k}\ , \tag{8.67}
$$

where L_{xx} is now given by (8.63) and L_{yy} by (8.31) as before.

In introducing the concept of mass operators it has been noted (Sects. 5.5 and 8.3.1) that an equivalent finite difference representation can be constructed by lumping the mass operators. Thus, in (8.62) the mass operators would be

$$
M_x^{\mathrm{fd}} = \{0, 0.5, 0\} \quad \text{and} \quad M_y^{\mathrm{fd}} = \{0, 1, 0\}^T\ . \tag{8.68}
$$

The resulting form of (8.62) is identical with (8.67). Thus, for bilinear Lagrange interpolation on rectangular finite elements, it is possible to implement Neumann boundary conditions with the Galerkin finite element method by introducing an additional set of points $T_{j+1,k}$ and $T_{j,k+1}$ given by (8.53) and to apply an interior formula, such as (8.29), throughout. It should be stressed that although this approach is expedient for coding efficiency, it should only be used *after* the equivalence has been demonstrated.

Typical results using the finite element method with Neumann boundary $\partial \bar{T}/\partial x(1, y, t) = 0$ and $\partial \bar{T}/\partial y(x, 1, t) = 80$, are shown in Table 8.3. The trends are similar to those with Dirichlet boundary conditions except that the errors are generally larger when Neumann boundary conditions are present. This same trend was apparent with the finite difference method.

8.5 Method of Fractional Steps

For the implicit methods described in this chapter the overall strategy has been to discretise and then to manipulate or modify the resulting algebraic equations to generate the 'one-dimensional' algorithms like (8.27 and 28).

An alternative strategy to the above is to split the governing equation, for example (8.1) into a pair of equations, each of which is locally one dimensional. Thus, (8.1) is replaced by

$$
0.5\frac{\partial \bar{T}}{\partial t} - \alpha_y\frac{\partial^2 \bar{T}}{\partial y^2} = 0 \quad \text{and} \tag{8.69}
$$

$$0.5\frac{\partial \bar{T}}{\partial t}-\alpha_x\frac{\partial^2 \bar{T}}{\partial x^2}=0 \ . \tag{8.70}$$

The equations are discretised and solved sequentially at each time-step. This class of methods was developed by Soviet mathematicians and is described in detail by Yanenko (1971) and Marchuk (1974). Mitchell and Griffiths (1980, pp. 70–74) refer to this class of methods as locally one-dimensional methods.

An explicit implementation of (8.69, 70) is

$$T_{j,k}^{n+1/2}=(1+\alpha_y \Delta t\, L_{yy})\, T_{j,k}^n \quad \text{and} \tag{8.71}$$

$$T_{j,k}^{n+1}=(1+\alpha_x \Delta t\, L_{xx})\, T_{j,k}^{n+1/2} \ . \tag{8.72}$$

This scheme coincides with (8.8 and 9) if $\alpha=\alpha_x=\alpha_y$. The algorithm (8.71, 72) has a truncation error of $O(\Delta t, \Delta x^2, \Delta y^2)$ and, if $\Delta x=\Delta y$, is stable for $s\le 0.5$.

A Crank-Nicolson (implicit) implementation of (8.69 and 70) is

$$(1-0.5\alpha_y \Delta tL_{yy})\, T_{j,k}^{n+1/2}=(1+0.5\,\alpha_y\,\Delta tL_{yy})\, T_{j,k}^n \tag{8.73}$$

and

$$(1-0.5\alpha_x \Delta tL_{xx})\, T_{j,k}^{n+1}=(1+0.5\,\alpha_x \Delta tL_{xx})\, T_{j,k}^{n+1/2} \ . \tag{8.74}$$

Equations (8.73, 74) lead to tridiagonal systems of equations along y and x gridlines respectively; consequently, the solution can be advanced in time economically using the Thomas algorithm. The scheme (8.73, 74) is second-order accurate in time and space with appropriate boundary condition specification and unconditionally stable in two and three dimensions.

Equations (8.73 and 74) can be combined into a single composite scheme by eliminating $T_{j,k}^{n+1/2}$ as long as the L_{xx} and L_{yy} operators commute. That is, the same formula is produced by $L_{xx}L_{yy}\, T_{j,k}^{n+1/2}$ as by $L_{yy}L_{xx}\, T_{j,k}^{n+1/2}$. The resulting composite scheme coincides with the ADI composite scheme produced by eliminating $T_{j,k}^*$ from (8.12 and 13).

However, a major difference occurs in the treatment of boundary conditions. For the method of fractional steps applied in the two-dimensional domain shown in Fig. 8.1 a Dirichlet boundary condition on $x=1$, $T(1, y, t)=b(y, t)$ is implemented, traditionally, at the intermediate time-level as

$$T_{NX,k}^{n+1/2}=b^{n+1/2} \ , \tag{8.75}$$

and similarly for boundary conditions on other boundaries. This treatment effectively reduces the accuracy of the overall scheme to first order in time.

Dwoyer and Thames (1981) examine the problem of correctly implementing boundary conditions in conjunction with a two-dimensional transport equation (Sect. 9.5). Mitchell and Griffiths (1980) show that the correct boundary condition on $x=1$, when using (8.71 and 72), is

$$T_{NX,k}^{n+1/2}=(1+\alpha_y \Delta tL_{yy})\, b_k^n \ . \tag{8.76}$$

This suggests that, when using (8.73, 74), compatible intermediate Dirichlet boundary conditions on $x = 1$ may be obtained by solving

$$(1 - 0.5\,\alpha_y \Delta t L_{yy})\, T_{NX,k}^{n+1/2} = (1 + 0.5\,\alpha_y \Delta t L_{yy})\, b_k^n \ . \tag{8.77}$$

It may be noted that the method of fractional steps does not provide an economical algorithm in terms of the corrections $\Delta T_{j,k}^{n+1}$, as was possible with approximate factorisation (Sect. 8.2.2 and following). Also, the method of fractional steps does not provide a direct evaluation of the steady-state residual (8.25), which is important when solving steady problems with a pseudotransient formulation (Sect. 6.4).

8.6 Closure

Multidimensional explicit methods are appropriate if small time-steps are necessary to obtain a sufficiently accurate solution. In addition it is often easier to apply a multidimensional explicit method efficiently on a parallel-processing computer.

For multidimensional parabolic partial differential equations, e.g., the diffusion equation, implicit schemes are often more effective than explicit schemes, primarily due to their inherently more stable behaviour, particularly if the accuracy of the spatial solution is more critical than that of the temporal solution. Additionally, an implicit formulation provides more flexibility for constructing higher-order schemes (Mitchell and Griffiths 1980, p. 61).

To retain the economy of the Thomas algorithm with a multidimensional implicit formulation, it is necessary to introduce some form of directional splitting. The recommended construction is to cast the equations as a linear system for the correction $\Delta T_{j,k}^{n+1}$ and to introduce an approximate factorisation, e.g. (8.22), which permits a multistage algorithm to be applied with each stage requiring the solution of a tridiagonal system of equations.

Approximate factorisation is effective with both the finite difference and finite element methods. The appearance of directional mass operators in the finite element approximate factorisation algorithm (8.39, 40) provides a means of obtaining a spatially more accurate scheme; that is, by choosing $\delta = \frac{1}{12}$ in (8.44). The higher accuracy is achieved with both Dirichlet (Sect. 8.3) and Neumann (Sect. 8.4) boundary conditions.

However, to maintain second-order temporal accuracy special attention must be given to the implementation of the boundary conditions for the intermediate solution correction $\Delta T_{j,k}^*$ that arises with the approximate factorisation construction. Although the finite difference and finite element methods handle Neumann boundary conditions in conceptually different ways, the form of the discretised equations are often structurally equivalent.

The splitting (approximate factorisation) techniques developed in this chapter are applicable, with minor modification, to the two-dimensional transport equation (Sect. 9.5), the two-dimensional Burgers' equations (Sect. 10.4) and the equations governing various classes of fluid flow, particularly when the Navier-Stokes equations are to be solved, e.g. Sects. 17.2.1, 17.3.3, 18.3 and 18.4.

8.7 Problems

Two-Dimensional Diffusion Equation (Sect. 8.1)

8.1 Apply the von Neumann stability analysis to (8.7) and show that the scheme is stable if $0 < s \le 0.5$.

8.2 For $\Delta x = \Delta y$ deduce what truncation error (8.7) has if $s = 1/6$.

8.3 Modify program TWDIF to implement (8.7) and verify the theoretical results of Problems 8.1 and 8.2.

Multidimensional Splitting Methods (Sect. 8.2)

8.4 Apply the von Neumann stability analysis to a three-dimensional equivalent of the ADI scheme (8.14 and 15) and the approximate factorisation scheme (8.23 and 24). Show that the three-dimensional ADI scheme is conditionally stable but that the approximate factorisation scheme is unconditionally stable.

8.5 Modify program TWDIF to implement the ADI scheme (8.14 and 15) and compare the accuracy and economy with that of the approximate factorisation scheme $ME = 1$, $\gamma = 0$, $\beta = 0.5$.

8.6 Repeat the comparison shown in Table 8.2 for $s_x = s_y = 0.5$ and 1.5, and explain the results. For $s_x = s_y = 0.5$ compare the results with the scheme developed in Problem 8.3.

Splitting Schemes and the Finite Element Method (Sect. 8.3)

8.7 Modify the program developed in Problem 8.5 to implement ADI-FEM (8.35, 36) and compare the accuracy with that of the solutions presented in Table 8.2.

8.8 Show that the scheme (8.43, 44) with $\gamma = 0$, $\beta = 0.5$ is equivalent to the scheme

$$[1 - (0.5\alpha\Delta t - \delta\Delta x^2)L_{xx}]T^*_{j,k} = [1 + (0.5\alpha\Delta t + \delta\Delta y^2)L_{yy}]T^n_{j,k} ,$$

$$[1 - (0.5\alpha\Delta t - \delta\Delta y^2)L_{yy}]T^{n+1}_{j,k} = [1 + (0.5\alpha\Delta t + \delta\Delta x^2)L_{xx}]T^*_{j,k} .$$

For the special case $\delta = 1/12$, this scheme coincides with the spatially fourth-order scheme of Mitchell and Fairweather (1964).

8.9 Replace (8.48) with $\Delta T^*_{NX,k}$ evaluated at $t_{n+1/2}$ from (8.41) in program TWDIF. Compare the solution accuracy with that given in Table 8.2.

Neumann Boundary Conditions (Sect. 8.4)

8.10 Apply program TWDIF with Neumann boundary conditions to obtain solutions for the methods given in Table 8.3 for the cases $s_x = s_y = 0.5$ and 1.5. Discuss the results.

8.11 Modify the program developed in Problem 8.3 to obtain the solution with Neumann boundary conditions comparable to the case considered in Table 8.3 with $s_x = s_y = 0.5$.

8.12 Modify the ADI program developed in Problem 8.5 to solve the same test case as Table 8.3 is based on. Compare the solution accuracy with that given in Table 8.3.

Method of Fractional Steps (Sect. 8.5)

8.13 Modify the program developed in Problem 8.5 to implement (8.73 and 74) and test on the case considered in Table 8.2.

8.14 Modify the program developed in Problem 8.12 to implement the Dirichlet boundary condition on $x = 1$ using (8.77) and compare the accuracy of the solutions with that of the conventional method of fractional steps.

9. Linear Convection-Dominated Problems

For most flow problems the motion of the fluid is an important determining factor of the overall flow behaviour. In the equations governing fluid dynamics (Chap. 11) the fluid motion is characterised by the velocity components u, v, w in the x, y, z (Cartesian coordinates) directions. In the one-dimensional x-momentum equation

$$\varrho\left(\frac{\partial u}{\partial t}+u\frac{\partial u}{\partial x}\right)+\frac{\partial p}{\partial x}=\mu\frac{\partial^2 u}{\partial x^2} \tag{9.1}$$

the velocity component u appears in the inertia term $(\partial u/\partial t + u\partial u/\partial y)$ and the viscous diffusion term $\mu(\partial^2 u/\partial x^2)$. The other terms are density (ϱ), pressure (p) and (constant) viscosity (μ).

Previously, in considering the diffusion equation, we examined the behaviour of terms like $\partial^2 u/\partial x^2$. Now we consider convective terms, like $u\partial u/\partial x$, and how they can be best handled, computationally.

The convective term has two separate features that must be taken into account. To begin with it contains first derivatives in the spatial coordinate. If a symmetric three-point formula is used to represent the term $\partial u/\partial x$, non-physical oscillations develop in the solution if the viscous term is small compared with the convective term. This behaviour is related to the dispersion-like influences (Sect. 9.2) in the truncation error. For steady flow this phenomenon will be examined in Sect. 9.3.

If asymmetric algebraic formulae are used to represent $\partial u/\partial x$ then, although the smoothness of the resulting solution is often improved, the accuracy of the representation of $\partial u/\partial x$ is typically one order lower than the corresponding representation obtained with a symmetric algebraic formula spanning the same number of gridpoints. For low-order asymmetric formulae this can introduce terms in the truncation error that are of comparable magnitude to important physical terms being represented.

This aspect will be considered in Sect. 9.2 for the pure convection equation, in Sect. 9.3 for the steady convection-diffusion equation and in Sect. 9.4 for the transport equation. The transport equation is formed from (9.1) by linearising the convective term (i.e. replacing $u\partial u/\partial x$ with $\varepsilon\partial u/\partial x$), dropping the pressure gradient term $\partial p/\partial x$ and assuming that ε and ϱ are known.

The second important feature of the convective term is that it is nonlinear in the dependent variable. For supersonic inviscid flow (Sect. 11.6.1) the nonlinear nature of the convective terms permits shock waves to appear. The nonlinear nature of the convective terms will be discussed, primarily in relation to Burgers' equation, in Chap. 10.

9.1   One-Dimensional Linear Convection Equation

To examine the problems of including convection the following linear equation will be considered:

$$\frac{\partial \bar{T}}{\partial t} + u \frac{\partial \bar{T}}{\partial x} = 0 \ , \tag{9.2}$$

where u is the known velocity and $\bar{T}$ is a passive scalar, e.g. temperature. Equation (9.2) is hyperbolic (Sect. 2.2) in character. Equation (9.2) can be interpreted as a model for the convective part of the energy equation (11.44).

If u is constant and positive then a general solution of (9.2) can be written as

$$\bar{T}(x,\, t) = F(x - ut) \ , \tag{9.3}$$

where the initial condition is given by

$$\bar{T}(x,\, 0) = F(x) \tag{9.4}$$

and $F(x)$ is known. If $F(x)$ is specified over the complete x range, $-\infty \le x \le \infty$, then the solution at some specific location (x_1, t_1) in the (x, t) plane is equal to the solution at $x_1 - ut_1$ at time $t = 0$, i.e.

$$\bar{T}(x_1 - t_1) = F(x_1 - ut_1) = \bar{T}(x_1 - ut_1,\, 0) \ .$$

This is illustrated in Fig. 9.1. The solution $\bar{T}$ is constant along lines such as AB, which is a characteristic for this equation (Fig. 2.5).

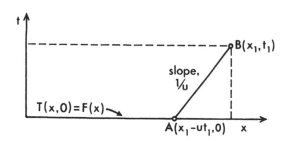

Fig. 9.1. Dependence of the solution on the initial data

9.1.1   FTCS Scheme

The simplest algorithm for the diffusion equation is the FTCS scheme (7.5). The corresponding forward time, centred space (FTCS) finite difference representation for (9.2) is

$$\frac{T_j^{n+1} - T_j^n}{\Delta t} + \frac{u(T_{j+1}^n - T_{j-1}^n)}{2\Delta x} = 0 \ . \tag{9.5}$$

Table 9.1. Algebraic (discretised) schemes for the convection equation $\dfrac{\partial \bar{T}}{\partial t} + u\dfrac{\partial \bar{T}}{\partial x} = 0$

Scheme	Algebraic form	Truncation error^a (E) (leading terms)	Amplification factor G ($\theta = m\pi\Delta x$)	Stability restrictions	Remarks
FTCS	$\dfrac{\Delta T_j^{n+1}}{\Delta t} + uL_x T_j^n = 0$	$Cu(\Delta x/2)\dfrac{\partial^2 T}{\partial x^2}$ $+u\left(\dfrac{\Delta x^2}{6}\right)(1+2C^2)\dfrac{\partial^3 T}{\partial x^3}$	$1 - iC\sin\theta$	unstable	$C = u\dfrac{\Delta t}{\Delta x}$ $L_x = \dfrac{1}{2\Delta x}\{-1,0,1\}$
Upwind	$\dfrac{\Delta T_j^{n+1}}{\Delta t} + u\dfrac{(T_j^n - T_{j-1}^n)}{\Delta x} = 0$	$-u\left(\dfrac{\Delta x}{2}\right)(1-C)\dfrac{\partial^2 T}{\partial x^2}$ $+u\left(\dfrac{\Delta x^2}{6}\right)(1-3C+2C^2)\dfrac{\partial^3 T}{\partial x^3}$	$1 - C(1-\cos\theta) - iC\sin\theta$	$C \leqq 1$	$\Delta T_j^{n+1} = T_j^{n+1} - T_j^n$
Leapfrog	$\dfrac{T_j^{n+1} - T_j^{n-1}}{2\Delta t} + uL_x T_j^n = 0$	$u\left(\dfrac{\Delta x^2}{6}\right)(1-C^2)\dfrac{\partial^3 T}{\partial x^3}$	$-iC\sin\theta \pm (1-C^2\sin^2\theta)^{\ddagger}$	$C \leqq 1$	

Table 9.1. (cont.)

Scheme	Algebraic form	Truncation error^a (E) (leading terms)	Amplification factor G ($\theta = m\pi\Delta x$)	Stability restrictions	Remarks
Lax-Wendroff	$\dfrac{\Delta T_j^{n+1}}{\Delta t} + uL_x T_j^n - 0.5uC\Delta x L_{xx} T_j^n$ $= 0$	$u\left(\dfrac{\Delta x^2}{6}\right)(1-C^2)\dfrac{\partial^3 T}{\partial x^3}$ $+ uC\left(\dfrac{\Delta x^3}{8}\right)(1-C^2)\dfrac{\partial^4 T}{\partial x^4}$	$1 - iC\sin\theta - 2C^2\sin^2\left(\dfrac{\theta}{2}\right)$	$C \leqq 1$	$L_{xx} = \left\{\dfrac{1}{\Delta x^2}, -\dfrac{2}{\Delta x^2}, \dfrac{1}{\Delta x^2}\right\}$
Crank-Nicolson	$\dfrac{\Delta T_j^{n+1}}{\Delta t} + uL_x\left(\dfrac{T_j^n + T_j^{n+1}}{2}\right) = 0$	$u\left(\dfrac{\Delta x^2}{6}\right)(1+0.5C^2)\dfrac{\partial^3 T}{\partial x^3}$	$\dfrac{(1-0.5iC\sin\theta)}{(1+0.5iC\sin\theta)}$	None	
Three-level fully implicit	$\dfrac{3}{2}\dfrac{\Delta T_j^{n+1}}{\Delta t} - \dfrac{1}{2}\dfrac{\Delta T_j^n}{\Delta t} + uL_x T_j^{n+1} = 0$	$u\left(\dfrac{\Delta x^2}{6}\right)(1+2C^2)\dfrac{\partial^3 T}{\partial x^3}$	$\dfrac{1 \pm \frac{1}{3}i(3 + i8C\sin\theta)^{\ddagger}}{2\left(1 + i\dfrac{2C}{3}\sin\theta\right)}$	None	
Linear F.E.M./ Crank-Nicolson	$M_x\dfrac{\Delta T_j^{n+1}}{\Delta t} + uL_x\left(\dfrac{T_j^n + T_j^{n+1}}{2}\right) = 0$	$C^2 u\left(\dfrac{\Delta x^2}{12}\right)\dfrac{\partial^3 T}{\partial x^3}$	$\dfrac{(2+\cos\theta - 1.5iC\sin\theta)}{(2+\cos\theta + 1.5iC\sin\theta)}$	None	$M_x = \left\{\dfrac{1}{6}, \dfrac{2}{3}, \dfrac{1}{6}\right\}$

^{a} The truncation error (E) has been expressed in terms of Δx and x-derivatives as in the modified equation approach (Sect. 9.2.2). Thus the algebraic scheme is equivalent to $\partial T/\partial t + u\partial T/\partial x + E(T) = 0$.

As an algorithm (9.5) can be written

$$T_j^{n+1} = T_j^n - 0.5C(T_{j+1}^n - T_{j-1}^n) \; , \tag{9.6}$$

which is consistent with (9.2) with a truncation error of $O(\Delta t, \Delta x^2)$. In (9.6), C is called the Courant number and is defined by

$$C = u \frac{\Delta t}{\Delta x} \; . \tag{9.7}$$

Application of the von Neumann stability analysis to (9.6) produces an amplification factor G for the mth component of the initial error distribution that is given by

$$G = 1 - iC\sin\theta \; . \tag{9.8}$$

Clearly, $|G| \geq 1$ for all θ so that (9.6) is unconditionally unstable. It may be recalled that the equivalent (FTCS) scheme was conditionally stable for the diffusion equation. It follows, from the lack of stability, that the FTCS scheme is of no practical use for pure convection problems. For completeness the properties of the FTCS scheme applied to (9.2) are summarised in Table 9.1.

9.1.2 Upwind Differencing and the CFL Condition

An alternative scheme is obtained by introducing a backward difference formula for $\partial \bar{T}/\partial x$ assuming that u is positive. Thus, (9.2) is discretised by

$$\frac{T_j^{n+1} - T_j^n}{\Delta t} + \frac{u(T_j^n - T_{j-1}^n)}{\Delta x} = 0 \; , \tag{9.9}$$

which as an algorithm, becomes

$$T_j^{n+1} = (1-C)T_j^n + CT_{j-1}^n \; . \tag{9.10}$$

For negative u (9.10) is replaced by $T_j^{n+1} = (1-|C|)T_j^n + |C| \, T_{j+1}^n$. Equation (9.10) indicates that the solution T_j^{n+1}, is determined by information upwind of node (j, n). Hence, (9.10) will be referred to as an upwind scheme (Table 9.1). Numerical solutions using the upwind scheme are provided in Sect. 9.1.5.

The von Neumann stability analysis applied to (9.10) gives the amplification factor shown in Table 9.1. Stable solutions are obtained if

$$C = u \frac{\Delta t}{\Delta x} \leq 1 \; . \tag{9.11}$$

The inequality $C \leq 1$ is referred to as the Courant-Friedrichs-Lewy (CFL) condition. This condition applies, generally, to explicit schemes for hyperbolic partial differential equations. Physically, the CFL condition indicates that a particle of fluid should not travel more than one spatial step-size Δx in one time-step Δt.

For the particular case $C=1$, (9.10) gives $T_j^{n+1}=T_{j-1}^n$, which is the exact solution of (9.2); this can be seen from Fig. 9.1.

A Taylor series expansion, about node (j, n), of the exact solution substituted into (9.10) produces the result

$$\frac{\partial \bar{T}}{\partial t}+u\frac{\partial \bar{T}}{\partial x}+0.5\Delta t\frac{\partial^2 \bar{T}}{\partial t^2}-0.5u\Delta x\frac{\partial^2 \bar{T}}{\partial x^2}+O(\Delta t^2, \Delta x^2)=0 \ . \tag{9.12}$$

Thus, (9.10) is consistent with (9.2) but has a truncation error, whose leading terms are $O(\Delta t, \Delta x)$. From (9.2)

$$\frac{\partial \bar{T}}{\partial t}=-u\frac{\partial \bar{T}}{\partial x} \quad \text{and} \quad \frac{\partial^2 \bar{T}}{\partial t^2}=u^2\frac{\partial^2 \bar{T}}{\partial x^2} \ ,$$

so that (9.12) can be written as

$$\frac{\partial \bar{T}}{\partial t}+u\frac{\partial \bar{T}}{\partial x}-0.5u\Delta x(1-C)\frac{\partial^2 \bar{T}}{\partial x^2}+O(\Delta t^2, \Delta x^2)=0 \ .$$

If (9.10) is interpreted as having a truncation error of $O(\Delta t^2, \Delta x^2)$, (9.10) is seen to be consistent with

$$\frac{\partial \bar{T}}{\partial t}+u\frac{\partial \bar{T}}{\partial x}-\alpha'\frac{\partial^2 \bar{T}}{\partial x^2}=0 \ , \tag{9.13}$$

instead of (9.2). Therefore, the use of the two-term, upwind finite difference representation for $\partial \bar{T}/\partial x$, combined with the forward difference formula for $\partial \bar{T}/\partial t$, is seen to introduce an artificial (numerical) diffusivity $\alpha'=0.5u\Delta x(1-C)$. Clearly, the artificial diffusivity is zero when $C=1$; this is not surprising since the solution given by (9.10) then coincides with the exact solution.

It might appear that setting Δt to give $C=1$ would avoid the artificial diffusivity problem. Although this is true for the linear convection equation, it is not possible to ensure $C=1$ everywhere for nonlinear equations, like Burgers' equation (Sect. 10.1), where u (in C) is spatially varying.

9.1.3 Leapfrog and Lax-Wendroff Schemes

Following the same approach as for the diffusion equation we seek a more accurate representation of (9.2) by introducing centred differences in both time and space. This gives the leapfrog method

$$\frac{T_j^{n+1}-T_j^{n-1}}{2\Delta t}+\frac{u(T_{j+1}^n-T_{j-1}^n)}{2\Delta x}=0 \ , \tag{9.14}$$

which as an algorithm becomes

$$T_j^{n+1} = T_j^{n-1} - C(T_{j+1}^n - T_{j-1}^n) \ . \tag{9.15}$$

Equation (9.15) is consistent with (9.2) with a truncation error of $O(\Delta t^2, \Delta x^2)$. The application of the von Neumann stability analysis to (9.15) produces the amplification factor G shown in Table 9.1. It can be seen that $|G| = 1.0$ if $C \le 1$ but $|G| > 1.0$ for some values of θ if $C > 1$. Thus, if the CFL condition is satisfied, the leapfrog scheme is neutrally stable. This is desirable since it implies no damping of the solution as time advances; which is what the physical process of convection (9.2) also implies.

It can be seen that the solution T_j^{n+1} given by (9.15) leaps over the solution at T_j^n; hence, the descriptive name: leapfrog method. However, the structure of (9.15) implies the grid is treated like a chessboard; the solution on black 'squares' is completely independent of the solution on red 'squares'. This can lead to two 'split' solutions. Roache (1972, p. 60) discusses strategies, such as averaging the solution at time-levels n and $n+1$, to overcome this problem.

Haltiner and Williams (1980, p. 111) have analysed the leapfrog scheme (9.15) applied to a problem with a harmonic initial condition, i.e. (9.4) becomes $F(x) = \exp(ijm\Delta x)$ where m is the wave number (Sect. 9.2). Haltiner and Williams obtain an analytic solution of (9.15) which consists of two waves or modes associated with the fact that three time-levels are linked by (9.15). One mode corresponds to the physical solution and the other is a spurious computational mode which propagates in the opposite direction to the physical solution and whose magnitude alternates in sign at each time step.

For the linear convection equation, the computational mode can be suppressed by an appropriate choice of the additional level of initial conditions, i.e. T^{n-1} as well as T^n. However, for nonlinear problems the "computational" mode, inherent in the leapfrog scheme, may develop and aggravate the separation of the solutions at consecutive time steps referred to above. Strategies such as averaging essentially suppress the "computational mode" with little effect on the physical mode.

Being a three-level scheme the leapfrog method requires the use of an alternative method for the first time-step. The leapfrog method is recommended (Morton, 1971) for hyperbolic problems like the convection equation (9.2); this may be contrasted with the unstable nature of the (equivalent) Richardson scheme for the diffusion equation.

The Lax-Wendroff method has been a very effective (and popular) algorithm for solving the equations that govern inviscid, compressible flow (Sects. 10.2 and 11.6.1). For the convection equation (9.2), the Lax-Wendroff scheme coincides with Leith's method (Roache 1972, p. 77). These methods are based on constructing a second-order representation for the term $\partial \bar{T}/\partial t$ from a Taylor series expansion. Thus,

$$\frac{\partial \bar{T}}{\partial t} \approx \frac{T_j^{n+1} - T_j^n}{\Delta t} - 0.5 \Delta t \frac{\partial^2 \bar{T}}{\partial t^2}$$

$$= \frac{T_j^{n+1} - T_j^n}{\Delta t} - 0.5 \Delta t u^2 \frac{\partial^2 \bar{T}}{\partial x^2} \ .$$

Introducing the centred finite difference expression for $\partial^2 \bar{T}/\partial x^2$ gives the Lax-Wendroff representation of (9.2) as

$$T_j^{n+1} = T_j^n - 0.5C(T_{j+1}^n - T_{j-1}^n) + 0.5C^2(T_{j-1}^n - 2T_j^n + T_{j+1}^n) . \qquad (9.16)$$

The Lax-Wendroff scheme is consistent with (9.2) with a truncation error of $O(\Delta t^2, \Delta x^2)$ and is stable if the CFL condition $C \leq 1.0$ is satisfied (Table 9.1). Numerical solutions using the Lax-Wendroff scheme are provided in Sect. 9.1.5.

9.1.4 Crank-Nicolson Schemes

The Crank-Nicolson scheme is very effective when applied to the one-dimensional diffusion equation. In this section, Crank-Nicolson finite difference and finite element schemes are applied to the one-dimensional convection equation (9.2).

The Crank-Nicolson finite difference scheme can be written

$$\frac{T_j^{n+1} - T_j^n}{\Delta t} + u(0.5L_x T_j^n + 0.5L_x T_j^{n+1}) = 0 , \qquad (9.17)$$

where $L_x T_j = (T_{j+1} - T_{j-1})/2\Delta x$. Equation (9.17) produces the following tridiagonal algorithm:

$$-0.25CT_{j-1}^{n+1} + T_j^{n+1} + 0.25CT_{j+1}^{n+1} = 0.25CT_{j-1}^n + T_j^n - 0.25CT_{j+1}^n , \qquad (9.18)$$

which can be solved efficiently using the Thomas algorithm (Sect. 6.2.2). The Crank-Nicolson scheme (9.17) is consistent with (9.2) and has a truncation error of $O(\Delta t^2, \Delta x^2)$. A von Neumann stability analysis produces the amplification factor G shown in Table 9.1. It is clear that the Crank-Nicolson scheme is unconditionally stable.

The application of the Galerkin finite element method with linear interpolation in space to (9.2) produces the following system of ordinary differential equations:

$$M_x\left[\frac{dT}{dt}\right]_j + uL_x T_j = 0 , \qquad (9.19)$$

where the mass operator $M_x = \{\frac{1}{6}, \frac{2}{3}, \frac{1}{6}\}$. Introducing Crank-Nicolson differencing for dT/dt gives the scheme

$$M_x\left[\frac{T_j^{n+1} - T_j^n}{\Delta t}\right] + uL_x[0.5(T_j^n + T_j^{n+1})] = 0 , \qquad (9.20)$$

which may be compared with (9.17). The mass operator can be generalised, as in (8.44) to

$$M_x \equiv \{\delta, 1-2\delta, \delta\} , \qquad (9.21)$$

where $\delta = \frac{1}{6}$ produces the finite element form and $\delta = 0$ produces the finite difference form. Substituting (9.21) into (9.20) produces the tridiagonal algorithm

$$(\delta - 0.25C)\, T_{j-1}^{n+1} + (1 - 2\delta)\, T_j^{n+1} + (\delta + 0.25C)\, T_{j+1}^{n+1}$$

$$= (\delta + 0.25C)\, T_{j-1}^n + (1 - 2\delta)\, T_j^n + (\delta - 0.25C)\, T_{j+1}^n \; . \tag{9.22}$$

This algorithm is executed in program TRAN (Fig. 9.8).

The finite element form (9.20) has a truncation error of $O(\Delta t^2, \Delta x^4)$. Thus, for sufficiently small time-steps that the solution error is dominated by the spatial discretisation, the Crank-Nicolson finite element method produces more accurate results than the Crank-Nicolson finite difference method. The second-order truncation error shown in Table 9.1 comes from the modified equation treatment (Sect. 9.2.2) of the Δt^2 term. For C and u of $O(1)$, this is the dominant term in the truncation error. The Crank-Nicolson finite element scheme (9.20) is unconditionally stable (Table 9.1).

Typical implicit schemes, considered for the diffusion equation, are unconditionally stable when applied to (9.2), as indicated in Table 9.1. However, the use of implicit schemes for hyperbolic equations is not necessarily advantageous. The use of an implicit scheme leads to a system of coupled equations at a given time-level $n + 1$. Therefore, a perturbation, e.g. due to round-off, introduced at one node (j, n) affects the solution at all other nodes $(j, n+1)$ at the next time-level. Physically, this behaviour corresponds to a parabolic partial differential equation (Sect. 2.3.2) like the diffusion equation.

In contrast, disturbances to solutions governed by hyperbolic partial differential equations (Sect. 2.2.2) are propagated at finite speeds. Using implicit schemes with hyperbolic partial differential equations typically produces inaccurate solutions if Δt (or equivalently C) is large. If Δt is small, i.e. $C \leq 1.0$ for the convection equation, there is no stability advantage in using implicit methods, although there may be an accuracy advantage, as with the Crank-Nicolson finite element scheme.

9.1.5 Linear Convection of a Truncated Sine Wave

To illustrate the behaviour of the various schemes for obtaining solutions to (9.2) we consider the propagation of a sine wave. Thus, (9.2) is to be solved subject to the initial condition

$$\bar{T}(x, 0) = \sin(10\pi x) \qquad \text{for } 0 \leq x \leq 0.1 \; , \tag{9.23}$$

$$= 0 \qquad \text{for } 0.1 < x \leq 1.0 \; ,$$

with boundary conditions

$$\bar{T}(0, t) = 0 \quad \text{and} \quad T(1, t) = 0 \; ,$$

The exact solution, up to $t = 0.9/u$, is

$$\bar{T} = 0 \qquad 0 \leq x \leq ut \; ,$$

$$= \sin[10\pi(x - ut)] \qquad ut < x \leq ut + 0.1 \; , \tag{9.24}$$

$$= 0 \qquad ut + 0.1 < x \leq 1.0 \; .$$

For a value $u = 0.1$ the exact solutions at $t = 0$ and 8 are shown in Fig. 9.2. The sine wave propagates with no reduction in amplitude at a speed $u = 0.1$.

The various schemes described in Sects. 9.1.2–4 are implemented in program TRAN (Fig. 9.8), with $\alpha = 0$. Program TRAN computes solutions to the one-dimensional transport equation and is described in Sect. 9.4.3. For the present problem the exact solution (9.24) is evaluated in subroutine EXSOL (Fig. 9.9).

Computational solutions have been obtained with 41 equally spaced points in the interval $0 \leq x \leq 1.0$ and with a Courant number $C = 0.8$. The output from program TRAN at $t = 8.0$ (i.e. after 40 time-steps) for the upwind scheme (9.10) is shown in Fig. 9.2. The computational solution is smooth, but compared with the exact solution, appears to have 'melted' (or diffused) away; this is consistent with the introduction of artificial viscous terms, as in (9.13).

The solution for the Lax-Wendroff scheme (9.16) is shown in Fig. 9.3 for values $t = 8.0$ and $C = 0.8$. It may be recalled that the structure of the Lax-Wendroff scheme is equivalent to the FTCS scheme (which is unstable) with additional diffusive terms added, where

$$\alpha_{add} = 0.5 Cu \Delta x \ .$$

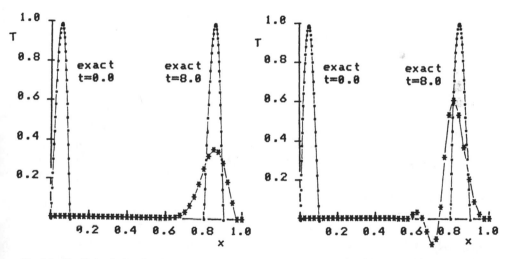

Fig. 9.2. Upwind solution for the convection equation with $C = 0.8$

Fig. 9.3. Lax-Wendroff solution for the convection equation with $C = 0.8$

The Lax-Wendroff solution is characterised by a primary wave of reduced amplitude travelling more slowly than the exact solution and the appearance of ancillary waves in the 'wake' of the primary wave. If the initial condition (9.23) were given a Fourier representation and the individual components (modes) propagated and reconstructed to generate the Lax-Wendroff solution at $t = 8.0$, the 'wake' would be consistent with the short wavelength components travelling more slowly than the dominant component. The trailing wave pattern is referred to as a 'dispersion' wake. Dispersion will be discussed in Sect. 9.2.

It may be noted that the upwind and Lax-Wendroff schemes reproduce the exact solution if $C = 1.0$. But the implicit schemes, such as the finite difference Crank-Nicolson scheme, do not. The Crank-Nicolson finite difference solution at $t = 8.0$ with $C = 0.8$ is shown in Fig. 9.4. The solution indicates a primary wave which is travelling more slowly than the exact solution and a substantial dispersion wake.

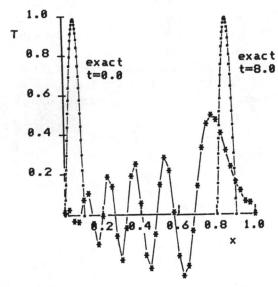

Fig. 9.4. Crank-Nicolson finite difference solution for the convection equation with $C = 0.8$

The corresponding Crank-Nicolson finite element solution is slightly more accurate than the solution shown in Fig. 9.4; the maximum amplitude is larger and the dispersion wake is smaller. However, at $C = 0.8$ and $u = 0.1$ the errors are dominated by the Δt^2 terms. For $C = 0.1$ and larger values of u the fourth-order spatial accuracy becomes more important and the Crank-Nicolson finite element scheme is considerably more accurate than the other schemes.

The flow patterns shown in Figs. 9.2–4 will be reconsidered at the end of Sect. 9.2. It should be emphasised that a relatively coarse mesh ($\Delta x, \Delta t$) and a relatively 'difficult' initial condition have been used to highlight some of the problems in computing convection-dominated flows. However, it is still true that obtaining computational solutions to the (pure) diffusion equation is considerably easier than when convection is present.

9.2 Numerical Dissipation and Dispersion

In fluid dynamics many phenomena are governed by systems of equations that are "hyperbolic" in character; they contain little or no dissipation. Solutions are characterised by wave-trains that propagate with little or no loss of amplitude. It is important that numerical schemes do not introduce non-physical dissipation, for example as in (9.10) and Fig. 9.2. It is also important that the numerical schemes do not alter the speed at which waves propagate; that is, the numerical schemes should not introduce artificial dispersion, producing the result shown in Fig. 9.4.

The solution for the propagation of a plane wave that is subjected to both dissipation and dispersion can be written as

$$\bar{T} = Rl T_{amp}\, e^{-p(m)t}\, e^{im[x-q(m)]t} \ , \tag{9.25}$$

where T_{amp} is real and positive, m is the wave-number which is related to the spatial wavelength λ by $\lambda = 2\pi/m$. In (9.25), $p(m)$ determines how rapidly the amplitude of the wave attenuates and $q(m)$ is the propagation speed of the wave.

If the motion of a plane wave represented by (9.25) is governed by the linear convection equation (9.2), $p(m)$ and $q(m)$ take the values

$$p(m) = 0 \quad \text{and} \quad q(m) = u \ ,$$

i.e. waves of any wavelength are propagated at the same speed with no damping.

It is instructive to consider two equations that are closely related to the convection equation (9.2); these are

$$\frac{\partial \bar{T}}{\partial t} + u\frac{\partial \bar{T}}{\partial x} - \alpha\frac{\partial^2 \bar{T}}{\partial x^2} = 0 \quad \text{and} \tag{9.26}$$

$$\frac{\partial \bar{T}}{\partial t} + u\frac{\partial \bar{T}}{\partial x} + \beta\frac{\partial^3 \bar{T}}{\partial x^3} = 0 \ . \tag{9.27}$$

Equation (9.26) is the transport equation which will be considered in Sect. 9.4. Equation (9.27) is the linear Korteweg de Vries equation. For plane waves governed by (9.26), the solution (9.25) has the parameter values

$$p(m) = \alpha m^2 \quad \text{and} \quad q(m) = u \ , \tag{9.28}$$

i.e. the amplitude of the wave is attenuated by the diffusive term but the wave propagation speed is unaffected. Since $m = 2\pi/\lambda$, short wavelengths are attenuated much more rapidly than long wavelengths. For plane waves governed by (9.27)

$$p(m) = 0 \quad \text{and} \quad q(m) = u - \beta m^2 \ , \tag{9.29}$$

i.e. the amplitude of the wave is unaltered but the wave propagates at a speed depending on its wavelength. If waves of more than one wavelength are present they propagate at different speeds, i.e. they disperse. The change in wave propagation speed is more pronounced for short wavelengths (large m). Thus, if β is positive short wavelengths travel much more slowly.

We can now formally define dissipation as the attenuation of the amplitude of plane waves and dispersion as the propagation of plane waves of different wave-number m at different speeds $q(m)$. Through the role of the higher-order derivatives in (9.26 and 27) we can also associate positive dissipation with the appearance of even-ordered spatial derivatives multiplied by coefficients of alternating sign. Similarly, a reduction in wave propagation speed can be associated with the appearance of odd-ordered spatial derivatives multiplied by coefficients of alternating sign.

In examining the consistency (Sect. 4.2) of the discretised equation, e.g. (9.9), it is apparent that the computational algorithm, e.g. (9.10), is equivalent to the original partial differential equation plus the truncation error, on a finite grid. But the truncation error consists, typically, of successive higher even- and odd-ordered derivatives. The link between dissipation, dispersion and higher-order derivatives in the truncation error will be pursued in Sect. 9.2.2.

9.2.1 Fourier Analysis

Quantitative information can be obtained about the numerical dissipation and dispersion introduced by a particular computational algorithm by comparing Fourier representations of the exact and numerical solutions. Thus, for the specific example considered in Sect. 9.1.5, a Fourier representation of the initial condition (9.23) would be

$$\bar{T}(x, 0) = \sum_{m=-\infty}^{\infty} T_m e^{imx} \ .$$

Using the separation of variables approach (Sect. 2.5.2) the exact solution at any intermediate time is

$$\bar{T}(x, t) = \sum_{m=-\infty}^{\infty} T_m e^{im(x-ut)} \ , \tag{9.30}$$

which will reconstruct the exact solution (9.24). The Fourier components in (9.30) all convect with the same velocity u and are not subject to any reduction in amplitude since there is no diffusive term in (9.2).

In a parallel manner a Fourier representation can be introduced for the solution of the computational algorithm, e.g. the upwind scheme (9.10), in the form

$$T(x, t) = \sum_{m=-\infty}^{\infty} T_m e^{-p(m)t} e^{im[x-q(m)t]} \ , \tag{9.31}$$

and since the algebraic equation of interest (9.10) is linear the components of the Fourier representation can be considered independently.

During one time-step Δt the mth component of the approximate solution will have decreased in amplitude by $\exp[-p(m)\Delta t]$ and will have advanced a distance $q(m)\Delta t$. During the same time-step the mth component of the exact solution will have maintained constant amplitude while advancing a distance $u\Delta t$. If the computational algorithm is to introduce no artificial dissipation or dispersion, $p(m)$ must be zero and $q(m)$ must equal u.

Expressions for $p(m)$ and $q(m)$ can be obtained by substituting (9.31) into (9.10) and constructing the amplitude ratio of the mth component of the Fourier representation for a single time-step, i.e.

$$G_m = \frac{T_m e^{-p(m)(t+\Delta t)} e^{im[x-q(m)(t+\Delta t)]}}{T_m e^{-p(m)t} e^{im[x-q(m)t]}}$$

$$= e^{-[p(m)+imq(m)]\Delta t} .$$

(9.32)

But from the substitution into (9.10), the amplitude ratio is

$$G_m = 1 + C[\cos(m\Delta x) - 1] - iC\sin(m\Delta x) ,$$

(9.33)

which, not surprisingly, is precisely the same expression (see Table 9.1) as is obtained in assessing the von Neumann stability of the upwind scheme (9.10). From (9.32, 33),

$$|G|_m = e^{-p(m)\Delta t} = 1 - 4C(1-C)\sin^2(0.5\, m\Delta x) .$$

(9.34)

For a stable solution $p(m) \geq 0$, so instability may be associated with negative dissipation. But (9.34) also gives a measure of how dissipative the particular algorithm (9.10) will be. The dissipation parameter $p(m)$ depends both on C and on the wave-number parameter $m\Delta x$. Attenuation is largest for $m\Delta x = \pi$, i.e. for the shortest wavelengths.

The wave propagation speed $q(m)$ is related to the phase ϕ_m of the amplification ratio G_m. From (9.32, 33)

$$\phi_m = -mq(m)\Delta t$$

$$= \tan^{-1} \frac{-C\sin(m\Delta x)}{1 + C[\cos(m\Delta x) - 1]} ,$$

(9.35)

ϕ_m is the change of phase experienced by the mth component during one time-step Δt. For the exact solution,

$$\phi_{ex} = -mu\Delta t = -Cm\Delta x .$$

Thus,

$$\frac{\phi_m}{\phi_{ex}} = \frac{q(m)}{u} = \left(\frac{-1}{Cm\Delta x}\right) \tan^{-1} \left(\frac{-C\sin(m\Delta x)}{1 + C[\cos(m\Delta x) - 1]}\right) .$$

(9.36)

The propagation speed (or equivalently, the phase change) is accurate for small values of $m\Delta x$ but the error grows as $m\Delta x \to \pi$.

If $0 < C < 0.5$, $q(m) < u$ for $m\Delta x$ large.

If $0.5 < C < 1.0$, $q(m) > u$ for $m\Delta x$ large.

Consideration of the different computational algorithms in Sect. 9.1 leads to different expressions on the right hand side of (9.33). In turn different expressions for $p(m)$ and $q(m)$ are obtained, through (9.32).

By introducing a Fourier representation for both the exact and approximate solutions, and considering the amplification ratio of the mth component, it is possible to obtain expressions for the dissipation and dispersion in the numerical scheme for any value of $m\Delta x$. Morton (1971, pp. 244–245) has used this approach for the numerical schemes considered in Sect. 9.1.

The shortest wavelength that a discrete grid can represent is $\lambda = 2\Delta x$. This corresponds to $m\Delta x = \pi$. Long wavelengths correspond to $m\Delta x \to 0$. If the numerical scheme introduces any dissipation or dispersion it is expected, from (9.26–29), that short wavelengths will be much more seriously affected than long wavelengths. From a practical viewpoint, it is preferable to model the long wavelength behaviour (small $m\Delta x$) accurately and to check that any spurious short wavelength behaviour is physically or numerically attenuated. This strategy is also consistent with the analysis of Sect. 3.4.2.

9.2.2 Modified Equation Approach

We have seen that dissipation and dispersion can be associated with the occurrence of second and third spatial derivatives, as in (9.26 and 27). This suggests that an examination of the truncation error may also permit the dissipative and dispersive tendencies of a particular computational algorithm to be identified. However, the truncation error must be interpreted in a special way.

It is assumed that $L(\bar{T}) = 0$ represents the differential equation and $L_a(T) = 0$ represents the algebraic equation, e.g. (9.9). Therefore, the previous development of the truncation error E_j^n can be written

$$L(\bar{T}) = L_a(\bar{T}) - E_j^n(\bar{T}) = 0 \ . \tag{9.37}$$

Now, the following expression for the truncation error must be obtained:

$$L_a(T) = L(T) + E(T) = 0 \ , \tag{9.38}$$

where the algebraic equation has been expanded as a Taylor series in the approximate solution. Equation (9.38) is manipulated so that $E(T)$ contains only spatial derivatives. For this purpose (9.38) is repeatedly differentiated to eliminate pure and cross derivatives in t. This is in contrast to the situation for (9.37) where $L(\bar{T}) = 0$ may be used to simplify the truncation error.

The equation $L(T) + E(T) = 0$ is a differential equation with an infinite number of terms. This is called the modified equation (Warming and Hyett 1974) and is the differential equation that the algebraic equation would solve if sufficient boundary conditions were available. The modified equation can be used to demonstrate consistency and the order of accuracy of the computational algorithm, as before. In the modified equation odd-order derivatives are associated with dispersion and even-ordered derivatives are associated with dissipation. Therefore, by considering the lowest odd- and even-order terms in the modified equation we can deduce the dissipative and dispersive properties of the algebraic scheme, but only for long wavelengths. This is because the modified equation is based on a Taylor expansion that assumes small Δt and Δx. However, as noted above, it is the long wavelength behaviour that is generally of greater interest.

Using the modified equation approach it is possible to show that the Lax-Wendroff scheme (9.16) applied to (9.2) has as the leading terms in the truncation error

$$\left(\frac{\Delta x^2}{6}\right)u(1-C^2)\frac{\partial^3 T}{\partial x^3}+\left(\frac{\Delta x^3}{8}\right)uC(1-C^2)\frac{\partial^4 T}{\partial x^4}\ .$$

As noted previously, the Lax-Wendroff scheme is consistent with (9.2) to $O(\Delta t^2, \Delta x^2)$. The above truncation error indicates second-order dispersion and third-order dissipation. The modified equation approach has been used to generate the truncation errors shown in Tables 7.1, 9.1 and 9.3, so that the numerical dissipation and dispersion associated with long wavelengths (small $m\Delta x$) can be assessed.

The modified equation analysis has the advantage that it extends to nonlinear equations, whereas Fourier analysis (Sect. 9.2.1) is only applicable to linear equations. Klopfer and McRae (1983) use a modified equation analysis to construct generalised Lax-Wendroff schemes for one-dimensional unsteady propagation of shocks, which are governed by the Euler equations (10.40).

For the equations governing fluid dynamics (Chap. 11) the analysis of computational algorithms using the modified equation approach requires a substantial amount of algebraic manipulation. Warming and Hyett recommend a systematic tabulation to keep track of individual terms. It is expected that symbolic manipulation using a computer, e.g. via MACSYMA, would be very effective in this area. The possibilities for symbolic manipulation in computational fluid dynamics are discussed by Steinberg and Roache (1985).

9.2.3 Further Discussion

A consideration of the truncation errors shown in Table 9.1 for the upwind, Lax-Wendroff and Crank-Nicolson schemes applied to the example at the end of Sect. 9.1 is facilitated if attention is focussed on the Fourier components that make up the solution. The results shown in Figs. 9.2–4 and Table 9.1 indicate that excessive numerical dissipation effectively obliterates the small dispersion errors for the upwind scheme and attenuates some of the dispersion errors in the Lax-Wendroff solution. The Crank-Nicolson schemes are dominated by dispersion errors which act to significantly slow down the propagation speed of individual Fourier components. Although the Crank-Nicolson schemes are stable for increasing Courant number the dispersion errors grow like C^2 (Table 9.1). For small C, and hence small Δt, the dispersion error of the Crank-Nicolson finite element scheme (Table 9.1) is considerably smaller than that of the other schemes considered in Sect. 9.1.

The use of Fourier analysis and the modified equation analysis permits the construction of more accurate schemes designed to reduce dissipation and dispersion errors. This can be illustrated with the Taylor-Galerkin (Donea 1984) finite element method. In this method (9.2) is first discretised in time to give

$$\frac{T^{n+1} - T^n}{\Delta t} + u \frac{\partial T}{\partial x} = 0 . \tag{9.39}$$

However, a more accurate representation of $\partial \bar{T}/\partial t$ at time level n can be written

$$\frac{\partial \bar{T}}{\partial t}\bigg|^n = \frac{T^{n+1} - T^n}{\Delta t} - 0.5 \Delta t \frac{\partial^2 \bar{T}}{\partial t^2}\bigg|^n - \frac{\Delta t^2}{6} \frac{\partial^3 \bar{T}}{\partial t^3}\bigg|^n - O(\Delta t^3) .$$

From (9.2),

$$\frac{\partial^2 \bar{T}}{\partial t^2} = u^2 \frac{\partial^2 \bar{T}}{\partial x^2} \quad \text{and} \quad \frac{\partial^3 \bar{T}}{\partial t^3} = u^2 \frac{\partial^2}{\partial x^2} \frac{\partial \bar{T}}{\partial t} .$$

Therefore a more accurate [than (9.39)] semi-discretisation of (9.2) is given by

$$\left[1 - \frac{u^2 \Delta t^2}{6} \frac{\partial^2}{\partial x^2}\right] \frac{T^{n+1} - T^n}{\Delta t} + u \frac{\partial T}{\partial x} - 0.5 u^2 \Delta t \frac{\partial^2 T}{\partial x^2} = 0 . \tag{9.40}$$

The application of a uniform-grid, linear interpolation Galerkin finite element formulation (Sect. 5.4) to (9.40) gives

$$\left[M_x - \frac{C^2}{6} \Delta x^2 L_{xx}\right](T_j^{n+1} - T_j^n) + C \Delta x L_x T_j^n - 0.5 C^2 \Delta x^2 L_{xx} T_j^n = 0 , \tag{9.41}$$

which is a generalisation of the Lax-Wendroff scheme (9.16). The mass operator $M_x = \{\frac{1}{6}, \frac{2}{3}, \frac{1}{6}\}$, arises from the finite element formulation. The term $(C^2/6) L_{xx}$ $(T_j^{n+1} - T_j^n)$ appears due to retaining additional terms in the Taylor series expansion of $\partial \bar{T}/\partial t$. Equation (9.41) has a truncation error of $O(\Delta t^3, \Delta x^3)$, which may be contrasted with the second-order truncation error of the Lax-Wendroff scheme (9.16). Donea (1984) has examined the dissipation, dispersion behaviour via a Fourier analysis (Sect. 9.2.1) and finds that (9.41) is superior to the Lax-Wendroff scheme for all wavelengths and for Courant numbers in the interval $0 < C < 1$.

The above strategy can be generalised by adding in controlled (anti-)dissipation and (anti-)dispersion as part of the Taylor series construction. Baker and Kim (1987) develop such a formulation and call it the Taylor-Weak statement (TWS) of the Galerkin finite element formulation. The TWS algorithm has four disposable constants. Specific choices of these parameters reproduce 17 well-known finite difference and finite element methods. An optimum choice can be made, in principle, for the linear convection equation (9.10). However, for more general nonlinear hyperbolic conservation equations, e.g. the Euler equations (10.40), an optimum choice will be solution dependent, although desirable dissipation, dispersion attributes can still be obtained in a gross sense.

We may infer the general guideline that first-order formulae for derivatives should be avoided. A first-order representation for the nth derivative in the governing equation will generate spatial derivatives of order $(n+1)$ and higher in the modified equation (which is the equivalent governing equation actually solved). This is particularly serious for the convective terms $(n=1)$, since the introduction of spurious second or third derivatives can change the character of the solution significantly.

9.3 Steady Convection-Diffusion Equation

For many flow problems dissipative mechanisms are only significant in a narrow layer, typically adjacent to a boundary. Computational solutions obtained with grids appropriate to the main flow region are often oscillatory when the true solution changes rapidly across the boundary layer. The steady convection-diffusion equation is a useful model equation with which to illustrate the phenomenon.

This equation can be written in one dimension as

$$u\frac{d\bar{T}}{dx} - \alpha\frac{d^2\bar{T}}{dx^2} = 0 \ . \tag{9.42}$$

It represents the steady-state balance between convection and diffusion; time plays no role at all. If (9.42) is combined with boundary conditions

$$\bar{T}(0) = 0 \quad \text{and} \quad \bar{T}(1) = 1.0 \tag{9.43}$$

the exact solution is obtained in the interval $0 \leq x \leq 1$ as

$$\bar{T} = [e^{ux/\alpha} - 1.0]/[e^{u/\alpha} - 1.0] \ , \tag{9.44}$$

which is shown in Fig. 9.5. The solution is characterised by a uniform distribution of T combined with a boundary layer adjacent to $x = 1.0$.

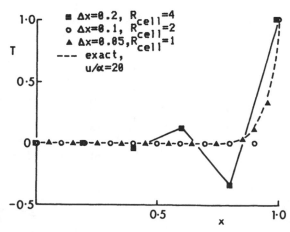

Fig. 9.5. Cell Reynolds number influence on the solution of the convection diffusion equation

9.3.1 Cell Reynolds Number Effects

If centred three-point finite difference expressions are substituted for the derivatives in (9.42), the following algorithm is obtained:

$$-(1+0.5R_{cell})T_{j-1}+2T_j-(1-0.5R_{cell})T_{j+1}=0 , \tag{9.45}$$

where $R_{cell}=u\Delta x/\alpha$, i.e. R_{cell} is a Reynolds (strictly Peclet) number based on the characteristic length Δx.

A Taylor series expansion of (9.45) about the jth node indicates that it is consistent with (9.42) with a truncation error of $O(\Delta x^2)$. Equation (9.45) is implicit but when equations are obtained for all nodes a tridiagonal system of equations is obtained which can be solved economically (Sect. 6.2.2).

For this relatively simple case, it is also possible to write an exact solution of (9.45) as

$$T_j=A_0+B_0\left[\frac{1+0.5R_{cell}}{1-0.5R_{cell}}\right]^j , \tag{9.46}$$

where A_0 and B_0 are chosen to satisfy the boundary conditions (9.43). For the case $u/\alpha=20$ and $\Delta x=0.05$, 0.1 and 0.2, solutions to (9.46) are shown in Fig. 9.5. The solution for $\Delta x=0.05$, which is equivalent to $R_{cell}=1$, is in reasonable agreement with the exact solution (9.44). The solution with the intermediate grid, $R_{cell}=2$, is accurate except in the boundary layer adjacent to $x=1$. However, the solution on the coarsest grid, $R_{cell}=4$, is not only inaccurate but also oscillatory.

From (9.46) it can be seen that oscillatory solutions are avoided if $R_{cell}\leqq2$. It is also instructive to relate the solution behaviour shown in Fig. 9.5 to the eigenvalues of the matrix of equations (9.45) when combined with boundary conditions (9.43). This system has the form

$$\underline{A}\,T=B , \tag{9.47a}$$

i.e.

$$\begin{bmatrix} b & c & & & & \\ a & b & c & & & \\ & & \cdot & & & \\ & & & \cdot & & \\ & & & & \cdot & \\ & & & a & b & c \\ & & & & a & b \end{bmatrix} \begin{bmatrix} T_2 \\ T_3 \\ \cdot \\ \cdot \\ \cdot \\ T_{J-2} \\ T_{J-1} \end{bmatrix} = \begin{bmatrix} -aT_1 \\ 0 \\ \cdot \\ \cdot \\ \cdot \\ 0 \\ -cT_J \end{bmatrix} , \tag{9.47b}$$

where $a=-(1+0.5R_{cell})$, $b=2$ and $c=-(1-0.5R_{cell})$. The eigenvalues of the above tridiagonal matrix are related to a, b and c as

$$\lambda_j=b+2(ac)^{1/2}\cos\left(\frac{j\pi}{J-1}\right) , \quad j=1, 2, \ldots, J-2 . \tag{9.48}$$

It can be seen that the condition $ac \geqq 0$ is required for the eigenvalues to be real. Substituting for a and c gives the condition

$$(1 + 0.5\,R_{cell})(1 - 0.5\,R_{cell}) \geqq 0 \quad \text{or} \quad R_{cell} \leqq 2 \ . \tag{9.49}$$

Thus, for this example, the oscillatory solutions coincide with the occurrence of complex eigenvalues.

For $\underline{A}$, in (9.47a) of general form, bounds on the eigenvalues λ are given by the Gershgorin circle theorem (Jennings 1977, p. 35)

$$|\lambda - a_{ii}| \leqq \sum_{j \neq i}^{N} |a_{ij}| \ .$$

That is, the eigenvalues of $\underline{A}$ lie in the union of circles associated with each row of $\underline{A}$. The centre of each circle is given by the diagonal entry a_{ii} and the radius by the sum of the magnitudes of the off-diagonal elements in each row. Carey and Sepehrnoori (1980) use the Gershgorin circle theorem to analyse (9.47a) for different values of R_{cell}. In the range $0 < R_{cell} \leqq 2$ the centre of the Gershgorin circle lies at $\{-2\alpha/\Delta x^2, 0\}$ in the complex plane with radius $2\alpha/\Delta x^2$, and all eigenvalues lie on the negative real axis. As R_{cell} increases, i.e. as α decreases, the radius reduces and the centre of the circle moves closer to the origin. At $R_{cell} = 2$ the centre is at $(-u/\Delta x, 0)$ with radius $u/\Delta x = 2\alpha/\Delta x^2$. For $R_{cell} > 2$ the radius remains constant at $u/\Delta x$ but the centre migrates to the origin and the eigenvalues are complex. When $R_{cell} = \infty$, the circle centre is at the origin and the eigenvalues are purely imaginary.

The limitation $R_{cell} \leqq B$ for non-oscillatory solutions, applies to other methods, such as the finite element method. Finlayson (1980, p. 241) provides values of B for finite element and orthogonal collocation methods. However, B is typically less than 10.

It should be emphasised that oscillatory solutions *may* occur if $R_{cell} > B$. Whether they actually do occur also depends on the flow geometry and imposition of boundary conditions. For example, decelerating flow ahead of an obstruction often produces an oscillatory solution if $R_{cell} > B$. However, oscillation-free solutions for uniform or accelerating flow can be obtained even though $R_{cell} > B$.

As might be expected, the replacement of the centred finite difference representation for $\partial \bar{T}/\partial x$ in (9.42) with the upwind finite difference expression $(T_j - T_{j-1})/\Delta x$ prevents the oscillations from occurring. The upwind scheme produces the following algorithm in place of (9.45):

$$-(1 + R_{cell})T_{j-1} + 2(1 + 0.5\,R_{cell})T_j - T_{j+1} = 0 \ . \tag{9.50}$$

Consideration of (9.48) shows that the upwind scheme has real eigenvalues for all values of R_{cell}. The solution for the upwind scheme with $u/\alpha = 20$ and $\Delta x = 0.2$ is shown in Fig. 9.6. As expected, the solution is not oscillatory but is not very accurate either.

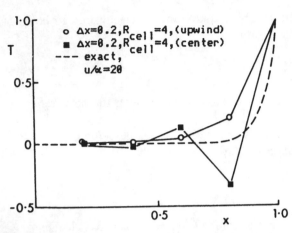

Fig. 9.6. Upwind solution of the convection diffusion equation

A Taylor series expansion of (9.50) indicates that it is consistent with (9.42) to $O(\Delta x)$ only. If (9.50) is treated as being accurate to $O(\Delta x^2)$ it is found to be consistent with the equation

$$u\frac{d\bar{T}}{dx} - \alpha(1 + 0.5R_{\text{cell}})\frac{d^2\bar{T}}{dx^2} = 0 \; , \tag{9.51}$$

i.e. the use of the first-order upwind difference formula has introduced an artificial diffusivity $0.5R_{\text{cell}}\alpha$. This may be compared with the artificial diffusion introduced by the upwind scheme for the convection equation, see (9.13). For accurate solutions a consideration of (9.51) would suggest the practical limit

$$0.5R_{\text{cell}} \ll 1 \; ,$$

which is more restrictive than (9.49).

If a Neumann boundary condition is introduced in place of the Dirichlet boundary condition at $x = 1.0$ the character of the solution is not radically altered. For large values of u/α the centred-difference scheme (9.45) produces solutions with even greater oscillations than those in Fig. 9.6.

9.3.2 Higher-Order Upwind Scheme

The very oscillatory behaviour of the three-point centred difference representation for $d\bar{T}/dx$ in (9.42) and the very dissipative nature of the two-point upwind scheme (9.50) suggests that a four-point representation of $d\bar{T}/dx$ will be necessary to obtain satisfactory results. For u positive the following representation would be appropriate:

$$\frac{dT}{dx} \approx aT_{j-2} + bT_{j-1} + cT_j + T_{j+1} \; . \tag{9.52}$$

Following the general procedure of Sect. 3.2.2 the following scheme is obtained:

$$\frac{dT}{dx}=\frac{T_{j+1}-T_{j-1}}{2\Delta x}+\frac{q(T_{j-2}-3T_{j-1}+3T_j-T_{j+1})}{3\Delta x}+O(\Delta x^2) \ . \tag{9.53}$$

For u negative (9.53) is replaced by

$$\frac{dT}{dx}=\frac{T_{j+1}-T_{j-1}}{2\Delta x}+\frac{q(T_{j-1}-3T_j+3T_{j+1}-T_{j+2})}{3\Delta x}+O(\Delta x^2) \ .$$

Equation (9.53) has been deliberately written as a modification to the three-point centred finite difference representation. The parameter q controls the size of the modification.

A Taylor series expansion about node j indicates that

$$T_{j-2}-3T_{j-1}+3T_j-T_{j+1}\equiv\left[-\Delta x^3\frac{d^3T}{dx^3}+0.5\Delta x^4\frac{d^4T}{dx^4}+\dots\right]_j \ .$$

That is, the modification can be used to counteract specific terms in the Taylor series expansion for the complete equation. In particular, the choice $q=0.5$ eliminates the $\Delta x^2 d^3 T/dx^3$ term and makes (9.53) of $O(\Delta x^3)$. Based on the association of wiggles with dispersion and with odd-ordered derivatives (Sect. 9.2) it can be anticipated that (9.53) with $q=0.5$ combined with the previous discretisation of $d^2\bar{T}/dx^2$, will produce a more accurate solution, with fewer wiggles, than (9.45).

Using (9.53) gives the following implicit algorithm after discretising (9.42):

$$\frac{q}{3}R_{cell}\,T_{j-2}-[1+(q+0.5)\,R_{cell}]\,T_{j-1}+(2+q\,R_{cell})\,T_j$$

$$-\left[1+\left(\frac{q}{3}-0.5\right)R_{cell}\right]T_{j+1}=0 \ , \tag{9.54}$$

which reduces to (9.45) if $q=0$. The structure of (9.54) introduces the complication of a quadridiagonal matrix to be solved. This is achieved in an efficient manner by using the generalised Thomas algorithm (Sect. 6.2.4) with $f=0$. The required coding, additional to subroutines BANFAC (Fig. 6.18) and BANSOL (Fig. 6.19), is indicated in Fig. 9.8, lines 126–133.

Equation (9.54) is combined with boundary conditions (9.43) to produce a computational solution to be compared with (9.44). Typical results on a coarse grid, $\Delta x=0.2$, $R_{cell}=4$, for various values of q are shown in Table 9.2. The case $q=0$ coincides with the three-point centred scheme (9.43), which produces the oscillatory solution plotted in Figs. 9.5 and 9.6.

The case $q=0.5$ is considerably more accurate although still slightly oscillatory. Increasing q is seen to produce a smoother solution but a more diffuse solution similar to the solution produced by the two-point upwind scheme (9.50) shown in

Table 9.2. Coarse grid solutions of (9.42, 43) using (9.54)

x	0.	0.2	0.4	0.6	0.8	1.0
Exact solution	0.	0.0000	0.0000	0.0003	0.0018	1.0000
$q=0$	0.	0.0164	−0.0328	0.1148	−0.3279	1.0000
$q=0.5$	0.	0.0000	−0.0003	0.0058	−0.0760	1.0000
$q=1.0$	0.	0.0000	0.0002	0.0036	0.0597	1.0000
$q=1.5$	0.	0.0004	0.0032	0.0222	0.1491	1.0000

Fig. 9.6. On a finer grid the case $q=0.5$ produces the most accurate solutions (not shown).

The equivalent ordinary differential equation actually solved by (9.54) can be obtained by expanding as a Taylor series about node j. The result is

$$\frac{dT}{dx} - \frac{\Delta x}{R_{\text{cell}}} \frac{d^2T}{dx^2} + (1-2q)\frac{\Delta x^2}{6}\frac{d^3T}{dx^3}$$

$$+ \left(2q - \frac{1}{R_{\text{cell}}}\right)\frac{\Delta x^3}{12}\frac{d^4T}{dx^4} + (1-10q)\frac{\Delta x^4}{120}\frac{d^5T}{dx^5} \ldots = 0 \ . \tag{9.55}$$

For practical problems with typically small values of α/u, R_{cell} is $O(1)$ or larger. This implies that the formal truncation error of representing the individual derivatives in (9.42) may have different influences on the solution accuracy of (9.54). Thus for $q \neq 0.5$, the third term in (9.55) is likely to have the dominant influence on the solution error. This suggests for problems with small diffusivities α it is appropriate to represent the convective terms, like $u\,dT/dx$ in (9.42), more accurately than the diffusive terms.

The results shown in Table 9.2 are consistent with (9.55). An increasing q value is seen to introduce positive dissipation. However, it should be remembered (Sect. 9.2) that (9.55) is more relevant to the long wavelength behaviour. Equation (9.55) indicates that (9.54) with $q=0.5$ and $R_{\text{cell}}=1.0$ is formally fourth order. However, for nonlinear convective problems R_{cell} will depend on the local solution so that it will not be possible to achieve fourth-order accuracy throughout the computational domain. The particular choice $q=0.375$ has been analysed by Leonard (1979) and is considered in relation to a two-dimensional problem in Sect. 17.1.5.

It is apparent from the analysis of Sects. 9.2 and 9.3 that odd-ordered spatial derivatives in the truncation error are associated with wiggles in the solution whereas even-ordered spatial derivatives are associated with dissipation or unstable growth depending on the sign of the multiplying coefficient. The tendency to cause wiggles without loss of amplitude will be referred to as a dispersion-like phenomenon even though the concept of dispersion is strictly only meaningful for unsteady flows. If a given steady flow is interpreted as the culmination of a transient flow (with steady boundary conditions) after a very large time, any dispersion-like wiggles can be thought of as emanating from the balance between dispersion effects and the imposition of boundary conditions.

9.4 One-Dimensional Transport Equation

The one-dimensional transport equation can be written

$$\frac{\partial \bar{T}}{\partial t} + u\frac{\partial \bar{T}}{\partial x} - \alpha\frac{\partial^2 \bar{T}}{\partial x^2} = 0 \ , \tag{9.56}$$

where $\bar{T}$ is a passive scalar (e.g. temperature) which is being convected with a known velocity $u(x, t)$ and being diffused (Fig. 1.7). If $\bar{T}$ is the temperature then α is the thermal diffusivity. To examine the behaviour of computational solutions of (9.56) we will assume that u and α are constants. Like the diffusion equation (Chap. 7) the transport equation is strictly parabolic (Sect. 2.3) and requires the same type of boundary and initial conditions, for example (7.2–4), as the diffusion equation.

However, when u/α becomes large the first two terms in (9.56) would be expected to dominate. Then the transport equation will demonstrate behaviour similar to the convection equation (9.2). It may be recalled that exact solutions of the convection equation are typically wave motions that propagate with no damping (reduction in amplitude). Therefore, for large values of u/α in the transport equation (9.56), solutions may be expected to demonstrate wave-like motion that is slightly damped.

From a consideration of the convection diffusion equation, which represents a steady-state limit of the transport equation, we may also expect spatial wave-like motions to appear in approximate solutions of (9.56) if symmetric three-point algebraic representations are used for the convective term with large values of u/α.

In this section we examine some of the algebraic schemes (FTCS, DuFort-Frankel, etc.), that have been investigated previously for the diffusion (Chap. 7) and convection (Sect. 9.1) equations, to see if the simultaneous appearance of convection and diffusion introduces any new difficulties.

9.4.1 Explicit Schemes

The FTCS scheme applied to (9.56) produces the algebraic equation

$$\frac{T_j^{n+1} - T_j^n}{\Delta t} + \frac{u(T_{j+1}^n - T_{j-1}^n)}{2\Delta x} - \frac{\alpha(T_{j-1}^n - 2T_j^n + T_{j+1}^n)}{\Delta x^2} = 0 \ , \tag{9.57}$$

which, as an algorithm, can be written

$$T_j^{n+1} = (s + 0.5C)\,T_{j-1}^n + (1 - 2s)\,T_j^n + (s - 0.5C)\,T_{j+1}^n \ , \tag{9.58}$$

where $s = \alpha\Delta t/\Delta x^2$ and $C = u\Delta t/\Delta x$.

A Taylor series expansion about the (j, n)-th node indicates that (9.58) is consistent with (9.56) with a truncation error of $O(\Delta t, \Delta x^2)$. If (9.58) is treated as being of $O(\Delta t^2, \Delta x^2)$ it is consistent with

$$\frac{\partial \bar{T}}{\partial t} + u\frac{\partial \bar{T}}{\partial x} - \alpha\frac{\partial^2 \bar{T}}{\partial x^2} + \frac{\Delta t}{2}\frac{\partial^2 \bar{T}}{\partial t^2} = 0 , \tag{9.59}$$

i.e. an additional term, previously interpreted as the leading term in the truncation error, is included in the partial differential equation that (9.58) is consistent with. Following the modified equation approach (Sect. 9.2), time derivatives can be eliminated in favour of spatial derivatives to give

$$\frac{\partial T}{\partial t} + u\frac{\partial T}{\partial x} - (\alpha - \alpha')\frac{\partial^2 T}{\partial x^2} - \left(\alpha u\Delta t + u^3\frac{\Delta t^2}{3} - u\frac{\Delta x^2}{6}\right)\frac{\partial^3 T}{\partial x^3}$$

$$+ \left(0.5\,\alpha^2\,\Delta t - \alpha u^2\,\Delta t^2 + 0.25\,u^4\,\Delta t^3 - \alpha\frac{\Delta x^2}{12} + u^2\frac{\Delta t\,\Delta x^2}{6}\right)\frac{\partial^4 T}{\partial x^4} = 0 , \tag{9.60}$$

where $\alpha' = u^2\,\Delta t/2$. Thus, it can be seen that the use of a first-order time differencing scheme in the transport equation introduces first-order dissipation and dispersion. For large values of u/α the artificial diffusion term $\alpha'\partial^2 \bar{T}/\partial x^2$ can easily be comparable to the physical diffusion term unless the magnitude of Δt is limited. Unless $\partial^4 T/\partial x^4$ is large, the last term in (9.60) is unlikely to be significant. Therefore, an accuracy limitation on Δt is

$$\Delta t \ll \frac{2\alpha}{u^2} \quad \text{or} \quad C^2 \ll 2s \quad \text{or} \quad R_{\text{cell}}\left(=\frac{C}{s}\right)\ll\frac{2}{C} . \tag{9.61}$$

Clearly, there is a strong incentive to use second-order time differencing when solving the transport equation.

It may be recalled that forward time differencing combined with centred spatial differencing produces an algorithm that is conditionally stable for the diffusion equation and always unstable for the convection equation. For the transport equation the von Neumann stability analysis produces the value for the amplification factor G shown in Table 9.3. For stable solutions, $|G|\leq 1$ implies the following restrictions (Noye 1983, p. 215):

$$0\leq C^2\leq 2s\leq 1 . \tag{9.62}$$

Equation (9.62) includes the 'diffusive' limit (Sect. 7.1.1). Strictly, (9.62) allows solutions for which $R_{\text{cell}} = u\Delta x/\alpha = C/s > 2.0$, so that oscillatory solutions may be expected. Also, from (9.61), such solutions would be inaccurate.

If a centred time difference expression is combined, as in the Richardson scheme (7.8), with a centred spatial difference expression to represent (9.56), the resulting algorithm is unconditionally unstable unless $s = 0$ which degenerates to the leapfrog scheme (9.15) for the convection equation (9.2).

The DuFort-Frankel scheme, applied to the transport equation, is

$$\frac{T_j^{n+1}-T_j^{n-1}}{2\Delta t}+u\frac{(T_{j+1}^n-T_{j-1}^n)}{2\Delta x}$$

$$-\frac{\alpha[T_{j-1}^n-(T_j^{n-1}+T_j^{n+1})+T_{j+1}^n]}{\Delta x^2}=0 \ . \tag{9.63}$$

The von Neumann analysis indicates that if the Courant number $C\leq 1$ there is no restriction on s. A Taylor series expansion about the (j, n)-th node indicates that, in contrast to the diffusion equation, it is necessary to ensure that $\Delta t\ll\Delta x\,(C^2\ll 1$ in Table 9.3) if (9.63) is to be consistent with (9.56). This is a very severe restriction.

If an upwind difference expression for $\partial T/\partial x$ replaces the centred difference representation in (9.57) the result, as an algorithm, can be written, for positive u,

$$T_j^{n+1}=(s+C)\,T_{j-1}^n+(1-2s-C)\,T_j^n+s\,T_{j+1}^n \ . \tag{9.64}$$

A Taylor series expansion indicates that (9.64) is consistent with (9.56) to $O(\Delta t, \Delta x)$. The truncation error for the upwind scheme shown in Table 9.3 indicates that an artificial diffusivity

$$\alpha'=0.5u\Delta x(1-C)$$

has been introduced. This same term appeared for the convection equation (Table 9.1). For the upwind scheme to generate accurate solutions of the transport equation it is required that

$$\alpha'\ll\alpha \quad\text{or}\quad R_{\text{cell}}\ll\frac{2}{1-C} \ .$$

The application of the von Neumann stability analysis to (9.64) gives the value for the amplification factor G shown in Table 9.3. The necessary condition for stability, $|G|\leq 1$, leads to the requirement

$$2s+C\leq 1 \ , \tag{9.65}$$

which is equivalent to

$$\Delta t\leq\frac{0.5\Delta x^2/\alpha}{1+0.5R_{\text{cell}}}, \tag{9.66}$$

which is considerably more restrictive than the limit for the diffusion equation (Table 7.1).

The major restrictions for the schemes considered so far are associated with the need to obtain accurate solutions when the derivatives are discretised with first-order approximations. The Lax-Wendroff scheme (9.16) provides a means of achieving second-order accuracy in both time and space. For the transport

Table 9.3. Algebraic (discretised) schemes for the transport equation $\partial \bar{T}/\partial t + u\partial \bar{T}/\partial x - \alpha\partial^2 \bar{T}/\partial x^2 = 0$

Scheme	Algebraic form	Truncation error^a (leading terms)	Amplification factor G $(\theta = m\pi\Delta x)$	Stability Restrictions	Remarks
FTCS	$\dfrac{\Delta T_j^{n+1}}{\Delta t} + uL_x T_j^n - \alpha L_{xx}T_j^n = 0$	$Cu(\Delta x/2)\dfrac{\partial^2 T}{\partial x^2}$ $-[C\alpha\Delta x - u(\Delta x^2/6)(1+2C^2)]\dfrac{\partial^3 T}{\partial x^3}$	$1 - 2s(1-\cos\theta) - iC\sin\theta$	$0 \leqq C^2 \leqq 2s \leqq 1$	$R_{\text{cell}} \ll 2/C$ for accuracy
Upwind	$\dfrac{\Delta T_j^{n+1}}{\Delta t} + u\dfrac{(T_j^n - T_{j-1}^n)}{\Delta x} - \alpha L_{xx}T_j^n = 0$	$-u(\Delta x/2)(1-C)\dfrac{\partial^2 T}{\partial x^2} - [C\alpha\Delta x$ $-u(\Delta x^2/6)(1-3C+2C^2)]\dfrac{\partial^3 T}{\partial x^3}$	$1 - (2s+C)(1-\cos\theta) - iC\sin\theta$	$C + 2s \leqq 1$	$R_{\text{cell}} \ll 2/(1-C)$ for accuracy
DuFort-Frankel	$(T_j^{n+1} - T_j^{n-1})/2\Delta t + uL_x T_j^n$ $-\dfrac{\alpha}{\Delta x^2}\{T_{j-1}^n - (T_j^{n-1} + T_j^{n+1})$ $+ T_{j+1}^n\} = 0$	$\alpha C^2 \dfrac{\partial^2 T}{\partial x^2} + (1-C^2)[u\Delta x^2/6]\dfrac{\partial^3 T}{\partial x^3}$ $-2\alpha^2 C^2/u\,\dfrac{\partial^3 T}{\partial x^3}$	$\dfrac{B \pm [B^2 - 8s(1+2s)]^{\frac{1}{2}}}{(2+4s)}$ where $B = 1 + 4s\cos\theta - i2C\sin\theta$	$C \leqq 1$	$C^2 \ll 1$ for accuracy

Scheme	Algebraic scheme	Modified equation terms	Amplification factor	Stability	Oscillations
Lax-Wendroff $\rule{0pt}{0pt}$	$\dfrac{\Delta T_j^{n+1}}{\Delta t}+uL_xT_j^n-\alpha^*L_{xx}T_j^n=0$ where $\alpha^*=\alpha+0.5uC\Delta x$	$-[C\alpha\,\Delta x-u(\Delta x^2/6)(1-C^2)]\dfrac{\partial^3 T}{\partial x^3}$ $+[C^2/u(\Delta x/2)-\alpha\Delta x^2/12]$ $-uC(\Delta x^3/8)(C^2-1)]\dfrac{\partial^4 T}{\partial x^4}$	$1-2s^*(1-\cos\theta)-iC\sin\theta$ where $s^*=\alpha^*\,\Delta t/\Delta x^2$	$0\leqq C^2\leqq 2s^*\leqq 1$	$R_{\text{cell}}\leqq 2$ to avoid spatial oscillations
Crank-Nicolson	$\dfrac{\Delta T_j^{n+1}}{\Delta t}$ $+\{uL_x-\alpha L_{xx}\}\left\{\dfrac{T_j^n+T_j^{n+1}}{2}\right\}=0$	$u(\Delta x^2/6)(1+0.5C^2)\dfrac{\partial^3 T}{\partial x^3}$ $-\alpha(\Delta x^2/12)(1+3C^2)\dfrac{\partial^4 T}{\partial x^4}$	$\dfrac{1-s(1-\cos\theta)-\mathrm{i}\,0.5C\sin\theta}{1+s(1-\cos\theta)+\mathrm{i}\,0.5C\sin\theta}$	None	$R_{\text{cell}}\leqq 2$ to avoid spatial oscillations
Three-level fully implicit	$\dfrac{3}{2}\dfrac{\Delta T_j^{n+1}}{\Delta t}-\dfrac{1}{2}\dfrac{\Delta T_j^n}{\Delta t}$ $+\{uL_x-\alpha L_{xx}\}\,T_j^{n+1}=0$	$u(\Delta x^2/6)(1+2C^2)\dfrac{\partial^3 T}{\partial x^3}$ $-\alpha(\Delta x^2/12)(1+12C^2)\dfrac{\partial^4 T}{\partial x^4}$	$\dfrac{1\pm\frac{1}{3}\mathrm{i}[3+16s(1-\cos\theta)+\mathrm{i}8C\sin\theta]^{\ddagger}}{2(1+\frac{4}{3}[2s(1-\cos\theta)+\mathrm{i}C\sin\theta])}$	None	$R_{\text{cell}}\leqq 2$ to avoid spatial oscillations
Linear F.E.M./ Crank-Nicolson	$M_x\dfrac{\Delta T_j^{n+1}}{\Delta t}+uL_x\left\{\dfrac{T_j^n+T_j^{n+1}}{2}\right\}$ $-\alpha L_{xx}\left\{\dfrac{T_j^n+T_j^{n+1}}{2}\right\}=0$	$uC^2(\Delta x^2/12)\dfrac{\partial^3 T}{\partial x^3}$ $+\alpha(\Delta x^2/12)(1-3C^2)\dfrac{\partial^4 T}{\partial x^4}$	$\dfrac{2+3\cos\theta-3s(1-\cos\theta)-\mathrm{i}1.5C\sin\theta}{2+3\cos\theta+3s(1-\cos\theta)+\mathrm{i}1.5C\sin\theta}$	None	$R_{\text{cell}}\leqq 2$ to avoid spatial oscillations

a The algebraic scheme is equivalent to $\partial T/\partial t+u\,\partial T/\partial x-\alpha\,\partial^2 T/\partial x^2+E(T)=0$

$L_x=\dfrac{1}{2\Delta x}\{-1,0,1\},\ L_{xx}=\dfrac{1}{\Delta x^2}\{1,2,1\},\ M_x=\{\tfrac{1}{6},\tfrac{2}{3},\tfrac{1}{6}\},\ C=u\Delta t/\Delta x,\ s=\alpha\Delta t/\Delta x^2,\ R_{\text{cell}}=C/s=u\Delta x/\alpha$

equation (9.56) the Lax-Wendroff scheme can be interpreted as an FTCS scheme with a modified diffusivity, $\alpha^* = \alpha(1 + 0.5\,CR_{cell})$. As $R_{cell} \to \infty$ the truncation errors and stability restrictions revert to those given in Table 9.1.

9.4.2 Implicit Schemes

For the diffusion equation implicit schemes are effective in removing the stability restriction $s \leq 0.5$. Here, the most effective one-dimensional two-level scheme for the diffusion equation, the Crank-Nicolson scheme, will be applied to the transport equation. Strictly, the present scheme should be described as a trapezoidal scheme, since the original Crank-Nicolson scheme was developed for the diffusion equation. However, the labels Crank-Nicolson and trapezoidal will be used interchangeably to refer to any two-level scheme that evaluates the spatial derivatives symmetrically about the $(n + 1/2)$-th time-level.

For the transport equation the Crank-Nicolson scheme gives

$$\frac{T_j^{n+1} - T_j^n}{\Delta t} + 0.5u\left(\frac{(T_{j+1}^n - T_{j-1}^n)}{2\Delta x} + \frac{T_{j+1}^{n+1} - T_{j-1}^{n+1}}{2\Delta x}\right)$$

$$-0.5\alpha\left(\frac{T_{j-1}^n - 2T_j^n + T_{j+1}^n}{\Delta x^2} + \frac{T_{j-1}^{n+1} - 2T_j^{n+1} + T_{j+1}^{n+1}}{\Delta x^2}\right) = 0 \; , \tag{9.67}$$

which can be written as an algorithm

$$-(s + 0.5C)\,T_{j-1}^{n+1} + 2(1 + s)\,T_j^{n+1} - (s - 0.5C)\,T_{j+1}^{n+1}$$
$$= (s + 0.5C)\,T_{j-1}^n + 2(1 - s)\,T_j^n + (s - 0.5C)\,T_{j+1}^n \; . \tag{9.68}$$

A Taylor series expansion of (9.68) about the (j, n)-th node indicates that it is consistent with (9.56) with a truncation error of $O(\Delta t^2, \Delta x^2)$. Thus, the problem of artificial diffusion arising from first-order truncation errors is avoided.

A von Neumann stability analysis of (9.68) produces the expression for the amplification factor G shown in Table 9.3, with no restriction on the magnitude of C or s for stability. But for solutions to be spatially non-oscillatory, the restriction $R_{cell} \leq 2$ is required.

For the long wavelength components (small $m\Delta x$) of the solution the dissipation and dispersion properties of computational algorithms for the transport equation can be inferred from the truncation errors listed in Table 9.3. Generally, the implicit schemes behave well. The physical dissipation in the system reduces the gross dispersion errors associated with short wavelengths ($m\Delta x \to \pi$) that occur with the convection equation. In particular, the finite element Crank-Nicolson scheme, shown in Table 9.3, has small dissipation and dispersion errors for small values of C. To obtain information about the short wavelength components it is necessary to carry out a Fourier analysis (Sect. 9.2.1). This technique is applied to a convecting temperature front in the next section.

9.4.3 TRAN: Convection of a Temperature Front

In this section some of the schemes described in Sects. 9.4.1 and 9.4.2 will be applied to the problem of a temperature front convecting with a velocity u through a fluid of thermal diffusivity α. For large values of the cell Reynolds (strictly Peclet) number, $R_{cell} = u \Delta x / \alpha$, the front width is restricted to a few grid spacings (Fig. 9.11) and all of the schemes described in Sects. 9.4.1 and 9.4.2 produce oscillatory solutions primarily associated with dispersion errors (Sect. 9.2).

Consequently, two new schemes will be developed here that produce much better dispersion-related behaviour. Both schemes will be based on Crank-Nicolson time differencing. The first scheme is an extension of the Crank-Nicolson mass operator scheme (9.20) that was applied to the linear convection equation (Sect. 9.1.5). For the one-dimensional transport equation (9.56) this becomes

$$M_x\left(\frac{T_j^{n+1}-T_j^n}{\Delta t}\right)+(uL_x-\alpha L_{xx})0.5(T_j^n+T_j^{n+1})=0 \ , \tag{9.69}$$

where $M_x \equiv \{\delta, 1-2\delta, \delta\}$ and L_x and L_{xx} are the conventional three-point centred difference formulae. The parameter δ will be chosen to reduce dispersion errors. It is clear that (9.69) will produce a tridiagonal system of equations that can be solved efficiently. A von Neumann stability analysis indicates that (9.69) is stable as long as $\delta \leq 0.25$.

Using the modified equation approach (Sect. 9.2.2) indicates that (9.69) is solving the following equivalent equation, as far as the longwave behaviour is concerned,

$$\frac{\partial T}{\partial t}+u\frac{\partial T}{\partial x}-\alpha\frac{\partial^2 T}{\partial x^2}+u\Delta x^2\left(\frac{1}{6}+\frac{C^2}{12}-\delta\right)\frac{\partial^3 T}{\partial x^3}$$

$$-\alpha\Delta x^2\left(\frac{1}{12}+\frac{C^2}{4}-\delta\right)\frac{\partial^4 T}{\partial x^4}+\ldots=0 \ . \tag{9.70}$$

Formally, (9.69) is a second-order scheme. The choice $\delta = 1/6 + C^2/12$ is seen to suppress the lowest-order dispersion term in the truncation error. If, in addition, $C^2 < 0.5$, the lowest-order dissipation term in the truncation error introduces positive dissipation. However, the optimal choice for δ requires $C \leq 1.0$ for stability.

An alternative means of obtaining better dispersion behaviour is to use the four-point upwind finite difference Crank-Nicolson scheme

$$\frac{T_j^{n+1}-T_j^n}{\Delta t}+(uL_x^{(4)}-\alpha L_{xx})0.5(T_j^n+T_j^{n+1})=0 \ , \tag{9.71}$$

where, for $u \geq 0$

$$L_x^{(4)}T=0.5\frac{(T_{j+1}-T_{j-1})}{\Delta x}+\frac{q(T_{j-2}-3T_{j-1}+3T_j-T_{j+1})}{3\Delta x} \ . \tag{9.72}$$

Equation (9.72) is the four-point convective operator introduced in conjunction with the steady convection-diffusion equation, Sect. 9.3.2.

Incorporation of (9.72) into (9.71) produces a quadridiagonal system of equations to be solved at each time-step. This requires using the generalised Thomas algorithm (Sect. 6.2.4). In program TRAN the generalised Thomas algorithm is implemented by making an additional forward sweep (lines 126–133) to render the system of equations tridiagonal. For the current application the generalised Thomas algorithm requires approximately 80% more operations than the conventional Thomas algorithm. The scheme represented by (9.71, 72) is stable as long as $q \geq -3/R_{cell}$. This is not a practical restriction since generally only non-negative values of q are of interest.

The equivalent modified equation (Sect. 9.2.2) to (9.71) is

$$\frac{\partial T}{\partial t} + u \frac{\partial T}{\partial x} - \alpha \frac{\partial^2 T}{\partial x^2} + u \Delta x^2 \left(\frac{(1-2q)}{6} + \frac{C^2}{12} \right) \frac{\partial^3 T}{\partial x^3}$$

$$- u \frac{\Delta x^3}{R_{cell}} \left(\frac{1 - 2R_{cell}q}{12} + \frac{C^2}{4} \right) \frac{\partial^4 T}{\partial x^4} + \ldots = 0 . \tag{9.73}$$

In (9.73) $u\Delta x^3/R_{cell} = \alpha\Delta x^2$. For flows for which $R_{cell} \gg 1$ the dispersive term is seen to be the largest term in the truncation error. This term can be eliminated by the choice $q = 0.5 + 0.25C^2$. For this choice of q and large values of R_{cell} the lowest-order dissipative term introduces positive dissipation.

The above schemes, (9.69, 71), and the schemes developed in Sects. 9.4.1 and 9.4.2 are applied to the propagating temperature front problem, Fig. 9.7. At $t=0$ a sharp front is located at $x=0$. For subsequent time the front convects to the right with a speed u and its profile loses its sharpness under the influence of the thermal diffusivity α. Consequently, for a given value of t, the larger the value of $R_{cell} = u\Delta x/\alpha$, the sharper the profile of the front.

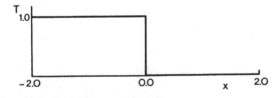

Fig. 9.7. Initial conditions for propagating temperature front

The governing equation for this problem is (9.56). For sufficiently small values of time the following are suitable boundary conditions:

$$T(-2, t) = 1.0 , \qquad T(2, t) = 0.0 . \tag{9.74}$$

An exact solution, by the separation-of-variables technique, is

$$\bar{T}(x, t) = 0.5 - \frac{2}{\pi} \sum_{k=1}^{N} \sin \left[(2k-1) \frac{\pi(x-ut)}{L} \right] \frac{\exp[-\alpha(2k-1)^2 \pi^2 t/L^2]}{2k-1} . \tag{9.75}$$

The various schemes, which are implemented in program TRAN, are indicated in Table 9.4. A listing of TRAN is provided in Fig. 9.8 and the main parameters are described in Table 9.5. The exact solution (9.75) is evaluated in the subroutine EXSOL (Fig. 9.9). Typical output from program TRAN, produced by the Lax-Wendroff scheme, is shown in Fig. 9.10.

Table 9.4. Various schemes implemented by program TRAN

ME	Description
1	FTCS scheme (9.58)
2	Lax-Wendroff scheme (9.58) with $s^* = s + 0.5C^2$
3	Explicit four-point upwind scheme (9.57, 72); EX-4PU
4	Crank-Nicolson, finite difference; $\delta = 0$, $q = 0$; CN-FDM
4	Crank-Nicolson, finite element; $\delta = 1/6$, $q = 0$; CN-FEM
4	Crank-Nicolson, mass operator; $\delta = 1/6 + C^2/12$, $q = 0$; CN-MO
4	Crank-Nicolson, four-point upwind; $\delta = 0$, $q = 0.5 + 0.25C^2$; CN-4PU

Table 9.5. Parameters used in program TRAN

Parameter	Description
ME	$= 1$, FTCS scheme (9.58)
ME	$= 2$, Lax-Wendroff scheme (9.16)
ME	$= 3$, explicit four-point upwind scheme (9.57, 72)
ME	$= 4$, general Crank-Nicolson scheme (9.69, 71)
JMAX	number of points in the interval $0 \leq x \leq 1$
NTIM	number of time steps
TIMAX	time of comparison with exact solution
JPR	$= 1$, propagating sine-wave;
	$= 2$, propagating temperature front
NEX	N in (9.75), used in EXSOL
LN	L in (9.75), used in EXSOL
C	Courant number (9.11)
U	u in (9.2, 56)
ALPH	thermal diffusivity α in (9.56)
S	$s = \alpha \Delta t / \Delta x^2$
RCEL	cell Reynolds number, $u \Delta x / \alpha$, Sect. 9.3
Q	q in (9.53, 72)
EM	δ in (9.69)
DT, DX, X	Δt, Δx, x
T	dependent variable (9.2, 56)
TEX	exact solution, (9.24 or 75)
AA, BB, CC	coefficients multiplying T_{j-1}^n, T_j^n and T_{j+1}^n for explicit schemes
AE, BE, CE	coefficients multiplying T_{j-1}^{n+1}, T_j^{n+1} and T_{j+1}^{n+1} for implicit schemes
A	elements in the tridiagonal matrix (9.18)
BANFAC	factorises $\underline{A}$ into upper triangular
BANSOL	solves $\underline{A}\,\boldsymbol{T} = \boldsymbol{R}$ for T_j^{n+1}
R	right-hand side of tridiagonal scheme (9.22)
TD	solution for T_j^{n+1} on return from BANSOL

```
1  C
2  C      TRAN SOLVES THE LINEAR TRANSPORT EQUATION USING
3  C      VARIOUS EXPLICIT AND IMPLICIT SCHEMES
4  C
5         DIMENSION R(65),T(65),TD(65),TEX(65),A(5,65)
6        1,D(4),X(65)
7         OPEN(1,FILE='TRAN.DAT')
8         OPEN(6,FILE='TRAN.OUT')
9         READ(1,1)ME,JMAX,NTIM,NEX,JPR,LN
10        READ(1,2)C,U,S,Q,EM
11      1 FORMAT(6I5)
12      2 FORMAT(2F5.2,3E10.3)
13 C
14        IF(JPR .EQ. 1)WRITE(6,3)
15        IF(JPR .EQ. 2)WRITE(6,4)
16      3 FORMAT(' PROPAGATING SINE-WAVE')
17      4 FORMAT(' PROPAGATING TEMP-FRONT')
18        JMAF = JMAX - 2
19        JMAP = JMAX - 1
20        AMP = JMAP
21        DX = 1.0/AMP
22        IF(JPR .EQ. 2)DX = 4./AMP
23        DT = C*DX/U
24        EL = LN
25        ALPH = S*DX*DX/DT
26        IF(ALPH .LT. 1.0E-10)ALPH = 1.0E-10
27        RCEL = U*DX/ALPH
28        QQ = Q*C/3.
29        IF(ME .LT. 3)QQ = 0.
30 C
31        MQ = 0
32        IF(ABS(QQ) .GT. 0.0001)MQ = 1
33        ATIM = NTIM
34        TIM = 0.
35        TIMAX = DT*ATIM
36 C
37        IF(ME .EQ. 1)WRITE(6,5)ME
38        IF(ME .EQ. 2)WRITE(6,6)ME
39        IF(ME .EQ. 3)WRITE(6,7)ME
40        IF(ME .EQ. 4)WRITE(6,8)ME
41      5 FORMAT(' ME =',I2,'    FTCS DIFFERENCING')
42      6 FORMAT(' ME =',I2,'    LAX-WENDROFF')
43      7 FORMAT(' ME =',I2,'    EXPLICIT 4PT UPWIND')
44      8 FORMAT(' ME =',I2,'    GENERAL CRANK-NICOLSON')
45        WRITE(6,9)JMAX,NTIM,C,U,DX,DT
46      9 FORMAT(' JMAX=',I3,'  NTIM=',I3,'  C=',F5.2,'  U=',F5.2,
47       1' DX=',F5.3,'  DT=',F5.3)
48        WRITE(6,10)S,ALPH,RCEL,Q,QQ
49        WRITE(6,11)NEX,LN,EM
50     10 FORMAT(' S=',F5.2,'  ALPH=',E10.3,'  RCEL=',F6.3,
51       1' Q=',F5.2,'  QQ=',F6.3)
52     11 FORMAT(' NEX=',I5,'  EL=',I5,'  EM=',E10.3)
53 C
```

Fig. 9.8. Listing of program TRAN

For a cell Reynolds number, $R_{cell} = 1.0$, solutions for the various explicit and implicit schemes are compared in Table 9.6. All methods produce smooth solutions with the Lax-Wendroff scheme being the most accurate explicit scheme. The most accurate implicit scheme is (9.71) with $q = 0.5$, which corresponds to the third-order treatment of $u\partial T/\partial x$.

For higher cell Reynolds numbers, results are presented for various Crank-Nicolson implicit schemes in Table 9.7 ($R_{cell} = 3.33$) and Table 9.8 ($R_{cell} = 100$). The Crank-Nicolson finite difference scheme serves as a reference. This solution is

```
54          IF(ME .GT. 3)GOTO 12
55          SS = S
56          IF(ME .EQ. 2)SS = S + 0.5*C*C
57          AA =  (0.5*C + SS) + 3.*QQ
58          BB = 1. - 2.*SS - 3.*QQ
59          CC = -0.5*C + SS + QQ
60          GOTO 13
61       12 AA = EM - 0.25*C - 0.5*S - 1.5*QQ
62          BB = 1.0 - 2.0*EM + S + 1.5*QQ
63          CC = EM + 0.25*C - 0.5*S - 0.5*QQ
64          AE = EM +0.25*C + 1.5*QQ + 0.5*S
65          BE = 1. - 2.*EM - 1.5*QQ - S
66          CE = EM - 0.25*C + 0.5*QQ + 0.5*S
67       13 WRITE(6,14)AA,BB,CC,AE,BE,CE
68       14 FORMAT(' AA=',F8.5,' BB=',F8.5,' CC=',F8.5,' AE=',F8.5,' BE=',
69          1F8.5,' CE=',F8.5,/)
70  C
71  C       INITIALISE T AND EVALUATE TEX
72  C
73          CALL EXSOL(JPR,JMAX,X,T,TEX,NEX,DX,U,ALPH,TIMAX,EL)
74  C
75          DO 16 J = 1,JMAX
76          DO 15 K = 1,5
77       15 A(K,J) = 0.
78       16 CONTINUE
79          WRITE(6,17)TIM
80       17 FORMAT(' INITIAL SOLUTION,  TIM =',F5.3)
81          WRITE(6,18)(X(J),J=1,JMAX)
82          WRITE(6,19)(T(J),J=1,JMAX)
83       18 FORMAT('   X=',12F6.3)
84       19 FORMAT('   T=',12F6.3)
85  C
86  C       MARCH SOLUTION IN TIME
87  C
88          DO 26 N = 1,NTIM
89          IF(ME .GT. 3)GOTO 21
90  C
91  C       EXPLICIT SCHEMES
92  C
93          D(1) = T(1)
94          D(2) = T(1)
95          D(3) = T(2)
96          DO 20 J = 2,JMAP
97          IF(ME .EQ. 3)D(4) = D(1)
98          D(1) = D(2)
99          D(2) = D(3)
100         D(3) = T(J+1)
101         T(J) = AA*D(1) + BB*D(2) + CC*D(3)
102         IF(ME .EQ. 3)T(J) = T(J) - QQ*D(4)
103      20 CONTINUE
104         GOTO 26
```

Fig. 9.8. (cont.) Listing of program TRAN

oscillatory showing an overshoot trailing the front which is particularly marked for $R_{cell} = 100$ (Table 9.8 and Fig. 9.11). The two low dispersion schemes (9.69, 71) are seen to be effective for $R_{cell} = 3.33$ in suppressing unphysical oscillations.

For $R_{cell} = 100$ the mass operator scheme with $\delta = \frac{1}{6} + C^2/12$ produces a sharp non-oscillatory temperature front. But the four-point upwind scheme produces a more spread out temperature front with a slightly oscillatory behaviour upwind and downwind of the front. In Fig. 9.11 grid-point values produced by the Crank-Nicolson mass operator (CN-MO) scheme are plotted only where they differ from the exact solution.

```
105  C
106  C      TRIDIAGONAL SYSTEM FOR IMPLICIT SCHEMES
107  C
108     21 IF(MQ .EQ. 1)DIM = 1.
109        DO 22 J = 2,JMAP
110        JM = J - 1
111        A(1,JM) = 0.5*QQ
112        A(2,JM) = AA
113        A(3,JM) = BB
114        A(4,JM) = CC
115        R(JM) = AE*T(JM) + BE*T(J) + CE*T(J+1)
116        IF(MQ .EQ. 1 .AND. J .GT. 2)DIM = T(JM-1)
117        IF(MQ .EQ. 1)R(JM) = R(JM) - 0.5*QQ*DIM
118     22 CONTINUE
119        R(1) = R(1) - A(2,1)*T(1)
120        IF(MQ .EQ. 0)GOTO 24
121        R(1) = R(1) - A(1,1)*T(1)
122        R(2) = R(2) - A(1,2)*T(1)
123  C
124  C      REDUCE A TO TRIDIAGONAL FORM
125  C
126        DO 23 JM = 3,JMAF
127        JMM = JM - 1
128        DUM = A(1,JM)/A(2,JMM)
129        A(2,JM) = A(2,JM) - A(3,JMM)*DUM
130        A(3,JM) = A(3,JM) - A(4,JMM)*DUM
131        A(1,JM) = 0.
132        R(JM) = R(JM) - R(JMM)*DUM
133     23 CONTINUE
134        A(1,1) = 0.
135        A(1,2) = 0.
136     24 A(2,1) = 0.
137        A(4,JMAF) = 0.
138  C
139        CALL BANFAC(A,JMAF,1)
140        CALL BANSOL(R,TD,A,JMAF,1)
141  C
142        DO 25 J = 2,JMAP
143     25 T(J) = TD(J-1)
144     26 CONTINUE
145  C
146        WRITE(6,27)TIMAX
147     27 FORMAT(' FINAL SOLUTION,    TIM =',F5.3)
148        WRITE(6,18)(X(J),J=1,JMAX)
149        WRITE(6,19)(T(J),J=1,JMAX)
150        WRITE(6,28)(TEX(J),J=1,JMAX)
151     28 FORMAT(' TEX=',12F6.3)
152        SUM = 0.
153        DO 29 J = 2,JMAP
154     29 SUM = SUM + (T(J) - TEX(J))**2
155        RMS = SQRT(SUM/(AMP-1.))
156        WRITE(6,30)RMS
157     30 FORMAT(' RMS ERR=',E12.5)
158        STOP
159        END
```

Fig. 9.8. (cont.) Listing of program TRAN

Further information about the relative performance of the two schemes (9.69 and 71) can be deduced by carrying out a Fourier analysis as in Sect. 9.2.1. The initial conditions are represented by a Fourier series and explicit expression are obtained from the discretised equation for the amplitude ratio G_m and the phase angle ϕ_m for each Fourier mode θ_m.

```
1          SUBROUTINE EXSOL(JPR,JMAX,X,T,TEX,NEX,DX,U,ALPH,TIMAX,EL)
2  C
3  C       SETS THE INITIAL T SOLUTION AND FINAL EXACT (TEX) SOLUTION
4  C
5          DIMENSION T(65),TEX(65),X(65)
6          JMAP = JMAX - 1
7          PI = 3.141592654
8          IF(JPR .EQ. 1)XST = 0.
9          IF(JPR .EQ. 2)XST = - 2.0
10         DO 1 J = 1,JMAX
11         AJ = J - 1
12         X(J) = XST + AJ*DX
13         T(J) = 0.
14       1 TEX(J) = 0.
15         IF(JPR .EQ. 2)GOTO 3
16 C
17 C       EXACT SOLUTION FOR PROPAGATING SINE-WAVE
18 C
19         JM = 0.1001/DX + 1.0
20         INC = U*TIMAX/DX + 0.001
21         DO 2 J = 1,JM
22         T(J) = SIN(10.*PI*X(J))
23         JP = J + INC
24       2 TEX(JP) = T(J)
25         RETURN
26 C
27 C       EXACT SOLUTION FOR PROPAGATING TEMP-FRONT
28 C
29       3 T(1) = 1.0
30         TEX(1) = 1.0
31         DO 5 J = 2,JMAP
32         IF(X(J) .LT. 0.)T(J) = 1.0
33         IF(ABS(X(J)) .LT. 1.0E-04)T(J) = 0.5
34         DEM = 0.
35         DO 4 K = 1,NEX
36         AK = 2*K - 1
37         DUM = AK*PI/EL
38         SNE = SIN(DUM*(X(J)-U*TIMAX))
39         DIM = - ALPH*DUM*DUM*TIMAX
40         IF(DIM .LT. -20.)GOTO 5
41       4 DEM = DEM + (SNE/AK)*EXP(DIM)
42       5 TEX(J) = 0.5 - 2.*DEM/PI
43         RETURN
44         END
```

Fig. 9.9. Listing of subroutine EXSOL

```
PROPAGATING TEMP-FRONT
ME = 2    LAX-WENDROFF
JMAX= 21   NTIM= 10   C=  .25   U=  .50   DX= .200   DT= .100
S=  .25   ALPH=  .100E+00   RCEL= 1.000   Q=  .00   QQ=  .000
NEX= 100   EL=  20   EM=  .000E+00
AA=  .40625 BB=  .43750 CC=  .15625 AE=  .00000 BE=  .00000 CE=  .00000

INITIAL SOLUTION,  TIM = .000
  X=-2.000-1.800-1.600-1.400-1.200-1.000 -.800 -.600 -.400 -.200  .000  .200
  X=  .400  .600  .800 1.000 1.200 1.400 1.600 1.800 2.000
  T= 1.000 1.000 1.000 1.000 1.000 1.000 1.000 1.000 1.000  .500  .000
  T=  .000  .000  .000  .000  .000  .000  .000  .000  .000
FINAL SOLUTION,    TIM =1.000
  X=-2.000-1.800-1.600-1.400-1.200-1.000 -.800 -.600 -.400 -.200  .000  .200
  X=  .400  .600  .800 1.000 1.200 1.400 1.600 1.800 2.000
  T= 1.000 1.000 1.000 1.000 1.000  .999  .997  .991  .973  .934  .862  .747
  T=  .593  .422  .261  .137  .059  .020  .005  .001  .000
TEX= 1.000 1.000 1.000 1.000 1.000 1.000  .998  .993  .978  .941  .868  .749
TEX=  .588  .412  .251  .132  .059  .022  .007  .002  .000
RMS ERR=  .44842E-02
```

Fig. 9.10. Typical output from program TRAN

Table 9.6. Temperature front solution at $t = 1.00$, for $R_{cell} = 1.0$, $C = 0.25$, $u = 0.50$, $\Delta x = 0.20$, $\Delta t = 0.05$

Scheme \ x	−0.40	−0.20	0.00	0.20	0.40	0.60	0.80	1.00	1.20	1.40	1.60	rms error
Exact	0.978	0.941	0.868	0.749	0.588	0.412	0.251	0.132	0.059	0.022	0.007	—
FTCS	0.981	0.947	0.877	0.760	0.598	0.414	0.245	0.121	0.047	0.014	0.003	0.006
Lax-Wendroff	0.973	0.934	0.862	0.747	0.593	0.422	0.261	0.137	0.059	0.020	0.005	0.004
EX-4PU, $q = 0.5$	0.976	0.940	0.872	0.762	0.609	0.431	0.258	0.123	0.040	0.005	−0.003	0.010
CN-FDM	0.981	0.946	0.870	0.742	0.575	0.398	0.245	0.135	0.067	0.030	0.012	0.006
CN-FEM	0.977	0.937	0.861	0.742	0.585	0.415	0.258	0.138	0.063	0.023	0.007	0.004
CN-4PU, $q = 0.5$	0.976	0.939	0.865	0.746	0.586	0.411	0.253	0.135	0.062	0.024	0.008	0.002

Table 9.7. Temperature front solution at $t = 1.00$ for $R_{cell} = 3.33$, $C = 1.00$, $u = 1.00$, $\Delta x = 0.10$, $\Delta t = 0.10$

x Scheme	0.50	0.60	0.70	0.80	0.90	1.00	1.10	1.20	1.30	1.40	1.50	rms error
Exact	0.979	0.949	0.890	0.793	0.658	0.500	0.342	0.207	0.110	0.051	0.021	—
CN-FDM	1.016	0.981	0.894	0.760	0.601	0.443	0.306	0.200	0.123	0.073	0.041	0.018
CN-4PU, $q = 0.75$	0.979	0.947	0.887	0.788	0.653	0.497	0.342	0.212	0.117	0.056	0.023	0.002
CN-MO, $\delta = 0.25$	0.972	0.950	0.883	0.795	0.660	0.504	0.343	0.206	0.109	0.052	0.022	0.002

Table 9.8. Temperature front solution at $t = 1.00$ for $R_{cell} = 100$, $C = 1.00$, $u = 1.00$, $\Delta x = 0.10$, $\Delta t = 0.10$

x Scheme	0.50	0.60	0.70	0.80	0.90	1.00	1.10	1.20	1.30	1.40	1.50	rms error
Exact	1.000	1.000	1.000	1.000	0.987	0.500	0.013	0.000	0.000	0.000	0.000	—
CN-FDM	1.217	1.253	1.093	0.825	0.553	0.335	0.187	0.098	0.048	0.022	0.010	0.113
CN-4PU, $q = 0.75$	1.010	1.050	1.056	0.958	0.741	0.466	0.223	0.065	-0.007	-0.024	-0.018	0.059
CN-MO, $\delta = 0.25$	1.000	1.000	1.000	0.998	0.954	0.501	0.045	0.002	0.000	0.000	0.000	0.013

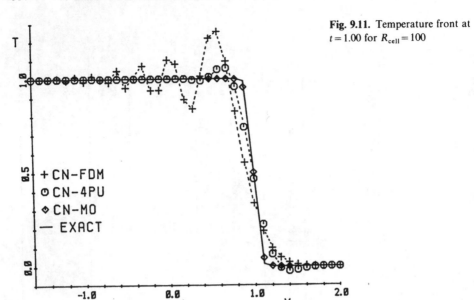

Fig. 9.11. Temperature front at $t = 1.00$ for $R_{cell} = 100$

For the Crank-Nicolson mass operator scheme (9.69) the following expressions are obtained:

$$G_m = \left(\frac{[1 - (2\delta + s)(1 - \cos \theta_m)]^2 + (0.5 \, C \sin \theta_m)^2}{[1 - (2\delta - s)(1 - \cos \theta_m)]^2 + (0.5 \, C \sin \theta_m)^2} \right)^{\frac{1}{2}} \quad \text{and} \tag{9.76}$$

$$\tan \phi_m = \frac{-C \sin \theta_m [1 - 2\delta(1 - \cos \theta_m)]}{[1 - 2\delta(1 - \cos \theta_m)]^2 - [0.5 \, C \sin \theta_m]^2 - [s(1 - \cos \theta_m)]^2} \tag{9.77}$$

The corresponding expressions for the exact solution are

$$G_{ex,m} = \exp[-C \Delta x^2 / R_{cell}] \quad \text{and} \quad \tan \phi_{ex,m} = \tan(-C\theta_m) \ . \tag{9.78}$$

For conditions corresponding to Tables 9.7 and 9.8 the ratio $G_m/G_{ex,m}$ and ϕ_m are presented in Tables 9.9 and 9.10 for various values of θ_m.

Table 9.9. Wavelength behaviour for $R_{cell} = 3.33$, $C = 1.0$, $\Delta x = 0.10$

Scheme	$\theta_m/\pi =$	0.05	0.25	0.50	0.75	1.00
CN-FDM	$G_m/G_{ex,m}$	0.996	0.858	0.619	0.389	0.251
CN-MO	$G_m/G_{ex,m}$	0.996	0.841	0.573	0.683	1.003
CN-4PU	$G_m/G_{ex,m}$	0.996	0.827	0.438	0.216	0.231
exact	$\phi_{ex,m}$	−9.00	−45.00	−90.00	−135.00	−180.00
CN-FDM	ϕ_m	−8.94	−39.19	−56.58	−49.09	0.00
CN-MO	ϕ_m	−9.00	−45.37	−100.21	−164.20	−180.00
CN-4PU	ϕ_m	−8.96	−40.65	−71.29	−130.14	−180.00

Table 9.10. Wavelength behaviour for $R_{cell}=100$, $C=1.0$, $\Delta x=0.10$

Scheme	$\theta_m/\pi=$	0.05	0.25	0.50	0.75	1.00
CN-FDM	$G_m/G_{ex,m}$	1.000	0.995	0.984	0.970	0.961
CN-MO	$G_m/G_{ex,m}$	1.000	0.994	0.980	0.967	1.000
CN-4PU	$G_m/G_{ex,m}$	1.000	0.958	0.674	0.277	0.010
exact	$\phi_{ex,m}$	-9.00	-45.00	-90.00	-135.00	-180.00
CN-FDM	ϕ_m	-8.94	-38.94	-53.13	-38.95	0.00
CN-MO	ϕ_m	-9.00	-45.00	-90.01	-135.08	-180.00
CN-4PU	ϕ_m	-8.96	-40.27	-61.30	-73.14	-180.00

For the Crank-Nicolson four-point upwind scheme (9.71) the amplitude ratio G_m and phase angle ϕ_m are

$$G_m=\left(\frac{[1-s^*(1-\cos\theta_m)]^2+\{0.5\,C\sin\theta_m[1+q(1-\cos\theta_m)/6]\}^2}{[1+s^*(1-\cos\theta_m)]^2+\{0.5\,C\sin\theta_m[1+q(1-\cos\theta_m)/6]\}^2}\right)^{\frac{1}{2}} \tag{9.79}$$

and

$$\tan\phi_m=\frac{-C\sin\theta_m[1+q(1-\cos\theta_m)/6]}{1-[s^*(1-\cos\theta_m)]^2-\{0.5C\sin\theta_m[1+q(1-\cos\theta_m)/6]\}^2}, \tag{9.80}$$

where $s^*=s+qC(1-\cos\theta_m)/3$. The amplitude ratio and phase angle variation with θ_m for $R_{cell}=3.33$ and 100 are shown in Tables 9.9 and 9.10, respectively.

For $R_{cell}=3.33$ (Table 9.9) the finite difference Crank-Nicolson scheme is seen to show poor agreement with the exact phase angle particularly for short wavelengths, $\theta_m\to\pi$. By contrast, the mass operator and four-point schemes show good phase agreement. Both schemes tend to be rather dissipative at intermediate wavelengths ($\theta_m\approx\pi/2$) and the four-point upwind scheme is dissipative for shorter wavelengths as well.

For $R_{cell}=100$ (Table 9.10) the Crank-Nicolson finite difference scheme maintains the amplitude for all wavelengths but introduces very significant phase errors for short wavelengths. This is consistent with the oscillatory temperature front solution shown in Fig. 9.11. The four-point upwind scheme shows some phase error for intermediate wavelengths but compensates for this by having significant attenuation of short wavelengths. The corresponding temperature front solution (Fig. 9.11) shows some oscillation and some smoothing out of the sharp temperature profile. The Crank-Nicolson mass operator scheme maintains the amplitude as well as the Crank-Nicolson finite difference scheme, but also achieves excellent phase behaviour. Consequently, the sharp profile of the temperature front (Fig. 9.11) is not unexpected.

Although the variable δ mass-operator scheme and the variable q four-point upwind scheme are both effective in reducing dispersion errors, their influence arises from the discretisation of different terms in the governing equation (9.56).

The mass operator is associated with the time derivative $\partial \bar{T}/\partial t$, whereas the four-point upwind differencing is associated with the convective term $u\partial \bar{T}/\partial x$. It follows that the optimal mass-operator strategy is not available for the one-dimensional steady convection-diffusion equation (Sect. 9.3) However, for multidimensional transport equations, mass operators appear in the finite element method associated with undifferentiated spatial directions. Consequently, the variable δ mass-operator mechanism for reducing dispersion-like behaviour is available for both steady and unsteady multidimensional transport equations (Sect. 9.5).

A similar Fourier analysis to the present is provided by Pinder and Gray (1977, pp. 150–162) for conventional finite difference and finite element schemes applied to the one-dimensional transport equation. A Fourier analysis of the Taylor-Galerkin formulation (Sect. 9.2.3) applied to the one-dimensional transport equation is provided by Donea et al. (1984).

In this section the various analytical techniques, von Neumann stability, modified equation approach and Fourier analysis, have been used to analyse some conventional schemes for convection-dominated problems and to design some better schemes.

9.5 Two-Dimensional Transport Equation

The additional complications of considering the multidimensional transport equation are comparable to those in progressing from the one-dimensional to multi-dimensional diffusion equation. In two dimensions, the transport equation can be written

$$\frac{\partial \bar{T}}{\partial t} + u \frac{\partial \bar{T}}{\partial x} + v \frac{\partial \bar{T}}{\partial y} - \alpha_x \frac{\partial^2 \bar{T}}{\partial x^2} - \alpha_y \frac{\partial^2 \bar{T}}{\partial y^2} = 0 \ . \tag{9.81}$$

The more severe stability restriction that occurred for explicit two-dimensional algorithms for the diffusion equation is also apparent for explicit representations of 9.81. For example, the two-dimensional version of the FTCS scheme (9.58) produces the algorithm

$$T_{j,k}^{n+1} = (s_x + 0.5C_x)\, T_{j-1,k}^n + (1 - 2s_x - 2s_y)\, T_{j,k}^n + (s_x - 0.5C_x)\, T_{j+1,k}^n$$

$$+ (s_y + 0.5C_y)\, T_{j,k-1}^n + (s_y - 0.5C_y)\, T_{j,k+1}^n \ , \quad \text{where} \tag{9.82}$$

$$s_x = \alpha_x \Delta t/\Delta x^2 \ , \quad C_x = u\Delta t/\Delta x \ , \quad s_y = \alpha_y \Delta t/\Delta y^2 \ , \quad C_y = v\Delta t/\Delta y \ .$$

A von Neumann stability analysis of (9.82) indicates that for a stable solution, the following restrictions apply (Hindmarsh et al. 1984)

$$(s_x + s_y) \leqq 0.5 \ , \quad \frac{C_x^2}{s_x} + \frac{C_y^2}{s_y} \leqq 2 \ . \tag{9.83}$$

If $s_x = s_y = s$ then $s \leqq 0.25$ for stability which is twice as severe as in one dimension.

The other major problem in multidimensions is the introduction of artificial diffusion, and particularly artificial cross-wind diffusion associated with first-order differencing schemes. As we have already seen the use of a first-order time differencing or upwind differencing for the convective terms introduces artificial diffusion (and dispersion) terms. In two dimensions the equivalent first-order differencing scheme introduces artificial diffusivities

$$\alpha'_x = -0.5u\,\Delta x(C_x - 1) \quad \text{and} \quad \alpha'_y = -0.5v\Delta y(C_y - 1) \ . \tag{9.84}$$

The errors introduced by these terms are particularly severe when the velocity vector is at $45°$ to either coordinate direction (de Vahl Davis and Mallinson 1976). This is discussed further in Sect. 9.5.3.

9.5.1 Split Formulations

Implicit algorithms for the two-dimensional transport equation avoid most of the stability difficulties of the explicit schemes but have the problem of achieving an efficient solution algorithm; just as for the diffusion equation.

Here, an examination is made of a typical split formulation for the two-dimensional transport equation. Application of the Galerkin finite element method (Sect. 5.3) with bilinear rectangular elements, to the transport equation (9.81) produces the result

$$M_x \otimes M_y \left[\frac{\partial T}{\partial t}\right]_{j,k} = [-uM_y \otimes L_x - vM_x \otimes L_y + \alpha_x M_y \otimes L_{xx}$$

$$+ \alpha_y M_x \otimes L_{yy}]\, T_{j,k} \ , \tag{9.85}$$

where $L_x \equiv (-1, 0, 1)/2\Delta x$, $L_y \equiv (1, 0, -1)^T/2\Delta y$ and $M_x = (\frac{1}{6}, \frac{2}{3}, \frac{1}{6})$ for a uniform grid. The operators L_{xx}, L_{yy} and M_y are given by (8.30 and 31).

Equation (9.85) may be compared with (8.29). The additional terms in (9.85) come from the convective terms. However, the directional mass operators in the convective terms behave in the same manner as for the diffusion terms. That is, for the finite element method, they spread the influence of the differential operators, L_x and L_y, in the normal direction. This follows from the tensor product $\otimes$ and the form of the operators M_x and M_y. Equation (9.85) can be generalised by defining

$$M_x = (\delta_x, 1 - 2\delta_x, \delta_x) \ , \quad M_y = (\delta_y, 1 - 2\delta_y, \delta_y)^T \ , \tag{9.86}$$

where δ_x and δ_y are parameters that may be chosen to improve the scheme just as δ in (9.21) was chosen in Sect. 9.4.3 to minimise dispersion errors. The choice, $\delta_x = \delta_y = 0$, produces a three-point finite difference scheme; the choice, $\delta_x = \delta_y = \frac{1}{6}$, gives the linear finite element scheme.

With a three-level representation for $\partial \bar{T}/\partial t$, (9.85) can be written as

$$M_x \otimes M_y \left[(1 + \gamma)\frac{\Delta T^{n+1}}{\Delta t} - \gamma \frac{\Delta T^n}{\Delta t}\right] = (1 - \beta)\,\text{RHS}^n + \beta\text{RHS}^{n+1} \ , \tag{9.87}$$

where

$$\Delta T^{n+1} = T^{n+1} - T^n \quad \text{and} \quad \Delta T^n = T^n - T^{n-1} \quad \text{and}$$

$$\text{RHS} = -[M_y \otimes (uL_x - \alpha_x L_{xx}) + M_x \otimes (vL_y - \alpha_y L_{yy})] T_{j,k} .$$

This same scheme was applied to a difference representation, (8.26), of the diffusion equation.

By introducing the same splitting as was applied to (8.29) the following two-stage algorithm for the two-dimensional transport equation is obtained:

$$\left[M_x + \left(\frac{\beta}{1+\gamma} \right) \Delta t (uL_x - \alpha_x L_{xx}) \right] \Delta T^*_{j,k} = \left(\frac{\Delta t}{1+\gamma} \right) \text{RHS}^n$$

$$+ \left(\frac{\gamma}{1+\gamma} \right) M_x \otimes M_y \Delta T^n_{j,k} , \qquad (9.88)$$

and

$$\left[M_y + \left(\frac{\beta}{1+\gamma} \right) \Delta t (vL_y - \alpha_y L_{yy}) \right] \Delta T^{n+1}_{j,k} = \Delta T^*_{j,k} . \qquad (9.89)$$

Clearly the inclusion of the convective terms, i.e. uL_x and vL_y, does not prevent the efficient splitting of (9.87) into a collection of tridiagonal subsystems of equations associated with each gridline.

The algorithm given by (9.88, 89) is consistent with (9.81) and has a truncation error of $O(\Delta t^2, \Delta x^2, \Delta y^2)$ for the two-level scheme, $\gamma = 0$, $\beta = 0.5$ and for the three-level scheme, $\gamma = 0.5$, $\beta = 1.0$. The same order of accuracy is achieved whether a linear finite element or a three-point finite difference scheme is used to evaluate the operators M_x, M_y, L_x, L_{xx}, L_y and L_{yy}. However, the coefficients in the truncation error are different in the different cases and the finite element formulation gives more accurate solutions.

A von Neumann stability analysis of (9.88 and 89) leads to a quadratic equation (for the three-level scheme) for the amplification factor G. Evaluation of this equation to assess the stability is most easily carried out numerically. The results indicate unconditional stability with $\beta \geq 0.5$ for both the finite element and finite difference formulations.

9.5.2 THERM: Thermal Entry Problem

In this section the split formulations developed in Sect. 9.5.1 will be applied to the problem of the flow of a 'cold' fluid entering a 'hot' two-dimensional duct with a very high aspect-ratio cross-section. Away from the side-walls of the duct the temperature distribution is effectively two dimensional, and governed by the steady form of the following energy equation:

$$\frac{\partial \bar{T}}{\partial t} + \frac{\partial (u\bar{T})}{\partial x} + \frac{\partial (v\bar{T})}{\partial y} - \alpha_x \frac{\partial^2 \bar{T}}{\partial x^2} - \alpha_y \frac{\partial^2 \bar{T}}{\partial y^2} = 0 , \qquad (9.90)$$

where $\alpha_x = \alpha_y = \alpha$, the thermal diffusivity. Equation (9.90) is expressed in unsteady form to facilitate a pseudo-transient solution strategy (Sect. 6.4). If the velocity components u and v are known (9.90) is a linear two-dimensional transport equation.

It is convenient to solve (9.90) in nondimensional form, after introduction of the nondimensional variables

$$T' = \frac{T - T_0}{T_w - T_0} , \quad u' = \frac{u}{u_m}, \quad v' = \frac{v\,\mathrm{Re}}{2.5u_m} , \quad t' = \frac{2.5u_m t}{a\,\mathrm{Re}} \tag{9.91}$$

$$x' = \frac{2.5x}{a\,\mathrm{Re}} , \quad y' = \frac{y}{a} , \quad \mathrm{Re} = \frac{4au_m}{v} \quad \text{and} \quad \mathrm{Pr} = \frac{k}{\mu Cp} ,$$

where T_0 and T_w are the entry and wall temperatures, u_m is the mean axial velocity and Re and Pr are the Reynolds and Prandtl numbers, respectively (Sect. 11.2.5). The particular choice for x' and the other nondimensional variables ensures that the temperature profile develops over approximately the same range of x' for different choices of Re. If the above nondimensionalisation is substituted into (9.90) the same equation results in terms of primed variables with

$$\alpha_x = \frac{10}{\mathrm{Pr}\,\mathrm{Re}^2} \quad \text{and} \quad \alpha_y = \frac{1.6}{\mathrm{Pr}} . \tag{9.92}$$

In the rest of this section primes will be dropped from nondimensional variables.

For the computational domain shown in Fig. 9.12, appropriate boundary conditions are

$$T(0, y) = 0 , \quad \text{on } x = 0 \quad \text{and} \quad \frac{\partial T}{\partial x} = 0 \quad \text{on } x = x_{max} \tag{9.93}$$

$$T(x, \pm 1) = 1 , \quad \text{on } y = \pm 1 .$$

Application of the group finite element formulation (Sect. 10.3 and Fletcher 1983) with linear interpolation on rectangular elements produces a system of ordinary differential equations

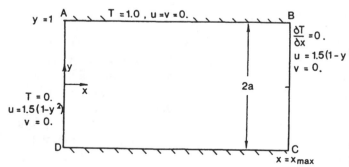

Fig. 9.12. Computational domain for thermal entry problem

$$M_x \otimes M_y \left[\frac{\partial T}{\partial t} \right]_{j,k} = \{ M_y \otimes [-L_x(uT) + \alpha_x L_{xx} T]$$

$$+ M_x \otimes [-L_y(vT) + \alpha_y L_{yy} T] \}_{j,k} , \qquad (9.94)$$

which differs from (9.85) only by the inclusion of u and v in the convective operators L_x and L_y. Consequently, the two-stage algorithm (9.88, 89) is implemented in program THERM (Fig. 9.13) with L_x and L_y operating on uT and vT, respectively.

Program THERM is written to operate on any prescribed velocity distribution. However, to facilitate the comparison of the temperature solution with a semi-analytic solution (Brown 1960) a fully developed velocity distribution is assumed

$$u = 1.5(1 - y^2) , \qquad v = 0 . \qquad (9.95)$$

To avoid the discontinuity in the solution at A (Fig. 9.12) the following boundary condition on $x = 0$ is implemented in THERM:

$$T(0, y) = y^{32} . \qquad (9.96)$$

This produces an inlet temperature profile that is zero except close to the corners $x = 0$, $y = \pm 1$.

The evaluation of RHS^n in (9.88) is carried out in subroutine RETHE (Fig. 9.14). The major parameters used in program THERM and subroutine RETHE are described in Table 9.11. Typical output from THERM is shown in Fig. 9.15 for the approximate factorisation finite difference method.

The centre-line solution is compared with the semi-analytic solution (Brown 1960) which is evaluated in subroutine TEXCL (Fig. 16.5; Vol. 2, p. 268). The semi-analytic solution is based on a reduced form of (9.90) which neglects the axial diffusion term $\alpha_x \partial^2 T / \partial x^2$. The justification for this is discussed in Sect. 16.1.4. It may be noted that the resulting semi-analytic solution is a good approximation to the solution of the present problem except close to the entrance $(x = 0)$ and for small values of Re. Because of the limitations on Brown's solution it will be referred to as semi-exact in program THERM and Table 9.12.

The centre-line solutions for a relatively coarse grid are shown in Table 9.12 and are plotted in Fig. 9.16. The rms errors shown in Table 9.12 are based on the centre-line solution and exclude the point $j = 2$ because of the possible doubtful relevance of the semi-exact solution close to the entrance $x = 0$.

Also shown in Table 9.12 and Fig. 9.16 are results obtained with the four-point upwind discretisation of $\partial(uT)/\partial x$ replacing L_x in (9.94) and the equivalent of (9.88) and (9.89). This is the same four-point upwind discretisation that was examined in Sects. 9.3.2 and 9.4.3. A consistent treatment of a two-dimensional problem implies that a four-point upwind discretisation should also be introduced for $\partial(vT)/\partial y$. This is not done in program THERM because v is zero in the current example and consequently, the term $\partial(vT)/\partial y$ has no influence on the solution. The four-point upwind scheme is not applied at $j = 2$ because of the difficulty of ascribing an appropriate value to $T_{0,k}$.

```
 1  C
 2  C      THERM APPLIES APPROXIMATION FACTORISATION TO SOLVE
 3  C      THE UNSTEADY THERMAL ENTRY PROBLEM  FOR T(X,Y)
 4  C
 5         DIMENSION T(41,42),DT(41,42),R(41,42),U(41,42),V(41,42),
 6        1EMX(3),EMY(3),B(5,65),RRT(65),DDT(65),CX(3),CY(3),TEX(42)
 7        2,ALF(10),DYFL(10),CXQ(4)
 8         COMMON CX,CXQ,CY,CCX,CCY,EMX,EMY,NX,NY,R,T,DT,U,V
 9         DATA  ALF/1.6815953,5.6698573,9.6682425,13.6676614,17.6673736,
10        121.6672053,25.6670965,29.6670210,33.6669661,37.6664327/
11         DATA DYFL/-0.9904370,1.1791073,-1.2862487,1.3620196,-1.4213257,
12        11.4704012,-1.5124603,1.5493860,-1.5823802,1.6122503/
13  C
14         OPEN(1,FILE='THERM.DAT')
15         OPEN(6,FILE='THERM.OUT')
16         READ(1,1)NX,NY,ME,ITMAX,GAM,BET,Q
17         READ(1,2)DTIM,EPS,RE,PR,XMAX
18       1 FORMAT(4I5,3F5.2)
19       2 FORMAT(3E10.3,2F5.2)
20  C
21         NXS = NX + 1
22         NXP = NX - 1
23         NYP = NY - 1
24         NYPP = NYP - 1
25         NYH = NY/2 + 1
26         ANX = NXP
27         DX = XMAX/ANX
28         ANY = NYP
29         DY = 2./ANY
30         ALX = 10./RE/RE/PR
31         ALY = 1.6/PR
32         CX(1) = -0.5/DX
33         CX(2) = 0.
34         CX(3) =  0.5/DX
35         CY(1) = -0.5/DY
36         CY(2) = 0.
37         CY(3) =  0.5/DY
38         CXQ(1) = Q/DX/3.
39         CXQ(2) = -3.*CXQ(1)
40         CXQ(3) = -CXQ(2)
41         CXQ(4) = -CXQ(1)
42         CCX = ALX/DX/DX
43         CCY = ALY/DY/DY
44         IF(ME .NE. 2)Q = 0.
45         EMX(1) = 0.
46         IF(ME .EQ. 3)EMX(1) = 1./6.
47         EMX(2) = 1. - 2.*EMX(1)
48         EMX(3) = EMX(1)
49         DO 3 J = 1,3
50       3 EMY(J) = EMX(J)
51  C
52         IF(ME .EQ. 1)WRITE(6,4)NX,NY,ME,ITMAX
53         IF(ME .EQ. 2)WRITE(6,5)NX,NY,ME,ITMAX
54         IF(ME .EQ. 3)WRITE(6,6)NX,NY,ME,ITMAX
55       4 FORMAT(' THERMAL ENTRY PROBLEM: 3PT FDM,  NX,NY =',2I3,'  ME =',
56        1I2,' ITMAX =',I3)
57       5 FORMAT(' THERMAL ENTRY PROBLEM: 4PT UPWIND FDM,  NX,NY =',2I3,
58        1' ME =',I2,'  ITMAX=',I3)
59       6 FORMAT(' THERMAL ENTRY PROBLEM: LIN FDM,  NX,NY =',2I3,'  ME =',
60        1I2,'  ITMAX =',I3)
61         WRITE(6,7)GAM,BET,DTIM,XMAX,PR,RE,Q
62       7 FORMAT(' GAM =',F5.2,'  BETA =',F5.2,'  DTIM =',F6.3,'  XMAX =',
63        1F5.2,'  PR =',F5.2,'  RE =',F5.1,'  Q =',F5.2,/)
64  C
65  C     GENERATE INITIAL SOLUTION
66  C
```

Fig. 9.13. Listing of program THERM

```
67              DO 9 K = 1,NY
68              AK = K - 1
69              Y = -1. + AK*DY
70              T(K,1) = Y**32
71              T(K,NXS) = 1.
72              U(K,1) = 1.5*(1. - Y*Y)
73              V(K,1) = 0.
74              DO 8 J = 2,NXS
75              U(K,J) = U(K,1)
76              V(K,J) = 0.
77              IF(J .EQ. NXS)GOTO 8
78              AJ = J - 1
79              T(K,J) = T(K,1) + AJ*DX*(T(K,NXS)-T(K,1))/XMAX
80           8 CONTINUE
81           9 CONTINUE
82              DO 12 J = 1,NXS
83              DO 10 K = 1,NY
84              DT(K,J) = 0.
85          10 R(K,J) = 0.
86              DO 11 K = 1,5
87          11 B(K,J) = 0.
88          12 CONTINUE
89              GAMH = GAM
90              GAM = 0.
91              BETH = BET
92              BET = 0.5
93              ITER = 0
94  C
95          13 DBT = BET*DTIM/(1.+GAM)
96              CXA = 0.5*DBT/DX
97              CYA = 0.5*DBT/DY
98              CCXA = DBT*CCX
99              CCYA = DBT*CCY
100 C
101             CALL RETHE(RMST,GAM,DTIM)
102 C
103             IF(RMST .LT. EPS)GOTO 24
104             IF(RMST .GT. 1.0E+04)GOTO 24
105 C
106 C           TRIDIAGONAL SYSTEMS IN THE X-DIRECTION
107 C
108             DO 19 K = 2,NYP
109             DO 15 J = 2,NX
110             JM = J - 1
111             B(2,JM) = EMX(1) - CCXA - CXA*U(K,JM)
112             B(3,JM) = EMX(2) + 2.*CCXA
113             B(4,JM) = EMX(3) - CCXA + CXA*U(K,J+1)
114             RRT(JM) = R(K,J)
115             IF(ME .NE. 2)GOTO 15
116             LST = 1
117             IF(J .EQ. 2)LST = 2
118             B(1,JM) = 0.
119             DO 14 L=LST,4
120             LJ = J + L - 3
121          14 B(L,JM) = B(L,JM) + CXQ(L)*U(K,LJ)*DBT
122          15 CONTINUE
123             B(2,1) = 0.
124             B(2,JM) = B(2,JM) + B(4,JM)
125             B(4,JM) = 0.
126             IF(ME .NE. 2)GOTO 17
127 C
128 C           REDUCE TO TRIDIAGONAL FORM
129 C
130             DO 16 JM = 3,NXP
131             JMM = JM - 1
132             DUM = B(1,JM)/B(2,JMM)
```

Fig. 9.13. (cont.) Listing of program THERM

```
133            B(2,JM) = B(2,JM) - B(3,JMM)*DUM
134            B(3,JM) = B(3,JM) - B(4,JMM)*DUM
135            B(1,JM) = 0.
136            RRT(JM) = RRT(JM) - RRT(JMM)*DUM
137      16 CONTINUE
138            B(1,2) = 0.
139    C
140      17 CALL BANFAC(B,NXP,1)
141            CALL BANSOL(RRT,DDT,B,NXP,1)
142    C
143            DO 18 J = 2,NX
144            JM = J - 1
145      18 R(K,J) = DDT(JM)
146      19 CONTINUE
147    C
148    C    TRIDIAGONAL SYSTEMS IN THE Y-DIRECTION
149    C
150            DO 22 J = 2,NX
151            DO 20 K = 2,NYP
152            KM = K - 1
153            B(2,KM) = EMY(1) - CCYA - CYA*V(KM,J)
154            B(3,KM) = EMY(2) + 2.*CCYA
155            B(4,KM) = EMY(3) - CCYA + CYA*V(K+1,J)
156      20 RRT(KM) = R(K,J)
157            B(2,1) = 0.
158            B(4,KM) = 0.
159    C
160            CALL BANFAC(B,NYPP,1)
161            CALL BANSOL(RRT,DDT,B,NYPP,1)
162    C
163    C    INCREMENT T
164    C
165            DO 21 K = 2,NYP
166            KM = K - 1
167            DT(K,J) = DDT(KM)
168      21 T(K,J) = T(K,J) + DDT(KM)
169      22 CONTINUE
170            DO 23 K = 2,NYP
171            T(K,NXS) = T(K,NXP)
172      23 DT(K,NXS) = DT(K,NXP)
173            ITER = ITER + 1
174            GAM = GAMH
175            BET = BETH
176            IF(ITER .LE. ITMAX)GOTO 13
177    C
178    C    COMPARE WITH SEMI-EXACT SOLUTION
179    C
180      24 SUM = 0.
181            DO 25 J = 2,NX
182            AJ = J - 1
183            X = AJ*DX
184    C
185            CALL TEXCL(X,TDX,PR,ALF,DYFL)
186    C
187            TEX(J) = TDX
188            IF(J .EQ. 2)GOTO 25
189            DIF = T(NYH,J) - TDX
190            SUM = SUM +  DIF*DIF
191      25 CONTINUE
192            RMS = SQRT(SUM/(ANX-1.0))
193            WRITE(6,26)ITER,RMST,RMS
194      26 FORMAT(' AFTER ',I3,' ITERATIONS,  RMS-RHS =',E10.3,
195         1'  RMS-ERR =',E10.3)
196            WRITE(6,27)(T(NYH,J),J=2,NX)
197            WRITE(6,28)(TEX(J),J=2,NX)
198      27 FORMAT(' C/L TEMP =',11F6.3)
199      28 FORMAT('  EX TEMP =',11F6.3)
200            STOP
201            END
```

Fig. 9.13. (cont.) Listing of program THERM

```
1
2          SUBROUTINE RETHE(RMST,GAM,DTIM)
3  C
4  C       EVALUATES RIGHT-HAND SIDE OF THE
5  C       TWO-DIMENSIONAL TRANSPORT EQUATION
6  C
7          DOUBLE PRECISION RTD,SUMT,AN,DSQRT
8          DIMENSION DT(41,42),T(41,42),R(41,42),U(41,42),V(41,42),
9         1DMX(3,3),DMY(3,3),EMX(3),EMY(3),CX(3),CY(3),CXQ(4)
10         COMMON CX,CXQ,CY,CCX,CCY,EMX,EMY,NX,NY,R,T,DT,U,V
11         NXP = NX - 1
12         NYP = NY - 1
13         DO 1 I = 1,3
14         DMX(I,1) = CCX*EMY(I)
15         DMX(I,2) = -2.*DMX(I,1)
16         DMX(I,3) = DMX(I,1)
17         DMY(1,I) = CCY*EMX(I)
18         DMY(2,I) = -2.*DMY(1,I)
19       1 DMY(3,I) = DMY(1,I)
20  C
21         SUMT = 0.
22         DO 6 J =2,NX
23         DO 5 K = 2,NYP
24         RTD = 0.
25         DO 4 N = 1,3
26         NK = K - 2 + N
27         DO 2 M = 1,3
28         MJ = J - 2 + M
29         DUM = DMX(N,M) + DMY(N,M) - CX(M)*EMY(N)*U(NK,MJ)
30        1 - CY(N)*EMX(M)*V(NK,MJ)
31       2 RTD = RTD + DUM*T(NK,MJ)*DTIM
32        1 + GAM*EMX(M)*EMY(N)*DT(NK,MJ)
33         MST = 1
34         IF(J .EQ. 2)MST = 2
35         DO 3 M = MST,4
36         MJ = J - 3 + M
37       3 RTD = RTD - CXQ(M)*EMY(N)*U(NK,MJ)*T(NK,MJ)*DTIM
38         IF(J .EQ. 2)RTD = RTD - CXQ(1)*EMY(N)*U(NK,1)*T(NK,1)*DTIM
39       4 CONTINUE
40         SUMT = SUMT + RTD*RTD
41       5 R(K,J) = RTD/(1. + GAM)
42       6 CONTINUE
43         AN = NXP*(NYP-1)
44         RMST = DSQRT(SUMT/AN)/DTIM
45         RETURN
46         END
```

Fig. 9.14. Listing of subroutine RETHE

It is clear that all methods are giving good agreement with the semi-exact solution, in view of the coarseness of the grid. The particularly close agreement of the linear finite element method is consistent with the steady-state modified equation which is actually being solved.

The relative behaviour of the mass operator construction and the four-point upwind discretisation for the convective terms can be compared by considering the steady-state modified equation for a scheme incorporating both features. This is

$$
\frac{\partial(uT)}{\partial x} + \frac{\partial(vT)}{\partial y} - \alpha_x \frac{\partial^2 T}{\partial x^2} - \alpha_y \frac{\partial^2 T}{\partial y^2}
$$

$$
+ (u\Delta x^2/6)\left\{ (1 - 2q_x - 6\delta_x)\frac{\partial^3 T}{\partial x^3} + [q_x + (6\delta_x - 0.5)/R_{\text{cell},x}]\frac{\partial^4 T}{\partial x^4} \right\} + \cdots
$$

$$
+ (v\Delta y^2/6)\left\{ (1 - 2q_y - 6\delta_y)\frac{\partial^3 T}{\partial y^3} + [q_y + (6\delta_y - 0.5)/R_{\text{cell},y}]\frac{\partial^4 T}{\partial y^4} \right\} + \cdots = 0 \,,
$$

$$
(9.97)
$$

Table 9.11. Parameters used in program THERM

Variable	Description
ME	= 1, approximate factorisation, three-point finite difference method, AF-FDM
	= 2, approximate factorisation, four-point upwind scheme, $q = 0.5$, AF-4PU
	= 3, approximate factorisation, linear finite element method, $\delta = 1/6$, AF-FEM
ITMAX	maximum number of iterations
GAM, BET	γ, β in (9.88, 89)
q	parameter controlling four-point upwind scheme, q_x in (9.97)
NX, NY	number of points in the x and y directions
DTIM	Δt
EPS	tolerance on the magnitude of RHS^n, (9.88)
RE, PR	Reynolds number, Prandtl number
XMAX	downstream extent of the computational domain
ALX, ALY	α_x, α_y, in (9.90, 92)
CX, CY	L_x, L_y in (9.88, 89)
CXQ	increment to L_x to generate $L_x^{(4)}$
EMX, EMY	M_x, M_y in (9.88, 89)
T; U, V	temperature; x and y velocity components
TEX	semi-exact centre-line temperature
TEXCL	calculates TEX for given x and Pr; requires ALF and DYFL; described in Sect. 16.1.4
DT	$\Delta T^n_{j,k}$ in (9.88)
DDT	$\Delta T^*_{j,k}$ and $\Delta T^{n+1}_{j,k}$ on return from BANSOL
B	quadridiagonal and tridiagonal matrix; left-hand sides of (9.88, 89)
R	RHS^n in (9.88), evaluated in RETHE; also used to store $\Delta T^*_{j,k}$ in (9.89)
RMS; RMST	$\| T_{C/L} - \bar{T}_{C/L} \|_{rms}$; $\| RHS^n \|_{rms}$

```
THERMAL ENTRY PROBLEM: 3PT FDM,  NX,NY = 11 11  ME = 1 ITMAX = 25
GAM = .50 BETA = 1.00 DTIM = .200 XMAX = 2.00 PR= .70 RE=100.0 Q = .00

AFTER 20 ITERATIONS,  RMS-RHS =  .857E-05  RMS-ERR =  .180E-01
C/L TEMP =  .459  .744  .881  .945  .975  .988  .995  .998  .999 1.000
 EX TEMP =  .493  .786  .909  .962  .984  .993  .997  .999  .999 1.000
```

Fig. 9.15. Typical output from program THERM

where q_x and q_y are the free parameters introduced into the four-point upwind discretisation of $\partial(uT)/\partial x$ and $\partial(vT)/\partial y$, respectively. The parameters δ_x and δ_y replace δ in the definitions of M_x and M_y, (9.94). The cell Reynolds numbers are defined as

$$R_{cell,x} = \frac{u \Delta x}{\alpha_x} \quad \text{and} \quad R_{cell,y} = \frac{v \Delta y}{\alpha_y} \ . \tag{9.98}$$

In the present example, $v = 0$ so that $R_{cell,y} = 0$. In contrast to the situation for the convection-diffusion equation, $R_{cell,x}$ is a function of position.

It is clear from (9.97) that either of the choices $q_x = q_y = 0.5$, $\delta_x = \delta_y = 0$, or $q_x = q_y = 0$, $\delta_x = \delta_y = \frac{1}{6}$ will eliminate the "dispersive" third-order derivatives. However, the ensuing treatment of the dissipative fourth-order derivatives is somewhat different,

Table 9.12. Centre-line solutions for thermal entry problem, 11×11 grid, $Re = 100$, $Pr = 0.70$, $\gamma = 0.5$, $\beta = 1.0$

Scheme	x 0.000	0.200	0.400	0.600	0.800	1.000	1.200	1.400	1.600	1.800	2.000	rms error
(semi-)exact	0.000	0.493	0.786	0.910	0.962	0.984	0.993	0.997	0.999	1.000	1.000	—
AF-FDM	0.000	0.459	0.744	0.881	0.945	0.975	0.988	0.995	0.998	0.999	1.000	0.018
AF-4PU, $q = 0.5$	0.000	0.452	0.754	0.898	0.959	0.984	0.994	0.998	0.999	1.000	1.000	0.011
AF-FEM	0.000	0.462	0.794	0.910	0.963	0.984	0.994	0.997	0.999	0.999	1.000	0.003

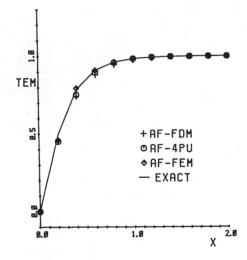

Fig. 9.16. Comparison of centre-line temperature distribution

particularly for large values of the cell Reynolds numbers. Both approaches introduce positive dissipation, but for the mass operator scheme the dissipation diminishes with increasing cell Reynolds number whereas for the four-point upwind scheme it does not. The results shown in Table 9.12 are consistent with (9.97).

9.5.3 Cross-Stream Diffusion

The use of two-point upwind differencing to represent convective terms introduces an artificial diffusivity $\alpha' = 0.5u\Delta x$ (Tables 9.1 and 9.3) for one-dimensional problems. In multi-dimensions the impact of artificial diffusivity is more usefully considered in relation to the local flow direction. Thus in two dimensions the streamwise and cross-stream artificial diffusivities are of greater physical interest than artificial diffusivities expressed relative to the coordinate directions.

The two-dimensional steady convection diffusion equation can be written (with $\alpha = \alpha_x = \alpha_y$ and u, v constant)

$$u\frac{\partial \bar{T}}{\partial x} + v\frac{\partial \bar{T}}{\partial y} - \alpha\left(\frac{\partial^2 \bar{T}}{\partial x^2} + \frac{\partial^2 \bar{T}}{\partial y^2}\right) = 0 \ . \tag{9.99}$$

It will be assumed that two-point upwind differencing is used to discretise $\partial \bar{T}/\partial x$ and $\partial \bar{T}/\partial y$ and that three-point centred differencing is used to discretise $\partial^2 \bar{T}/\partial x^2$ and $\partial^2 \bar{T}/\partial y^2$. A Taylor-series expansion of the resulting discrete equation indicates consistency to $O(\Delta x^2, \Delta y^2)$ with

$$u\frac{\partial T}{\partial x}+\frac{\partial T}{\partial y}-(\alpha+\alpha'_x)\frac{\partial^2 T}{\partial x^2}-(\alpha+\alpha'_y)\frac{\partial^2 T}{\partial y^2}=0 \ , \tag{9.100}$$

where $\alpha'_x=0.5u\Delta x$ and $\alpha'_y=0.5v\Delta y$. Equation (9.100) indicates the presence of artificial diffusivity associated with each coordinate direction and is clearly a direct extension of the one-dimensional result (9.51).

It is convenient to transform (9.100) based on a cartesian coordinate system, tangential to the local flow direction,

$$q\frac{\partial T}{\partial s}-\alpha\left(\frac{\partial^2 T}{\partial s^2}+\frac{\partial^2 T}{\partial n^2}\right)-\alpha'_s\frac{\partial^2 T}{\partial s^2}-\alpha'_{sn}\frac{\partial^2 T}{\partial s\partial n}-\alpha'_n\frac{\partial^2 T}{\partial n^2} \ , \tag{9.101}$$

where

$$\alpha'_s=0.5q(\Delta x\cos^3\alpha+\Delta y\sin^3\alpha) \ ,$$

$$\alpha'_{sn}=0.5q(\Delta y\sin\alpha-\Delta x\cos\alpha)\sin 2\alpha \ ,$$

$$\alpha'_n=0.5q(\Delta x\cos\alpha\sin^2\alpha+\Delta y\sin\alpha\cos^2\alpha) \ .$$

In (9.101) s and n denote directions tangential and normal to the local flow direction, respectively. In addition the uniform flow q makes an angle α with the x axis, so that

$$u=q\cos\alpha \quad \text{and} \quad v=q\sin\alpha \ . \tag{9.102}$$

The first three terms in (9.101) arise from the direct transformation of (9.99). The last three terms are artificial diffusion terms. For the case $\Delta x=\Delta y$ it is clear that the cross-stream artificial diffusivity α'_n is a maximum at $\alpha=45°$ but is zero if the flow is aligned with either the x or the y axis. In contrast the stream-wise artificial diffusivity is a maximum at $\alpha=0$ or $90°$ and falls to a minimum at $\alpha=45°$.

At first sight it might appear that streamwise artificial diffusion is of more concern since for $\Delta x=\Delta y$, $|\alpha'_s| \geq |\alpha'_n|$ for all α in the range $0\leq\alpha\leq90°$. However, for most flows, $\partial^2 T/\partial s^2$ is very small; an exception is the region adjacent to a stagnation point. In contrast for free shear layers or boundary layers (Chap. 16) $\partial^2 T/\partial n^2 \gg \partial^2 T/\partial s^2$ and $\partial^2 T/\partial n^2$ is of comparable magnitude to the convective term $q\,\partial T/\partial s$. Consequently cross-stream artificial diffusion may introduce significant errors if a low-order discretisation of the convective terms is utilised and the local flow direction does not coincide with the coordinate lines. Conversely methods that add purely streamwise artificial diffusion, e.g. Brooks and Hughes (1982), are effective in producing oscillation-free accurate solution for many types of flow problems.

Cross-stream artificial diffusion is examined theoretically and numerically by de Vahl Davis and Mallinson (1976), Raithby (1976) and Griffiths and Mitchell (1979).

9.6 Closure

The various computational techniques introduced in conjunction with the diffusion equation in Chap. 7 have been applied to a sequence of equations: a linear convection equation, steady convection diffusion equation and transport equation. This sequence can be interpreted as models, of varying degrees of complexity, for the equations (Chap. 11) governing flow problems. Related techniques will be applied to the fluid dynamic governing equations in Chaps. 14–18.

The examination of the model equations has drawn attention to the need to recognise and control numerical dissipation and dispersion. This is particularly troublesome when first-order algebraic representations are used with diffusion either absent or of a much smaller magnitude than convection. The second situation, of convection dominating diffusion, also permits non-physical oscillations to occur if symmetric three-point algebraic formulae are used to discretise the convective terms on a coarse grid. However, the introduction of mass operators, which are a feature of the finite element method, provide a mechanism for exercising some control over dispersion-like oscillations. Through the splitting constructions, introduced in Sect. 8.3, mass operators are available in steady and unsteady multidimensional problems.

An alternative means of controlling dispersion-like oscillations is provided by the introduction of a four-point upwind discretisation of the convective terms. This technique is effective although it may introduce additional higher-order dissipation. The four-point upwind discretisation can be incorporated into implicit algorithms but typically produces a quadridiagonal system of equations that can be solved by the Thomas algorithm after first reducing the system to tridiagonal form. The use of a four-point scheme also introduces the possible need of additional boundary conditions at inflow boundaries or the local reduction to a three-point scheme.

Comparison of the methods of analysis used in this chapter emphasises the cost-effectiveness of the modified equation approach (Sect. 9.2.2) for constructing improved algorithms. Because of the underlying Taylor series expansion this approach only provides information about the long wavelength behaviour.

The short wavelength behaviour requires a Fourier analysis (Sect. 9.2.1) which, in practice, also provides a von Neumann stability analysis (Sect. 4.3). However, although the modified equation approach extends directly to nonlinear schemes and equations, the Fourier analysis requires a local linearisation which may detract from its effectiveness for nonlinear flow problems (Chaps. 10, 11 and 14–18).

9.7 Problems

1-D Linear Convection Equation (Sect. 9.1)

9.1 Obtain the expression for the truncation error shown in Table 9.1 for the Lax-Wendroff scheme.

9.2 Apply the von Neumann stability analysis to the Crank-Nicolson mass operator form (9.22) and confirm that there is no stability restriction.

9.3 Apply program TRAN to the convecting sine wave problem (Sect. 9.1.5) using the Crank-Nicolson mass operator scheme and compare solutions with those shown in Figs. 9.2–4.

Numerical Dissipation and Dispersion (Sect. 9.2)

9.4 Apply the Fourier analysis, Sect. 9.2.1, to the one-dimensional convection equation for the following schemes:
(i) Upwind difference scheme,
(ii) Lax-Wendroff scheme,
(iii) Crank-Nicolson finite difference scheme,
with $C = 0.8$ and $\Delta x = 0.5$. Obtain results equivalent to Table 9.9; it is recommended that a computer program be written to do this. Analyze the solutions shown in Figs. 9.2–4 in relation to the Fourier analysis.

9.5 Apply the modified equation approach to:
(i) Upwind difference scheme,
(ii) Lax-Wendroff scheme,
(iii) Crank-Nicolson finite difference scheme,
for the one-dimensional convection equation and confirm the results given in Table 9.1. Compare the predicted behaviour with that of the Fourier analysis of Problem 9.4 and the results shown in Figs. 9.2–4.

9.6 Construct a generalised, three-level discretisation, equivalent to (8.26), of the one-dimensional convection equation. By considering the truncation error determine if there are optimal choices of γ and β for minimising dispersion and dissipation errors. Modify program TRAN to run this case and test on the convecting sine-wave problem.

Steady Convection-Diffusion Equation (Sect. 9.3)

9.7 Incorporate the centred difference scheme (9.45) and the upwind scheme (9.50) into a computer program and confirm the results shown in Figs. 9.5 and 9.6.

9.8 Implement the program developed in Problem 9.7 for $u/\alpha = 5$ and $\Delta x = 0.1$ with a Neumann boundary condition, $dT/dx = g$ at $x = 1.0$. Choose g to satisfy Eq. (9.44). Compare the solution with a corresponding Dirichlet boundary condition case.

9.9 Introduce symmetric five-point schemes for dT/dx and d^2T/dx^2, by minimising the truncation error with one adjustable parameter left in each expression. Apply this scheme to the convection diffusion equation with Dirichlet boundary conditions (9.43). Use the exact solution (9.44) to supply the additional boundary conditions required. Modify subroutines BANFAC and BANSOL to solve the resulting pentadiagonal matrix. Obtain solutions to compare with those given in Table 9.2 and determine whether specific choices of the two adjustable parameters will produce more accurate, but stable, solutions.

1-D Transport Equation (Sect. 9.4)

9.10 Apply the modified equation analysis to the Lax-Wendroff and Crank-Nicolson discretisations of the one-dimensional transport equation and confirm the results given in Table 9.3.

9.11 For the explicit four-point upwind scheme, i.e. (9.71) with T_j^n replacing $0.5(T_j^n + T_j^{n+1})$ in the spatial operator, determine an optimal value of q to reduce the size of the truncation error. Apply the von Neumann stability analysis to determine any stability limits. Run program TRAN with the optimum q, if stable, for conditions corresponding to Table 9.6 and compare results.

9.12 Modify program TRAN to implement the three-level fully implicit scheme (Table 9.3). Obtain solutions to the propagating temperature front problem (Sect. 9.4.3) for increasing cell Reynolds numbers and determine empirically the limit for oscillation-free solutions. Repeat for the Crank-Nicolson finite difference and finite element schemes.

2-D Transport Equation (Sect. 9.5)

9.13 By considering the truncation error of the Crank-Nicolson mass operator formulation of the two-dimensional transport equation (9.81) determine optimum values of δ_x and δ_y in (9.86) to minimise dispersion errors. Apply a von Neumann analysis to (9.88 and 89) with $\gamma = 0$ and $\beta = 0.5$ to establish whether the optimal values will produce stable solutions.

9.14 For program THERM compare AF-FDM; AF-4PU ($q = 0.5$) and AF-FEM for the number of iterations to reach convergence. Do this for the three marching algorithms, (a) $\gamma = 0$, $\beta = 0.5$, (b) $\gamma = 0$, $\beta = 1.0$, (c) $\gamma = 0.5$, $\beta = 1.0$, and for grids, 11×11 and 21×21.

9.15 Based on the results in Table 9.12, Problem 9.14 and any additional solutions from THERM compare the computational efficiency of AF-FDM, AF-4PU and AF-FEM.

10. Nonlinear Convection-Dominated Problems

In Chap. 9 *linear* convection-dominated problems are considered. It is demonstrated there that care is required in discretising, particularly the convective terms, to avoid introducing spurious higher-order derivatives (in the modified equation sense) that may be of comparable magnitude to physically important terms in the governing equation. This is a particular problem with two-point upwind differencing.

The identification of higher-order derivatives in the modified equation (Sect. 9.2.2) with the processes of dissipation and dispersion allows more accurate computational algorithms to be developed. For example, a four-point upwind scheme avoids excessive dispersion-related oscillations (Sects. 9.3.2 and 9.4.3) without introducing unacceptable dissipation, as is the case with the two-point upwind scheme.

For the energy equation (Sect. 11.2.4) the convective terms are linear if the velocity field is known. However the convective terms in the momentum equations (Sects. 11.2.2 and 11.2.3) are nonlinear. In this chapter the additional complications, associated with nonlinear convection, are considered. The one-dimensional Burgers' equation is a suitable model for this purpose since it possesses the same form of convective nonlinearity as the incompressible Navier-Stokes equations (Sect. 11.5) and has readily evaluated exact solutions for many combinations of initial and boundary conditions.

The convective nonlinearity must be suppressed in applying the von Neumann stability analysis. This is done by treating solution-dependent coefficients multiplying derivatives as being temporarily frozen. The modified equation approach to analysing nonlinear computational algorithms is applicable (Klopfer and McRae 1983) but the appearance of products of higher-order derivatives makes the construction of more accurate schemes less precise (Sect. 10.1.4) than was the case for linear equations (Sect. 9.4.3).

The nonlinear nature of the convective terms permits severe gradients in the dependent variables to develop. This suggests the use of a nonuniform grid locally refined in the vicinity of the rapid solution change. The influence of a nonuniform grid on the truncation error and solution accuracy are examined briefly in Sect. 10.1.5. This subject is treated at greater length in Chap. 12.

The governing equations for fluid dynamics (Chap. 11) usually occur as systems, as well as being nonlinear. The additional complications of constructing computational algorithms for systems of nonlinear equations are considered, in a preliminary fashion, in Sect. 10.2.

The Galerkin finite element method (Chap. 5) produces accurate and economical algorithms for linear convection-dominated problems (Sects. 9.1.4, 9.4.3 and 9.5). However, the treatment of convective nonlinearities is potentially uneconomical in multidimensions or if higher-order interpolation is used. The lack of economy can be overcome by adopting a *group* finite element formulation; this is described in Sect. 10.3.

The split schemes, developed in Sects. 8.2, 8.3 and 9.5, are extended to the two-dimensional Burgers' equations in Sect. 10.4.2. The two-dimensional Burgers' equations, like the one-dimensional form, have a readily available exact solution, at least in the steady-state limit, thereby permiting direct evaluation of the accuracy of specific computational algorithms (Sect. 10.4.3).

10.1 One-Dimensional Burgers' Equation

In one dimension, Burgers' equation (Burgers 1948) is

$$\frac{\partial \bar{u}}{\partial t} + \bar{u}\frac{\partial \bar{u}}{\partial x} - v\frac{\partial^2 \bar{u}}{\partial x^2} = 0 \ . \tag{10.1}$$

Burgers' equation is similar to the transport equation (9.56) except that the convective term is now nonlinear. This is the same form of nonlinearity as appears in the momentum equations, Sects 11.2.2 and 11.2.3. To emphasise the connection with the momentum equations, the diffusivity α in (9.56) has been replaced with the kinematic viscosity v in (10.1). Further discussion of Burgers' equation is provided by Fletcher (1983a) and Whitham (1974).

10.1.1 Physical Behaviour

If the 'viscous' term is dropped from (10.1) the result is the inviscid Burgers' equation

$$\frac{\partial \bar{u}}{\partial t} + \bar{u}\frac{\partial \bar{u}}{\partial x} = 0 \ , \tag{10.2}$$

which may be compared with (9.2). The nonlinearity in (10.2) allows discontinuous solutions to develop. The way that this can occur is illustrated schematically in Fig. 10.1. A wave is convecting from left to right and solutions for successive times $t = 0, t_1, t_2$ are indicated. Points on the wave with larger values of $\bar{u}$ convect faster and consequently overtake parts of the wave convecting with smaller values of $\bar{u}$. For (10.2) to have a unique solution (and a physically sensible result) it is necessary to postulate a shock (*ab* in Fig. 10.1) across which $\bar{u}$ changes discontinuously.

The comparable wave development for the 'viscous' Burgers' equation (10.1) is shown in Fig. 10.2. The effect of the viscous term $v\partial^2 \bar{u}/\partial x^2$ is twofold. First, it reduces the amplitude of the wave for increasing t (just as for the transport

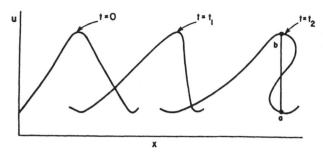

Fig. 10.1. Formation of a multivalued solution of the inviscid Burgers' equation (10.2)

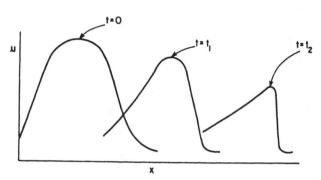

Fig. 10.2. Evolution of the solution of Burgers' equation (10.1)

equation). Second, it prevents multivalued solutions, as in Fig. 10.1, from developing.

The above features make Burgers' equation a very suitable model for testing computational algorithms for flows where severe gradients or shocks are anticipated. This role as a test-bed for computational algorithms is facilitated by the Cole-Hopf transformation (Cole 1951; Hopf 1950) which allows exact solutions of Burgers' equation to be obtained for many combinations of initial and boundary conditions. Examples of this use of Burgers' equation are provided by Fletcher (1983a). In Sect. 10.1.4 Burgers' equation is used to demonstrate various computational algorithms for convection-dominated flows.

There is an alternative way of handling the nonlinear convective term. This requires rewriting (10.1) in conservation or divergence form,

$$\frac{\partial \bar{u}}{\partial t} + \frac{\partial \bar{F}}{\partial x} - v \frac{\partial^2 \bar{u}}{\partial x^2} = 0 \ , \tag{10.3}$$

where $\bar{F} = 0.5\bar{u}^2$. The appearance of the groups of terms, like $\bar{F}$, arises naturally in the derivation of the equations governing fluid flow (Sect. 11.2.1) and is often directly modelled in computational algorithms, particularly for compressible flow (Sects. 14.2 and 18.3). The same idea is used in the group finite element formulation (Sect. 10.3).

The appearance of the nonlinear convective term in (10.2) also permits aliasing to occur. If an initial condition $\bar{u}(x, 0)$, associated with (10.2), is decomposed into its

Fourier components, as was done to generate (9.30), the subsequent evolution of the solution, which is governed by (10.2), will generate a rapid growth in the spectrum of wavelengths with significant amplitudes, through the product of the Fourier components implied by the nonlinear term $\bar{u}\partial\bar{u}/dx$.

When a solution is obtained on a grid of spacing Δx the smallest wavelength that can be resolved is $2\Delta x$. The energy associated with wavelengths shorter than $2\Delta x$ reappears associated with long wavelengths. This phenomenon is called aliasing (Hamming 1973, p. 505). Unfortunately the aliased shortwave contribution to the solution distorts the true longwave solution and may even cause instability (referred to as nonlinear instability) where very long time integrations are made, as in computational weather forecasting.

It may be recalled (Sect. 9.2) that if any dissipation, physical or computational, is present it attenuates the amplitude of the short wavelengths very significantly; in this case the errors introduced by aliasing are minimal. This is the situation for most engineering flow problems, which possess physical dissipation, typically. However, many classes of geophysical fluid motion are characterised by the convection of wave-trains with little or no physical dissipation. Then aliasing, and nonlinear instability, can be a problem.

10.1.2 Explicit Schemes

An FTCS finite difference representation of (10.1) is

$$\frac{u_j^{n+1}-u_j^n}{\Delta t}+\frac{u_j^n(u_{j+1}^n-u_{j-1}^n)}{2\Delta x}-\frac{v(u_{j-1}^n-2u_j^n+u_{j+1}^n)}{\Delta x^2}=0 \ . \tag{10.4}$$

The contribution u_j^n to the convective term is the local solution at the node (j, n). Since this is available directly, the various schemes (e.g. DuFort-Frankel, etc.), described for the linear transport equation (Sect. 9.4), can also be applied to Burgers' equation.

The concepts of truncation error and the modified equation are also relevant to nonlinear equations like Burgers' equation. In applying the von Neumann stability analysis the contribution u_j^n to the convective term is temporarily frozen, since the stability analysis is strictly only applicable to linear equations. Although the application of the von Neumann analysis to nonlinear equations cannot be rigorously justified it is found to be effective in practice.

An obvious FTCS representation of (10.3), equivalent to (10.4), is

$$\frac{u_j^{n+1}-u_j^n}{\Delta t}+\frac{F_{j+1}^n-F_{j-1}^n}{2\Delta x}-\frac{v(u_{j-1}^n-2u_j^n+u_{j+1}^n)}{\Delta x^2}=0 \ . \tag{10.5}$$

The algorithm formed from (10.5) is implemented in program BURG (Fig. 10.4) when ME = 1.

A potentially more accurate treatment of the convective term is provided by a four-point upwind discretisation. This was applied to the equivalent linear convective term in Sects. 9.3.2 and 9.4.3. The four-point upwind discretisation of $\partial F/\partial x$ is introduced by replacing $(F_{j+1}-F_{j-1})/2\Delta x$ in (10.5) with $L_x^{(4)}F$ where, for u positive,

$$L_x^{(4)}F=\frac{F_{j+1}-F_{j-1}}{2\Delta x}+\frac{q(F_{j-2}-3F_{j-1}+3F_j-F_{j+1})}{3\Delta x}.\qquad(10.6)$$

If u is negative (10.6) is replaced by

$$L_x^{(4)}F=\frac{F_{j+1}-F_{j-1}}{2\Delta x}+\frac{q(F_{j-1}-3F_j+3F_{j+1}-F_{j+2})}{3\Delta x}.\qquad(10.7)$$

Equations (10.6 and 7) have truncation errors of $O(\Delta x^2)$ for any choice of q, except $q=0.5$ when they are of $O(\Delta x^3)$. For the propagating shock problem considered in Sect. 10.1.4, the explicit four-point upwind scheme (10.5, 6) is implemented in program BURG (Fig. 10.4) when ME$=3$.

The equations that govern unsteady one-dimensional inviscid shock flows (Sects. 10.2 and 14.2) are similar in structure to the conservation form of the inviscid Burgers' equation, i.e.

$$\frac{\partial\bar{u}}{\partial t}+\frac{\partial\bar{F}}{\partial x}=0 ,\qquad(10.8)$$

where $\bar{F}=0.5\bar{u}^2$. Previously the Lax-Wendroff scheme (9.16) was applied to the linear counterpart, i.e. (9.2), of (10.8). Since $\bar{F}$ is nonlinear the replacement of $\partial^2\bar{u}/\partial t^2$ with an equivalent spatial derivative is more complicated than in Sect. 9.1.3. Thus from (10.8)

$$\frac{\partial^2\bar{u}}{\partial t^2}=-\frac{\partial^2\bar{F}}{\partial x\partial t}=-\frac{\partial}{\partial x}\left(\frac{\partial\bar{F}}{\partial t}\right) \quad\text{and}\quad \frac{\partial\bar{F}}{\partial t}=A\frac{\partial\bar{u}}{\partial t}=-A\frac{\partial\bar{F}}{\partial x} ,$$

where $A=\partial\bar{F}/\partial\bar{u}$. For Burgers' equation, $A=\bar{u}$. Making use of the above relations gives

$$\frac{\partial^2\bar{u}}{\partial t^2}=\frac{\partial}{\partial x}\left(A\frac{\partial\bar{F}}{\partial x}\right).\qquad(10.9)$$

With a suitable discretised form for $\partial(A\partial\bar{F}/\partial x)/\partial x$, the Lax-Wendroff scheme, applied to (10.8), is

$$u_j^{n+1}=u_j^n-0.5\frac{\Delta t}{\Delta x}(F_{j+1}^n-F_{j-1}^n)+0.5\left(\frac{\Delta t}{\Delta x}\right)^2[A_{j+\frac{1}{2}}(F_{j+1}^n-F_j^n)$$
$$-A_{j-\frac{1}{2}}(F_j^n-F_{j-1}^n)] ,\qquad(10.10)$$

where, for Burgers' equation, $A_{j+\frac{1}{2}} = u_{j+\frac{1}{2}} = 0.5(u_j + u_{j+1})$, etc. Equation (10.10) has a truncation error of $O(\Delta t^2, \Delta x^2)$ and is stable if $|u_{max}\Delta t/\Delta x| \leqq 1.0$.

Because of the explicit appearance of the Jacobian A at the half steps, $j-\frac{1}{2}$ and $j+\frac{1}{2}$, (10.10) is computationally less economical than the equivalent algorithm (9.16) for the linear convection equation. A more economical evaluation is obtained by replacing (10.10) with an equivalent two-stage algorithm

$$u^*_{j+\frac{1}{2}} = 0.5(u^n_j + u^n_{j+1}) - 0.5\frac{\Delta t}{\Delta x}(F^n_{j+1} - F^n_j) \quad \text{and} \tag{10.11}$$

$$u^{n+1}_j = u^n_j - \frac{\Delta t}{\Delta x}(F^*_{j+\frac{1}{2}} - F^*_{j-\frac{1}{2}}) . \tag{10.12}$$

Equation (10.11) generates an intermediate solution $u^*_{j+\frac{1}{2}}$, nominally at $(n+\frac{1}{2})\Delta t$, which allows $F^*_{j+\frac{1}{2}}$, etc., in (10.12) to be evaluated. The two-stage scheme (10.11, 12) is equivalent to (9.16) if $F = u$. For ease of coding, values of F at $j+\frac{1}{2}$ are stored in $F(j)$. This is done in program BURG (Fig. 10.4) for option $ME = 2$.

Historically (Richtmyer and Morton 1967) the two-stage algorithm (10.11, 12) has been successful for predicting inviscid compressible flow behaviour. Attempts to extend it to viscous flow, e.g. Thommen (1966), have been disadvantaged by the reduction in formal accuracy to $O(\Delta t, \Delta x^2)$. This gives accurate steady-state results but the time step is restricted by the CFL stability condition. For transient problems the time step must be restricted further to achieve acceptable accuracy. Thommen's extension of (10.11, 12) to discretise (10.3) can be written

$$u^*_{j+\frac{1}{2}} = 0.5(u^n_j + u^{n+1}_j) - 0.5\frac{\Delta t}{\Delta x}(F^n_{j+1} - F^n_j)$$
$$+ 0.5s[0.5(u^n_{j-1} - 2u^n_j + u^n_{j+1}) + 0.5(u^n_j - 2u^n_{j+1} + u^n_{j+2})] , \tag{10.13}$$

$$u^{n+1}_j = u^n_j - \frac{\Delta t}{\Delta x}(F^*_{j+\frac{1}{2}} - F^*_{j-\frac{1}{2}}) + s(u^n_{j-1} - 2u^n_j + u^n_{j+1}) , \tag{10.14}$$

where $s = v\Delta t/\Delta x^2$. This scheme is implemented in program BURG $(ME = 2)$. An examination of (10.13) indicates that four grid points are involved in evaluating the viscous term; this implies additional boundary conditions to evaluate u_{j+2}. Equations (10.13 and 14) produce stable solutions if

$$\Delta t(A^2\Delta t + 2v) \leqq \Delta x^2 . \tag{10.15}$$

Peyret and Taylor (1983, p. 56) indicate that a practical explicit criterion for a stable solution is $\Delta t \leqq \Delta x^2/(2v + |A|\Delta x)$.

Equations (10.13 and 14) can be interpreted as one member of the $S^\alpha_{\beta, \gamma}$ family introduced by Lerat and Peyret (1975). This family of methods is discussed by Peyret and Taylor (1983, pp. 53–57). The related S^α_β family for the inviscid Burgers' equation is discussed in Sect. 14.2.2. The parameter γ determines the proportional evaluation of the viscous term, in (10.13) centred at grid points j and $j+1$. For the Thommen scheme, $\alpha=\beta=\gamma=0.5$.

A general classification of difference schemes for the inviscid Burgers' equations (10.2) that includes the S^α_β family is provided by Yanenko et al. (1983). The classification also includes implicit schemes and covers the one-dimensional unsteady Euler equations (10.40, 41). Yanenko et al. describe and reference many schemes of Soviet origin that are not well-known in the Western literature.

10.1.3 Implicit Schemes

Applying implicit schemes to nonlinear equations such as Burgers' equation is not quite as straightforward as for linear equations (Chap. 9). A Crank-Nicolson implicit formulation of (10.3) is

$$\frac{\Delta u_j^{n+1}}{\Delta t}=-0.5L_x(F_j^n+F_j^{n+1})+0.5vL_{xx}(u_j^n+u_j^{n+1}) \ , \tag{10.16}$$

where $\Delta u_j^{n+1}=u_j^{n+1}-u_j^n$, $L_x=(-1, 0, 1)/2\Delta x$ and $L_{xx}=(1, -2, 1)/\Delta x^2$. To make use of the very efficient Thomas algorithm (Sect. 6.2.2) it is necessary to reduce (10.16) to a system of linear tridiagonal equations for the solution u_j^{n+1}. The appearance of the nonlinear implicit term F_j^{n+1} poses a problem.

However, this problem can be overcome in the same way that was used in introducing the solution correction term ΔT_j^{n+1} with the split schemes (Sect. 8.2). That is, a Taylor series expansion of F_j^{n+1} is made about the nth time-level. Thus

$$F_j^{n+1}=F_j^n+\Delta t\left[\frac{\partial F}{\partial t}\right]_j^n+0.5\Delta t^2\left[\frac{\partial^2 F}{\partial t^2}\right]_j^n+ \ \dots \ \ \text{or}$$

$$F_j^{n+1}=F_j^n+A\Delta u_j^{n+1}+O(\Delta t^2) \ ,$$

where $A=[\partial F/\partial u]_j^n=u_j^n$ for (10.3). Therefore from (10.16), the following tridiagonal algorithm can be constructed:

$$\frac{\Delta u_j^{n+1}}{\Delta t}=-0.5L_x(2F_j^n+u_j^n\Delta u_j^{n+1})+0.5vL_{xx}(u_j^n+u_j^{n+1}) \quad \text{or} \tag{10.17}$$

$$u_j^{n+1}+0.5\Delta t[L_x(u_j^n u_j^{n+1})-vL_{xx}u_j^{n+1}]=u_j^n+0.5v\Delta t L_{xx}u_j^n \ . \tag{10.18}$$

In forming (10.18) it is clear that $L_x F_j^n$ has cancelled out. In tridiagonal form (10.18) can be written as

$$a_j^n u_{j-1}^{n+1}+b_j^n u_j^{n+1}+c_j^n u_{j+1}^{n+1}=d_j^n \ , \tag{10.19}$$

where $a_j^n = -0.25(\Delta t/\Delta x)u_{j-1}^n - 0.5s$,

$$b_j^n = 1 + s \ ,$$

$$c_j^n = 0.25\frac{\Delta t}{\Delta x}u_{j+1}^n - 0.5s \ ,$$

$$d_j^n = 0.5su_{j-1}^n + (1-s)u_j^n + 0.5su_{j+1}^n \quad \text{and} \quad s = \frac{v\Delta t}{\Delta x^2} \ .$$

It is clear that the coefficients a_j^n and c_j^n are functions for u^n and so must be reevaluated at every time step. However the form of (10.19) permits direct use of the Thomas algorithm (Sect. 6.2.2).

The implicit scheme (10.18) has a truncation error of $O(\Delta t^2, \Delta x^2)$ and is unconditionally stable in the von Neumann sense. The linearised Crank-Nicolson scheme (10.17) can be extended to include the mass-operator form and the four-point upwind discretisation of the convective term (Sect. 9.4.3). Following (9.69 and 71), a generalised Crank-Nicolson discretisation of (10.3) is written

$$M_x\left(\frac{\Delta u_j^{n+1}}{\Delta t}\right) = -0.5L_x^{(4)}(2F_j^n + u_j^n\Delta u_j^{n+1}) + 0.5vL_{xx}(u_j^n + u_j^{n+1}) \ , \qquad (10.20)$$

where $M_x = \{\delta, 1-2\delta, \delta\}$ and $L_x^{(4)}F$ is given by (10.6) for u positive. The truncation error for (10.20) is $O(\Delta t^2, \Delta x^2)$. In quadridiagonal form (10.20) can be reduced to

$$e_j^n u_{j-2}^{n+1} + a_j^n u_{j-1}^{n+1} + b_j^n u_j^{n+1} + c_j^n u_{j+1}^{n+1} = d_j^n \ , \qquad (10.21)$$

where

$$e_j^n = \frac{q}{6}\frac{\Delta t}{\Delta x}u_{j-2}^n \ ,$$

$$a_j^n = -(0.25 + 0.5q)\frac{\Delta t}{\Delta x}u_{j-1}^n + \delta - 0.5s \ ,$$

$$b_j^n = 1 + 0.5q\frac{\Delta t}{\Delta x}u_j^n - 2\delta + s \ ,$$

$$c_j^n = \left(0.25 - \frac{q}{6}\right)\frac{\Delta t}{\Delta x}u_{j+1}^n + \delta - 0.5s \ ,$$

$$d_j^n = (\delta + 0.5s)u_{j-1}^n + (1 - 2\delta - s)u_j^n + (\delta + 0.5s)u_{j+1}^n \ .$$

The quadridiagonal system of equations (10.21) can be solved using the generalised Thomas algorithm, Sect. 6.2.4. In practice, the quadridiagonal system (10.21) is reduced to tridiagonal form (lines 134–144 in Fig. 10.4) and subsequently subroutines BANFAC and BANSOL are applied in the conventional manner. The generalised Crank-Nicolson algorithm (10.21) is implemented in program BURG when ME$=4$ or greater.

Stable solutions of (10.21) are obtained if $\delta \leq 0.25$ and $q \geq 0$. As might be expected from Sect. 9.4.3, specific choices of δ and q will reduce dispersion-related oscillations. This is pursued in Sect. 10.1.4.

10.1.4 BURG: Numerical Comparison

In this subsection the various explicit (Sect. 10.1.2) and implicit (Sect. 10.1.3) schemes will be applied to the propagation of a shock wave governed by the "viscous" Burgers' equation (10.3). At $t=0$ the shock wave is located at $x=0$. The initial conditions are

$$u_0(x) = \bar{u}(x, 0) = 1.0 \quad \text{for} \quad -x_{\max} \leq x \leq 0 \ , \tag{10.22}$$

$$u_0(x) = \bar{u}(x, 0) = 0 \quad \text{or} \quad 0 < x \leq x_{\max} \ .$$

The following boundary conditions are applied at $x = \pm x_{\max}$:

$$\bar{u}(-x_{\max}, t) = 1.0 \ , \quad \bar{u}(x_{\max}, t) = 0 \ . \tag{10.23}$$

For this combination of initial and boundary conditions, (10.3) has an exact solution given by

$$\bar{u} = \int\limits_{-\infty}^{\infty} [(x-\xi)/t] e^{-0.5 \mathrm{Re} G} d\xi \Big/ \int\limits_{-\infty}^{\infty} e^{-0.5 \mathrm{Re} G} d\xi \ , \tag{10.24}$$

where

$$G(\xi; x, t) = \int\limits_{0}^{\xi} u_0(\xi') d\xi' + \frac{0.5(x-\xi)^2}{t} \quad \text{and} \quad \mathrm{Re} = \frac{1}{\nu} \ .$$

Equation (10.24) is plotted in Fig. 10.3 for two values of Re. In using (10.24) as the

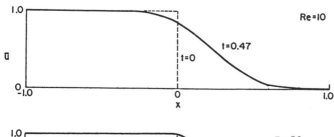

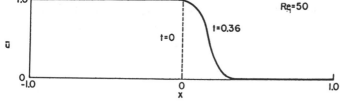

Fig. 10.3. Exact solution of Burgers' equation

```
1  C
2  C      BURG SOLVES BURGERS EQUATION FOR A PROPAGATING 'SHOCK'
3  C
4          DIMENSION R(65),U(65),UD(65),UE(65),A(5,65),
5        1,EF(4),X(65),AA(2),BB(2),CC(2)
6          OPEN(1,FILE='BURG.DAT')
7          OPEN(6,FILE='BURG.OUT')
8          READ(1,1)ME,JMAX,NTIM,XMAX,AD
9          READ(1,2)C,S,Q,EM
10       1 FORMAT(3I5,2E10.3)
11       2 FORMAT(F5.2,3E10.3)
12 C
13         JMAF = JMAX - 2
14         JMAP = JMAX - 1
15         AMP = JMAP
16         DX = 2.*XMAX/AMP
17         DT = C*DX
18         SA = AD*C*C
19         ALPH = S*DX*DX/DT
20         IF(ALPH .LT. 1.0E-10)ALPH = 1.0E-10
21         RCEL = DX/ALPH
22         QQ = Q*C/3.
23         IF(ME .LT. 3)QQ = 0.
24         MQ = 0
25         IF(ABS(QQ) .GT. 0.0001)MQ = 1
26 C
27         ATIM = NTIM
28         TIM = 0.
29         TIMAX = DT*ATIM
30 C
31         WRITE(6,3)
32       3 FORMAT(' PROPAGATING SHOCK WAVE(BURGERS EQUATION)')
33         IF(ME .EQ. 1)WRITE(6,4)ME
34         IF(ME .EQ. 2)WRITE(6,5)ME
35         IF(ME .EQ. 3)WRITE(6,6)ME
36         IF(ME .GE. 4)WRITE(6,7)ME
37       4 FORMAT(5H ME =,I2,3X,18H FTCS DIFFERENCING)
38       5 FORMAT(5H ME =,I2,3X,13H LAX-WENDROFF)
39       6 FORMAT(5H ME =,I2,3X,20H EXPLICIT 4PT UPWIND)
40       7 FORMAT(5H ME =,I2,3X,23H GENERAL CRANK-NICOLSON)
41         WRITE(6,8)JMAX,NTIM,C,DX,DT,XMAX
42       8 FORMAT(' JMAX=',I3,'  NTIM=',I3,'  C=',F5.2,
43        1'  DX=',F5.3,'  DT=',F5.3,'  XMAX=',F5.2)
44         WRITE(6,9)S,ALPH,RCEL,Q,EM,SA
45       9 FORMAT(' S=',F5.2,'  ALPH=',E10.3,'  RCEL=',F6.1,'  Q=',F5.2,
46        1' EM=',F6.3,'  SA=',E10.3)
47 C
48         IF(ME .GT. 3)GOTO 10
49         AA(1) =   0.5*C  + 3.*QQ
50         BB(1) =     - 3.*QQ
51         CC(1) = -0.5*C + QQ
52         IF(ME .NE. 2)GOTO 11
53         AA(1) = 0.
54         BB(1) = 0.5*C
55         CC(1) = -0.5*C
56         AA(2) = C
57         BB(2) = -C
58         CC(2) = 0.
```

Fig. 10.4. Listing of program BURG

exact solution it is necessary to limit the time and the smallest value of Re so that the boundary conditions (10.23) are consistent with the exact solution.

The various schemes described in Sects. 10.1.2 and 10.1.3 are implemented in program BURG (Fig. 10.4) according to the value of the parameter ME, as in Table 10.1. The choice $\delta = 1/6$, $q = 0$ produces the Crank-Nicolson, *group* finite element scheme. The group finite element formulation (Sect. 10.3) is necessary to discretise the dependent variable F explicitly. For solutions of (10.3) with small

```
59           GOTO 11
60     10 AA(1) = EM  + 0.5*S
61        BB(1) = 1.0 - 2.0*EM - S
62        CC(1) = EM + 0.5*S
63     11 WRITE(6,12)AA,BB,CC
64     12 FORMAT(' AA=',2F8.5,'  BB=',2F8.5,'   CC=',2F8.5,/)
65 C
66 C       INITIALISE U AND EVALUATE UEX
67 C
68        CALL EXSH(JMAX,X,U,UE,DX,XMAX.ALPH,TIMAX)
69 C
70        UD(1) = U(1)
71        UD(JMAX) = U(JMAX)
72        DO 14 J = 1,JMAX
73        DO 13 K = 1,5
74     13 A(K,J) = 0.
75     14 CONTINUE
76        WRITE(6,15)TIM
77     15 FORMAT(' INITIAL SOLUTION,  TIM =',F5.3)
78        WRITE(6,16)(X(J),J=1,JMAX)
79        WRITE(6,17)(U(J),J=1,JMAX)
80     16 FORMAT(' X=',12F6.3)
81     17 FORMAT(' U=',12F6.3)
82 C
83 C       MARCH SOLUTION IN TIME
84 C
85        DO 27 N = 1,NTIM
86        IF(ME .GT. 3)GOTO 21
87 C
88 C       EXPLICIT SCHEMES
89 C
90        IP = 1
91     18 IF(IP .EQ. 1)EF(1) = 0.5*U(1)*U(1)
92        IF(IP .EQ. 2)EF(1) = 0.5*UD(1)*UD(1)
93        EF(2) = EF(1)
94        EF(3) = EF(2)
95        IF(IP .EQ. 1)EF(4) = 0.5*U(2)*U(2)
96        IF(IP .EQ. 2)EF(4) = 0.5*UD(2)*UD(2)
97        DO 20 J = 2,JMAP
98        IF(ME .EQ. 3)EF(1) = EF(2)
99        EF(2) = EF(3)
100       EF(3) = EF(4)
101       IF(IP .EQ. 1)EF(4) = 0.5*U(J+1)*U(J+1)
102       IF(IP .EQ. 2)EF(4) = 0.5*UD(J+1)*UD(J+1)
103       DUM = S*(U(J-1) - 2.*U(J) + U(J+1))
104       IF(ME .NE. 2 .OR. IP .EQ. 2)GOTO 19
105       JP = J+1
106       IF(J .EQ. JMAP)JP = J
107       DUM = 0.5*DUM+0.5*S*(U(J)-2.*U(J+1)+U(JP+1))+0.5*(U(JP)-U(J))
108    19 UD(J) = U(J) + AA(IP)*EF(2) + BB(IP)*EF(3) + CC(IP)*EF(4) + DUM
109       IF(ME .EQ. 3)UD(J) = UD(J) - QQ*EF(1)
110    20 CONTINUE
111       IF(ME .NE. 2 .OR. IP .EQ. 2)GOTO 25
112       IP = IP + 1
113       GOTO 18
```

Fig. 10.4. (cont.) Listing of program BURG

values of the viscosity v it is desirable to add some artificial dissipation to the right-hand side of (10.16). The following expression is added:

$$0.5 v_a \Delta t L_{xx}(F_j^n + F_j^{n+1}) \ .$$

After linearising $L_{xx}F_j^{n+1}$, the modified form of (10.20) is implemented as

$$M_x u_j^{n+1} + 0.5\Delta t[L_x^{(4)}(u_j^n u_j^{n+1}) - v L_{xx} u_j^{n+1} - v_a \Delta t L_{xx}(u_j^n u_j^{n+1})]$$

$$= M_x u_j^n + 0.5 v \Delta t L_{xx} u_j^n \ . \tag{10.25}$$

```
114 C
115 C      TRIDIAGONAL SYSTEM FOR IMPLICIT SCHEMES
116 C
117    21 IF(MQ .EQ. 1)DIM = U(1)
118       DO 22 J = 2,JMAP
119       JM = J - 1
120       IF(JM .NE. 1)DIM = U(JM-1)
121       A(1,JM) = 0.5*QQ*DIM
122       A(2,JM) = EM - 0.5*S - (0.25*C + 1.5*QQ)*U(JM)
123       A(3,JM) = 1.0 - 2.0*EM + S + 1.5*QQ*U(J)
124       A(4,JM) = EM - 0.5*S + (0.25*C - 0.5*QQ)*U(J+1)
125       IF(ME .NE. 5 )GOTO 22
126       A(2,JM) = A(2,JM) - 0.5*SA*U(JM)
127       A(3,JM) = A(3,JM) + SA*U(J)
128       A(4,JM) = A(4,JM) - 0.5*SA*U(J+1)
129    22 R(JM) = AA(1)*U(JM) + BB(1)*U(J) + CC(1)*U(J+1)
130       R(1) = R(1) - A(2,1)*U(1)
131       IF(MQ .EQ. 0)GOTO 24
132       R(1) = R(1) - A(1,1)*U(1)
133       R(2) = R(2) - A(1,2)*U(1)
134 C
135 C      REDUCE A TO TRIDIAGONAL FORM
136 C
137       DO 23 JM = 3,JMAF
138       JMM = JM - 1
139       DUM = A(1,JM)/A(2,JMM)
140       A(2,JM) = A(2,JM) - A(3,JMM)*DUM
141       A(3,JM) = A(3,JM) - A(4,JMM)*DUM
142       A(1,JM) = 0.
143       R(JM) = R(JM) - R(JMM)*DUM
144    23 CONTINUE
145       A(1,1) = 0.
146       A(1,2) = 0.
147    24 A(2,1) = 0.
148       A(4,JMAF) = 0.
149 C
150       CALL BANFAC(A,JMAF,1)
151       CALL BANSOL(R,UD,A,JMAF,1)
152 C
153    25 DO 26 J = 2,JMAP
154       IF(ME .LE. 3)U(J) = UD(J)
155       IF(ME .GT. 3)U(J) = UD(J-1)
156    26 CONTINUE
157    27 CONTINUE
158 C
159       WRITE(6,28)TIMAX
160    28 FORMAT(' FINAL SOLUTION,    TIM =',F5.3)
161       WRITE(6,16)(X(J),J=1,JMAX)
162       WRITE(6,17)(U(J),J=1,JMAX)
163       WRITE(6,29)(UE(J),J=1,JMAX)
164    29 FORMAT('  UE=',12F6.3)
165       SUM = 0.
166       DO 30 J = 2,JMAP
167    30 SUM = SUM + (U(J) - UE(J))**2
168       RMS = SQRT(SUM/(AMP-1.))
169       WRITE(6,31)RMS
170    31 FORMAT(' RMS ERR=',E12.5)
171       STOP
172       END
```

Fig. 10.4. (cont.) Listing of program BURG

Table 10.1. Various schemes implemented in program BURG

ME	Description
1	FTCS scheme (10.5)
2	two-stage Lax-Wendroff scheme (10.13, 14)
3	Explicit four-point upwind scheme (10.5, 6)
4	Crank-Nicolson, fin. diff.; $\delta=0$, $q=0$ in (10.20); CN-FDM
4	Crank-Nicolson, fin. elem.; $\delta=1/6$, $q=0$ in (10.20); CN-FEM
4	Crank-Nicolson, mass oper.; $\delta=0.12$, $q=0$ in (10.20); CN-MO
4	Crank-Nicolson, 4pt. upwind; $\delta=0$, $q=0.5$ in (10.20); CN-4PU
5	generalised Crank-Nicolson plus additional dissipation

The parameter v_a is chosen empirically. Equation (10.25) is implemented when $ME=5$.

The various parameters used in program BURG are described in Table 10.2. The exact solution (10.24) is evaluated in subroutines EXSH (Fig. 10.5) and ERFC (Fig. 10.6). Typical output produced by program BURG for the FTCS scheme is shown in Fig. 10.7.

Table 10.2. Parameters used in program BURG

Parameter	Description
ME	$=1$, FTCS scheme (10.5)
	$=2$, two-stage Lax-Wendroff scheme (10.13, 14)
	$=3$, explicit four-point upwind scheme (10.5, 6)
	$=4$, generalised Crank-Nicolson (10.20)
	$=5$, generalised Crank-Nicolson plus additional dissipation
JMAX	number of points in the interval, $-x_{max} \leq x \leq x_{max}$
NTIM	number of time steps
TIMAX	time of comparison with exact solution
XMAX	x_{max} in (10.22)
AD	artificial dissipation, v_a in (10.25)
C	reference Courant number, $C=u(1)\Delta t/\Delta x$
ALPH	viscosity, v in (10.3)
S, SA	$s=v\Delta t/\Delta x^2$, $s_a=v_a C^2$ (Table 10.5)
RCEL	reference cell Reynolds number, $u(1)\Delta x/v$
Q	q in (10.6)
EM	δ in (10.20)
DT, DX, X	Δt, Δx, x
U	dependent variable, u in (10.3)
EF	F in (10.3)
UE	exact solution, $\bar{u}$ in (10.24)
AA, BB, CC	coefficients multiplying F_{j-1}^n, etc., for explicit schemes
	coefficients multiplying u_{j-1}^n, etc., for implicit schemes
A	elements in quadridiagonal matrix (10.21)
R	vector containing d_j^n in (10.21)
UD	intermediate explicit solution; first-stage Lax-Wendroff
UD	for u_j^{n+1} on return from BANSOL

```
1
2          SUBROUTINE EXSH(JMAX,X,U,UE,DX,XMAX,ALPH,TIMAX)
3   C
4   C      SETS THE INITIAL U SOLUTION AND FINAL EXACT (UEX) SOLUTION
5   C
6          DIMENSION U(65),UE(65),X(65)
7          JMAP = JMAX - 1
8          PI = 3.141592654
9          XST = -XMAX
10         DO 1 J = 1,JMAX
11         AJ = J - 1
12         X(J) = XST + AJ*DX
13         U(J) = 0.
14       1 UE(J) = 0.
15         U(1) = 1.0
16         UE(1) = 1.0
17  C
18         DO 2 J = 2,JMAP
19         IF(X(J) .LT. 0.)U(J) = 1.0
20         IF(ABS(X(J)) .LT. 1.0E-04)U(J) = 0.5
21         AJ = X(J)
22         XB = - AJ
23         XA = AJ - TIMAX
24         PQ = SQRT(0.25*PI)
25         PP = SQRT(PI*TIMAX*ALPH)
26         XB = XB*PQ/PP
27         XA = XA*PQ/PP
28  C
29  C      ERFC CALCULATES THE COMPLEMENTARY ERROR FUNCTION
30  C
31         CALL ERFC(XA,EXA)
32         CALL ERFC(XB,EXB)
33  C
34         SUMA = PP*EXA
35         SUMB = PP*EXB
36         DUM  = 0.5*(AJ-0.5*TIMAX)/ALPH
37         IF(DUM .GT. 20.)DUM = 20.
38         IF(DUM .LT.-20.)DUM =-20.
39         SUMC = EXP(DUM)
40         UE(J) = SUMA/(SUMA + SUMC*SUMB)
41       2 CONTINUE
42         RETURN
43         END
```

Fig. 10.5. Listing of subroutine EXSH

```
1
2          SUBROUTINE ERFC(X,ERC)
3   C
4   C      COMPLEMENTARY ERROR FUNCTION
5   C
6          B = ABS(X)
7          IF(B .LT. 4)GOTO 1
8          D = 1.0
9          GOTO 2
10       1 C = EXP(-B*B)
11         T = 1./(1. + 0.3275911*B)
12         D = 0.254829592*T - 0.284496736*T*T + 1.421413741*T*T*T
13         1 - 1.453152027*T*T*T*T + 1.061405429*T*T*T*T*T
14         D = 1. - D*C
15       2 IF(X .LT. 0.)D = -D
16         ERC = 1. - D
17         RETURN
18         END
```

Fig. 10.6. Listing of subroutine ERFC

```
PROPAGATING SHOCK WAVE(BURGERS EQUATION)
ME = 1    FTCS DIFFERENCING
JMAX= 21  NTIM= 20  C=  .25  DX= .200  DT= .050  XMAX= 2.00
S=  .25  ALPH=  .200E+00 RCEL=   1.0  Q=  .00 EM=  .000  SA=  .000E+00
AA=  .12500  .00000 BB=  .00000  .00000  CC= -.12500  .00000

INITIAL SOLUTION,  TIM = .000
X=-2.000-1.800-1.600-1.400-1.200-1.000 -.800 -.600 -.400 -.200  .000  .200
X=  .400  .600  .800 1.000 1.200 1.400 1.600 1.800 2.000
U= 1.000 1.000 1.000 1.000 1.000 1.000 1.000 1.000 1.000 1.000  .500  .000
U=  .000  .000  .000  .000  .000  .000  .000  .000  .000
FINAL SOLUTION,   TIM =1.000
X=-2.000-1.800-1.600-1.400-1.200-1.000 -.800 -.600 -.400 -.200  .000  .200
X=  .400  .600  .800 1.000 1.200 1.400 1.600 1.800 2.000
U= 1.000 1.000 1.000 1.000 1.000  .999  .997  .991  .975  .940  .869  .751
U=  .588  .407  .247  .132  .063  .027  .010  .003  .000
UE= 1.000 1.000 1.000 1.000 1.000  .999  .996  .989  .973  .937  .868  .753
UE=  .591  .409  .247  .132  .063  .027  .011  .004  .000
RMS ERR=  .14106E-02
```

Fig. 10.7. Typical output from program BURG

For a value of the viscosity, $v = 0.2$, a representative cell Reynolds number is $R_{cell} = u(1)\Delta x/v = 1.0$. This implies that the convective and dissipative terms in (10.3) are of the same order. Solutions for $R_{cell} = 1.0$, obtained with the various explicit and implicit schemes provided in Sects. 10.1.2 and 10.1.3, are given in Table 10.3. All methods give smooth results. The FTCS scheme is the most accurate explicit scheme and the Crank-Nicolson four-point upwind scheme with $q = 0.5$ is the most accurate implicit scheme.

Representative solutions for a viscosity $v = 0.03$ ($R_{cell} = 3.33$) are shown in Table 10.4. The Crank-Nicolson finite difference scheme produces a solution that indicates dispersion-like wiggles particularly behind the shock. Both the Crank-Nicolson finite element scheme and the Crank-Nicolson four-point upwind scheme produce smooth, accurate solutions for this case. The shock is spread over about nine grid points so that the computational algorithms can resolve the sharp gradient in u without difficulty.

If $v = 0.001$ ($R_{cell} = 100$) the shock is much sharper (Table 10.5 and Fig. 10.8) and is likely to cause dispersion-related oscillations.

In principle it is possible to construct low-dispersion schemes following the procedure in Sect. 9.4 of generating the modified equation (Sect. 9.2.2) and choosing q or δ in (10.20) to cancel the coefficient of the lowest-order odd derivative. The nonlinear nature of (10.3) causes many additional products to be generated which do not fit into the dispersive/dissipative pattern of Sect. 9.2.2. If an equivalent linear scheme is constructed by locally freezing the nonlinear coefficients, the following optimum values for δ and q are obtained:

$$\delta_{opt} = \frac{1}{6} + \frac{(u_j \Delta t/\Delta x)^2}{12} , \qquad (10.26)$$

$$q_{opt} = 0.5 + \frac{(u_j \Delta t/\Delta x)^2}{4} . \qquad (10.27)$$

Table 10.3. Propagating shock solution at $t = 1.00$, for $R_{cell} = 1.0$, $c = 0.25$, $\Delta x = 0.20$, $\Delta t = 0.05$

Scheme \ x	−0.40	−0.20	0.00	0.20	0.40	0.60	0.80	1.00	1.20	1.40	1.60	rms error
Exact	0.973	0.937	0.868	0.753	0.591	0.409	0.247	0.132	0.063	0.027	0.011	—
FTCS	0.975	0.940	0.869	0.751	0.588	0.407	0.247	0.132	0.063	0.027	0.010	0.0014
Lax-Wendroff	0.979	0.943	0.869	0.748	0.583	0.403	0.244	0.131	0.063	0.027	0.010	0.004
EX-4PU, $q = 0.5$	0.971	0.936	0.869	0.757	0.598	0.415	0.248	0.129	0.059	0.025	0.009	0.003
CN-FDM	0.975	0.939	0.867	0.747	0.582	0.403	0.246	0.135	0.067	0.030	0.013	0.003
CN-FEM	0.973	0.936	0.865	0.748	0.589	0.411	0.252	0.136	0.064	0.027	0.010	0.002
CN-4PU, $q = 0.5$	0.972	0.936	0.867	0.753	0.593	0.410	0.247	0.131	0.063	0.028	0.012	0.0007

Table 10.4. Propagating shock solution at $t = 1.00$ for $R_{cell} = 3.33$, $c = 1.00$, $\Delta x = 0.10$, $\Delta t = 0.10$

Scheme \ x	0.00	0.10	0.20	0.30	0.40	0.50	0.60	0.70	0.80	0.90	1.00	rms error
Exact	1.000	0.999	0.995	0.969	0.847	0.500	0.153	0.031	0.005	0.001	0.000	—
CN-FDM	0.998	0.995	1.020	1.021	0.814	0.438	0.157	0.043	0.011	0.003	0.001	0.015
CN-FEM	1.000	1.000	0.999	0.986	0.846	0.491	0.153	0.023	0.002	0.000	0.000	0.003
CN-4PU, $q = 0.5$	1.000	1.000	1.000	0.980	0.847	0.501	0.144	0.022	0.004	0.001	0.000	0.003

Table 10.5. Propagating shock solution at $t = 2.00$ for $R_{cell} = 100$, $c = 1.00$, $s_a = 0.25$, $\Delta x = 0.10$, $\Delta t = 0.10$

Scheme \ x	0.50	0.60	0.70	0.80	0.90	1.00	1.10	1.20	1.30	1.40	1.50	rms errors
Exact	1.000	1.000	1.000	1.000	1.000	0.500	0.000	0.000	0.000	0.000	0.000	—
CN-FDM	1.004	0.984	0.987	1.086	0.996	0.413	0.028	0.000	0.000	0.000	0.000	0.020
CN-MO, $\delta = 0.12$	1.000	1.001	0.995	1.010	1.011	0.522	−0.046	0.007	−0.001	0.000	0.000	0.009
CN-4PU, $q = 0.5$	1.000	1.000	1.000	1.009	0.983	0.572	−0.056	−0.007	0.000	0.000	0.000	0.015

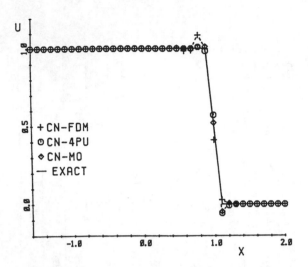

Fig. 10.8. Velocity distribution at $t = 2.00$ with $R_{cell} = 100$

Clearly these values are a natural extension of the optimal values derived in Sect. 9.4.3 for the convecting temperature front. However, the use of (10.26) or (10.27) are less successful than in Sect. 9.4.3, mainly due to the additional nonlinear terms not accounted for.

Consequently an alternative strategy has been adopted here. That is to choose δ and q empirically and to introduce sufficient artificial dissipation to eliminate most of the ripples without diffusing the shock front over too many grid points. The results shown in Table 10.5 and Fig. 10.8 indicate that this is a successful strategy and that even the Crank-Nicolson finite difference scheme is producing quite accurate solutions with judicious use of artificial dissipation.

However, this example is well suited to the use of artificial dissipation, since the exact solution is constant except at the shock. Consequently a large value of v_a in (10.25) has no influence on the solution except in the immediate vicinity of the shock. A better technique is to make v_a a function of the solution so that it only assumes a significant value at the shock. This technique is discussed in Sect. 14.2.3.

The one-dimensional Burgers' equation with a shock-like (i.e. thin) internal layer can be computed with spectral methods (Sect. 5.6). A direct application of the Galerkin spectral method produces a solution that demonstrates global oscillations for a nine-term Legendre polynomial solution at $v = 0.01$ (Fletcher 1984, pp. 194–198). Basdevant et al. (1986) obtain solutions with $v = 0.01/\pi$ for spectral tau and collocation (pseudospectral) methods (Sect. 5.6.3) based on Chebyshev polynomials. Without co-ordinate transformation it is found that 256 collocation points must be used to avoid oscillatory solutions. It is concluded that spectral methods, being global, are not as well suited to the calculation of thin internal layers as are local methods, such as finite difference and finite element methods.

10.1.5 Nonuniform Grid

When severe gradients occur either internally or adjacent to a boundary, more accurate solutions can be obtained by grid refinement. However, for a uniform grid throughout the computational domain, grid refinement can be computationally expensive, particularly in multidimensions.

An obvious alternative is to introduce a grid which is refined only in the parts of the computational domain where severe gradients are to be expected. The mechanics of generating such a grid are discussed in Chap. 13. Here the use of non-uniform grids for one-dimensional steady problems will be examined, primarily to contrast the expected improvement in the solution accuracy through local grid refinement with the possible reduction in the order of the truncation error.

The convection-diffusion equation (Sect. 9.3)

$$u\frac{d\bar{T}}{dx} - \alpha\frac{d^2\bar{T}}{dx^2} = 0 \ , \tag{10.28}$$

with u and α constant and boundary conditions $\bar{T}(0) = 0$, $\bar{T}(1) = 1$, produces a severe boundary gradient adjacent to $x = 1$ (Fig. 9.5) for large values of u/α.

Local grid refinement adjacent to $x = 1$ can be achieved by allowing the grid to grow geometrically. That is,

$$x_{j+1} - x_j = r_x(x_j - x_{j-1}) = r_x\Delta x_j \ , \tag{10.29}$$

where r_x is the grid growth factor. For the particular example shown in Fig. 9.5 a value of r_x less than one would be appropriate to place more grid points close to $x = 1.0$. Typically r_x is constant through all or a large part of the computational domain.

A general three-point representation of $d\bar{T}/dx$ on a nonuniform grid that produces the smallest truncation error can be obtained using the technique described in Sect. 3.2.2. A Taylor series expansion of such a three-point discretisation about x_j is

$$\frac{(T_{j+1} - T_j)/r_x + r_x(T_j - T_{j-1})}{(1 + r_x)\Delta x_j}$$

$$= \left[\frac{dT}{dx} + \frac{\Delta x_j^2}{6}r_x\frac{d^3T}{dx^3} + \left(\frac{\Delta x_j^3}{24}\right)r_x(r_x - 1)\frac{d^4T}{dx^4} + \ \cdots \ \right]_j \ , \tag{10.30}$$

where r_x is defined by (10.29) and $\Delta x_j = x_j - x_{j-1}$. This representation is of second order but introduces an additional third-order error that is not present for a uniform grid ($r_x = 1.0$).

Application of the Galerkin finite element method to (10.28) with linear interpolation (Sect. 5.3.1) produces a "three-point" representation of $d\bar{T}/dx$. A Taylor series expansion of the finite-element three-point representation of $d\bar{T}/dx$ is

$$\frac{T_{j+1} - T_{j-1}}{(1 + r_x)\Delta x_j} = \left[\frac{dT}{dx} + \frac{\Delta x_j}{2}(r_x - 1)\frac{d^2T}{dx^2} + \frac{\Delta x_j^2}{6}\frac{r_x^3 + 1}{r_x + 1}\frac{d^3T}{dx^3} + \ \cdots \ \right]_j \ . \tag{10.31}$$

This representation introduces a first-order dissipative term whose sign depends on whether the grid is geometrically growing or diminishing in the positive x direction.

Clearly if $(r_x - 1)$ or Δx_j are not small the influence of the term for large u/α may be significant compared with $\alpha d^2 \bar{T}/dx^2$ in (10.28).

A three-point representation of $d^2 T/dx^2$ in (10.28) that minimises the truncation error has the following Taylor series expansion about x_j:

$$
\left[\frac{T_{j+1} - T_j}{r_x} - (T_j - T_{j-1}) \right] \left(\frac{2}{(1 + r_x)\Delta x_j^2} \right)
$$

$$
= \left[\frac{d^2 T}{dx^2} + \frac{\Delta x_j}{3}(r_x - 1)\frac{d^3 T}{dx^3} + \frac{\Delta x_j^2}{12}\frac{r_x^3 + 1}{r_x + 1}\frac{d^4 T}{dx^4} + \cdots \right]_j . \tag{10.32}
$$

The left-hand side of (10.32) also results from a linear Galerkin finite element discretisation of $d^2 \bar{T}/dx^2$. It is apparent that the use of a nonuniform grid leads to the appearance of a first-order truncation error.

Substitution of the left-hand sides of (10.30 and 32) into (10.28) produces the tridiagonal scheme:

$$
-(0.5 R_{cell} r_x + 1) T_{j-1} + \left[0.5 R_{cell} \frac{r_x^2 - 1}{r_x} + \left(1 + \frac{1}{r_x} \right) \right] T_j
$$

$$
+ \left(\frac{0.5 R_{cell}}{r_x} - \frac{1}{r_x} \right) T_{j+1} = 0 . \tag{10.33}
$$

The substitution of the left-hand sides of (10.31 and 32) produces the slightly simpler tridiagonal scheme

$$
-(0.5 R_{cell} + 1) T_{j-1} + \left(1 + \frac{1}{r_x} \right) T_j + \left(0.5 R_{cell} - \frac{1}{r_x} \right) T_{j+1} = 0 , \tag{10.34}
$$

where $R_{cell} = u\Delta x_j/\alpha$. Solutions based on (10.33 and 34) have been obtained with $u/\alpha = 20$ and 11 points in the interval $0 \le x \le 1$ and are presented in Tables 10.6 and 10.7. For a uniform grid, $r_x = 1$, the cell Reynolds number $R_{cell} = 2$; this case is plotted in Fig. 9.5.

For the uniform-grid case the rms error is dominated by the local errors at the grid points close to $x = 1$. Reducing r_x is seen to reduce the rms error (Table 10.6) with solutions of (10.34) being considerably more accurate than those of (10.33). Since $r_x < 1.0$ the use of (10.34) introduces positive dissipation. When using (10.33) the choice $r_x = 0.80$ produces the minimum rms solution error. This corresponds to

Table 10.6. Variation of the rms error with r_x for solutions of (10.28) at $u/\alpha = 20$

rms error	r_x 0.70	0.80	0.90	1.00
(10.33)	0.0338	0.0276	0.0319	0.0455
(10.34)	0.0048	0.0080	0.0197	0.0455

Table 10.7. Solutions to (10.28) using a nonuniform grid $r_x = 0.8$, $u/\alpha = 20$

x	0.753	0.827	0.885	0.932	0.970	1.000	rms error
T, exact	0.0072	0.0312	0.1010	0.2584	0.5480	1.0000	—
T, (10.33)	0.0006	0.0111	0.0615	0.2050	0.5013	1.0000	0.0276
T, (10.34)	0.0035	0.0229	0.0879	0.2439	0.5376	1.0000	0.0080

a relatively uniform error distribution (Table 10.7). For growth factors $r_x < 0.80$, although the local solution error is smaller close to $x = 1.0$, it becomes larger at smaller values of x. For more complicated governing equations in multidimensions, grid growth factors in the range $0.8 \leq r_x \leq 1.2$ are usually appropriate.

The use of a nonuniform grid with a severe internal gradient can be illustrated using a modified form of Burgers' equation

$$\frac{\partial \bar{u}}{\partial t} + (\bar{u} - 0.5)\frac{\partial \bar{u}}{\partial x} - v\frac{\partial^2 \bar{u}}{\partial x^2} = 0 \ . \tag{10.35}$$

Compared with (10.1) an additional term, $-0.5\partial \bar{u}/\partial x$, has been introduced. For the initial and boundary conditions given by (10.22 and 10.23), the inclusion of the additional term is equivalent to allowing the grid to migrate with the shock so that the shock does not move relative to the grid. As a result, (10.35) has a particularly simple steady-state solution

$$\bar{u}_{ss} = 0.5\left[1 - \tanh\left(\frac{0.25x}{v}\right)\right] \ . \tag{10.36}$$

Decreasing v sharpens the gradient centred at $x = 0$. Equation (10.35) can be written in conservation form as

$$\frac{\partial \bar{u}}{\partial t} + \frac{\partial \bar{F}}{\partial x} - v\frac{\partial^2 \bar{u}}{\partial x^2} = 0 \ , \tag{10.37}$$

where $\bar{F} = 0.5(\bar{u}^2 - \bar{u})$. Since only steady-state solutions of (10.37) are required it is appropriate to introduce the fully implicit linearised two-level discretisation,

$$\frac{\Delta u_j^{n+1}}{\Delta t} = vL_{xx}(u_j^n + \Delta u_j^{n+1}) - L_x\left(F_j^n + \frac{dF}{du}\Delta u_j^{n+1}\right) , \tag{10.38}$$

where L_{xx} and L_x are given by the left-hand sides of (10.32), (10.30 and 31) and $dF/du = u - 0.5$.

The application of (10.31) to $\partial F/\partial x$ is an example of the group finite element formulation (Sect. 10.3), since F is a nonlinear group. The following tridiagonal system of equations for Δu_j^{n+1} is obtained:

$$\{1 + \Delta t[L_x(u - 0.5) - vL_{xx}]\}\Delta u_j^n = \Delta t(vL_{xx}u_j^n - L_x F_j^n) \ . \tag{10.39}$$

Equation (10.39) is integrated until Δu_j^{n+1} is negligibly small corresponding to a solution of $(vL_{xx}u_j - L_xF_j) = 0$.

For a shock centred at $x=0$ it is convenient to introduce a grid growing geometrically in both the positive and negative x directions from $x=0$. Solutions based on (10.30, 31) for $v=0.08$ with 11 points in the interval, $-x_{max} \leq x \leq x_{max}$, are shown in Tables 10.8 and 10.9. For $r_x = 1.0$ the solutions correspond to $R_{cell} = 5$ based on $u(1) = 1.0$.

Table 10.8. Variation of the rms errors with r_x for the solution of (10.39) with $v = 0.08$

rms error	r_x 1.00	1.10	1.20	1.30
(10.30)	0.0084	0.0093	0.0123	0.0161
(10.31)	0.0084	0.0090	0.0082	0.0068

Table 10.9. Solutions of (10.39) using a nonuniform grid, $r_x = 1.3$, with $v = 0.08$

x	−1.368	−0.882	−0.509	−0.221	0.000	0.221	0.509	0.882	1.368	rms error
u, exact	1.000	0.996	0.960	0.799	0.500	0.201	0.040	0.004	0.000	—
u, (10.30)	0.999	1.001	0.993	0.807	0.500	0.193	0.007	−0.001	0.001	0.0161
u, (10.31)	0.999	1.002	0.967	0.788	0.500	0.212	0.033	−0.002	0.001	0.0068

The errors shown in Table 10.8 indicate that the use of the discretisation for $\partial F/\partial x$ given by the left-hand side of (10.30) contributes to a reduction in accuracy with increasing r_x. In contrast, the use of (10.31) does cause a reduction in rms error for r_x sufficiently large. However, the improvement in accuracy is not very great. For $r_x = 1.3$ the two solutions are compared with the exact solution in Table 10.9. Both solutions are exact at $x=0.0$ and the major contributions to the rms error comes from the four points, $x = \pm 0.221$, ± 0.509.

The application of a nonuniform grid to "real" flow problems usually requires considerable numerical experimentation to arrive at the best choice for the grid growth factor, r_x, and the discretisation scheme. For unsteady problems with locally severe gradients it may be necessary to make the grid adapt to the local solution. That is, the region of local refinement may change with time. Such adaptive grid techniques (Thompson et al. 1985; Kim and Thompson 1990) are beyond the scope of this book.

Nonuniform grids may also be required where the computational boundary does not coincide with the local grid. Noye (1983) provides a summary of traditional techniques for handling this situation. A more modern approach to irregular boundaries is through the use of generalised curvilinear coordinates (Chap. 12), so that the computational boundary automatically coincides with a grid line.

10.2 Systems of Equations

The material presented in Sect. 10.1 and previous chapters is related to the occurrence of a *single* governing equation. However, fluid dynamic phenomena are described, usually, by systems of equations (Chap. 11). An exception is inviscid, incompressible flow which is governed by Laplace's equation (Sect. 11.3). The purpose of this section is to consider a system of governing equations and to apply to it representative discretisations that have already been applied to a single governing equation.

One-dimensional unsteady compressible inviscid flow is governed by a system of three equations: continuity (Sect. 11.2.1), x-momentum (Sect. 11.2.2) and energy (Sect. 11.2.4). After a suitable nondimensionalisation (Sect. 14.2.3) the system of equations can be written

$$\frac{\partial \bar{q}}{dt} + \frac{\partial \bar{F}}{\partial x} = 0 , \quad \text{where} \tag{10.40}$$

$$\bar{q} = \begin{pmatrix} \varrho \\ \varrho u \\ \dfrac{p}{\gamma(\gamma-1)} + 0.5\varrho u^2 \end{pmatrix} \quad \text{and} \quad \bar{F} = \begin{pmatrix} \varrho u \\ \varrho u^2 + \dfrac{p}{\gamma} \\ \left(\dfrac{p}{\gamma-1} + 0.5\varrho u^2\right)u \end{pmatrix} . \tag{10.41}$$

In (10.41) ϱ is the density, u is the velocity, p is the pressure and γ is the specific heats ratio.

Although ϱ, u and p are the dependent variables, discretisation is applied directly to q and F. This will be illustrated here for the two-stage Lax-Wendroff scheme (10.11, 12). Applied to (10.40) this gives

$$q^*_{j+1/2} = 0.5(q^n_j + q^n_{j+1}) - 0.5\frac{\Delta t}{\Delta x}(F^n_{j+1} - F^n_j) \quad \text{and} \tag{10.42}$$

$$q^{n+1}_j = q^n_j - \frac{\Delta t}{\Delta x}(F^*_{j+1/2} - F^*_{j-1/2}) . \tag{10.43}$$

At each stage of the solution development, ϱ, u and p are evaluated from q (10.41) so that the components of F can be determined.

Klopfer and McRae (1983) use the modified equation approach (Sect. 9.2.2) to correct the main (dispersion) error for the Lax-Wendroff scheme applied to the Euler equations, (10.40) modelling the one-dimensional shock tube problem (Problem 14.8). The resulting scheme is more accurate, less oscillatory but typically requires a smaller $\Delta t_{\max}$ for stable solutions.

Implicit schemes can also be constructed for systems of equations. Thus, a Crank-Nicolson treatment of (10.40) would be

$$q^{n+1}_j - q^n_j = -0.25\frac{\Delta t}{\Delta x}[(F^n_{j+1} - F^n_{j-1}) + (F^{n+1}_{j+1} - F^{n+1}_{j-1})] . \tag{10.44}$$

To produce a linear system of algebraic equations for Δq_j^{n+1} the nonlinear terms F^{n+1} are expanded as

$$F^{n+1} = F^n + \underline{A}\,\Delta q^{n+1} + \cdots \tag{10.45}$$

where $\underline{A}$ ($\equiv \partial F/\partial q$) is a 3×3 matrix evaluated from (10.41). The specific form of $\underline{A}$ can be deduced from (14.99). Substituting (10.45) into (10.44) and rearranging gives

$$-0.25\frac{\Delta t}{\Delta x}\underline{A}_{j-1}\Delta q_{j-1}^{n+1} + \underline{I}\Delta q_j^{n+1} + 0.25\frac{\Delta t}{\Delta x}\underline{A}_{j+1}\Delta q_{j+1}^{n+1}$$

$$= -0.5\frac{\Delta t}{\Delta x}(F_{j+1}^n - F_{j-1}^n) \ . \tag{10.46}$$

Equation (10.46) is a 3×3 block tridiagonal system of equations for Δq_j^{n+1} that can be solved by the block Thomas algorithm (Sect. 6.2.5). The 3×3 block size follows directly from the number of governing equations at each grid point. Having obtained Δq_j^{n+1} from (10.46) the solution at the nth time step is obtained from $q_j^{n+1} = q_j^n + \Delta q_j^{n+1}$.

The stability of the discretised scheme, e.g. (10.46), can be assessed using the von Neumann analysis (Sect. 4.3) as for scalar equations. Since the von Neumann analysis is only applicable to linear equations it is convenient to linearise (10.40) *before* discretisation.

That is

$$\frac{\partial \bar{q}}{\partial t} + \underline{A}\frac{\partial \bar{q}}{\partial x} = 0 \ ,$$

where $\underline{A} \equiv \partial F/\partial q$ as in (10.45). The von Neumann stability analysis proceeds in the manner described in Sect. 4.3.4 except that an amplification matrix $\underline{G}$ is produced in place of the amplification factor G in (4.33). For the Crank-Nicolson scheme (10.46) the result is

$$\underline{G} = \left(1 + 0.5\mathrm{i}\frac{\Delta t}{\Delta x}\underline{A}\sin\theta\right)^{-1}\left(1 - 0.5\mathrm{i}\frac{\Delta t}{\Delta x}\underline{A}\sin\theta\right) \ . \tag{10.47}$$

If λ_m are the eigenvalues of $\underline{G}$, stability requires that $|\lambda_m| \le 1.0$ for all m. The stability will depend on the local magnitude of the eigenvalues of $\underline{A}$ (14.32). Apart from the appearance of the Jacobian matrix $\underline{A}$ the structure of (10.47) is very similar to that for the Crank-Nicolson scheme applied to the linear convection equation (Table 9.1).

The system of equations (10.40) is considered at greater length in Sect. 14.2.3 (Vol. 2) where a program, SHOCK, is provided based on the two-stage Lax-Wendroff scheme (10.42, 43), and the MacCormack scheme (14.49, 50). A typical solution for a propagating shock wave is shown in Fig. 10.9, after 25 time steps. This

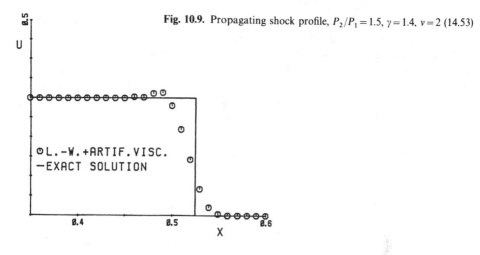

Fig. 10.9. Propagating shock profile, $P_2/P_1 = 1.5$, $\gamma = 1.4$, $v = 2$ (14.53)

solution is obtained with a uniform grid $\Delta x = 0.01$ in the interval $0 \le x \le 1.0$. At $t = 0$ the shock is located at $x = 0.5$; subsequently it propagates to the right with a nondimensional shock speed $= 1.195$.

Artificial dissipation has been added to (10.40) to produce a smoother solution. The form of the artificial dissipative terms is indicated by (14.53). Broadly it is similar to the term added to the one-dimensional Burgers' equation (10.3) to give (10.25). However, v, the equivalent of v_a in (10.25), is a function of the local solution so that it only has a significant value at the shock.

Comparing the results of Fig. 10.9 with Fig. 10.8 it is clear that the computed shock profile for the "real" governing equation is less sharply defined than for the one-dimensional Burgers' equation. This is partly due to the relative complexity of the equations and is partly a reflection of the relative accuracy of the Lax-Wendroff and Crank-Nicolson schemes.

10.3 Group Finite Element Method

The finite element method has been applied to the Sturm-Liouville problem in Sect. 5.4, the one-dimensional diffusion equation in Sect. 5.5.1, the two dimensional diffusion equation in Sect. 8.3 and the two-dimensional transport equation in Sect. 9.5.2. In all these applications the governing equations are linear.

However, the fluid dynamic governing equations (Chap. 11) typically contain nonlinear convective terms. In primitive variables the convective terms are quadratically nonlinear for incompressible flow and cubically nonlinear for compressible flow.

The conventional finite element method introduces a separate approximate solution, like (5.44) for each dependent variable appearing in the governing equation. Application of the Galerkin method produces large numbers of products of nodal values of dependent variables, particularly from the nonlinear convective terms. This feature contributes to an inherent lack of economy of the finite element method in multidimensions (Fletcher 1984, p. 243).

However, the contribution to the lack of economy from the nonlinear terms can be avoided by introducing the group finite element formulation (Fletcher 1979, 1983b). The group formulation is generally applicable to nonlinear problems but is particularly effective in dealing with convective nonlinearities.

The group formulation requires two steps:
i) the equations are cast in conservation form, for example (10.3),
ii) a single approximate solution is introduced for the group of terms appearing in the differential terms, for example $\bar{F}$ in (10.3).

10.3.1 One-Dimensional Group Formulation

The group finite element method will be illustrated for the one-dimensional Burgers' equation in conservation form (10.3). Separate approximate solutions, equivalent to (5.44), are introduced for $\bar{u}$ and $\bar{F}$ in (10.3) as

$$u = \sum_l \phi_l u_l \quad \text{and} \quad F = \sum_l \phi_l F_l , \tag{10.48}$$

where F_l represents F evaluated at the lth node. Substitution into (10.3) and evaluation of the weighted residual equation (5.5) based on the Galerkin weighting function produces a system of ordinary differential equations. On a uniform grid this system can be written

$$M_x \left[\frac{du}{dt} \right]_j + L_x F_j - \nu L_{xx} u_j = 0 , \tag{10.49}$$

where $M_x = (\frac{1}{6}, \frac{2}{3}, \frac{1}{6})$, $L_x = \{-1, 0, 1\}/2\Delta x$ and $L_{xx} = \{1, -2, 1\}/\Delta x^2$, for linear interpolating functions, ϕ_l, in (10.48). With Crank-Nicolson time discretisation, (10.49) produces an algorithm that is given by (10.20) with $\delta = \frac{1}{6}$ and $q = 0$.

If the expression $L_x F_j$ is expressed in terms of u_j the result can be written

$$L_x F_j = 0.5(u_{j-1} + u_{j+1}) \left\{ \frac{u_{j+1} - u_{j-1}}{2\Delta x} \right\} . \tag{10.50}$$

If the conventional finite element method is applied to (10.1) the discretisation of the convective term $u \partial u / \partial x$, on a uniform grid, is

$$u \frac{\partial u}{\partial x} \Rightarrow \frac{(u_{j-1} + u_j + u_{j+1})}{3} \left\{ \frac{u_{j+1} - u_{j-1}}{2\Delta x} \right\} . \tag{10.51}$$

The linear terms $\partial u / \partial t$ and $\partial^2 u / \partial x^2$ produce the same discretisation as in (10.49).

The discretised form generated by the conventional finite element method has been combined with Crank-Nicolson time differencing to produce the solution of the one-dimensional Burgers' equation shown in Fig. 10.10 as CN-FEM(C). The

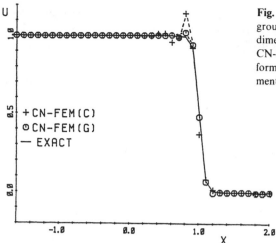

Fig. 10.10. Comparison of conventional and group finite element methods for the one-dimensional Burgers' equation at $t = 2.0$. CN-FEM(C) = conventional finite element formulation CN-FEM(G) = group finite element formulation

group finite element formulation is implemented by (10.20) with $\delta = \frac{1}{6}$ and $q = 0$. In Fig. 10.10 it is shown as CN-FEM(G).

The solutions shown in Fig. 10.10 are based on 41 equally-space points in the interval $-2.0 \leq x \leq 2.0$ and correspond to the boundary and initial conditions considered in Sect. 10.1.4. The cell Reynolds number, based on $u(1)$, is $R_{cell} = 5$. The conventional finite element solution is indicating a larger oscillation trailing the shock than the group finite element solution. In other parts of the computational domain the solutions are similar.

Numerical convergence behaviour of the conventional and group finite element methods applied to the one-dimensional Burgers' equation is compared by Fletcher (1983c). For linear, quadratic and cubic interpolation both formulations are of comparable accuracy. In one spatial dimension both formulations demonstrate similar economy.

10.3.2 Multidimensional Group Formulation

The group finite element formulation is demonstrated, here, on the two-dimensional Burgers' equations (10.57 and 58). The equivalent divergence vector form of these equations is

$$\frac{\partial \bar{q}}{\partial t} + \frac{\partial \bar{F}}{\partial x} + \frac{\partial \bar{G}}{\partial y} - \nu \left(\frac{\partial^2 \bar{q}}{\partial x^2} + \frac{\partial^2 \bar{q}}{\partial y^2} \right) - \bar{S} = 0 \ , \tag{10.52}$$

where

$$\bar{q} = \{u, v\} \ , \qquad \bar{F} = \{u^2, uv\} \ , \qquad \bar{G} = \{uv, v^2\} \ ,$$

$$\bar{S} = \left\{ \frac{0.5u(u^2 + v^2)}{\nu}, \frac{0.5v(u^2 + v^2)}{\nu} \right\} \ . \tag{10.53}$$

As well as the approximate solutions for u and v, additional approximate solutions are introduced for the groups u^2, uv, v^2 and the components of S. A typical approximate solution (for rectangular elements) is

$$uv = \sum_{l=1}^{4} (uv)_l \phi_l(\xi, \eta) \, , \tag{10.54}$$

where $\phi_l(\xi, \eta)$ are bilinear interpolating functions (5.59). The term $(uv)_l$ represents the evaluation of uv at the lth node.

Application of the Galerkin finite element method, as in Sect. 5.3, produces the discretised equations

$$M_x \otimes M_y \left[\frac{\partial q}{\partial t} \right]_{j,k} = \mathrm{RHS} \, , \quad \text{where} \tag{10.55}$$

$$\mathrm{RHS} = -M_y \otimes L_x F - M_x \otimes L_y G$$
$$+ v(M_y \otimes L_{xx} + M_x \otimes L_{yy})q + M_x \otimes M_y S \, . \tag{10.56}$$

The similarity in structure with (9.85) is noteworthy. This is a feature of the group formulation; that is, at the level at which the discretisation takes place, the equations are linear, although indeterminate. Substitution for the nodal groups in terms of the unknown nodal variables introduces the nonlinearity but also makes the system determinate.

The tensor products of the mass operator and difference operator, e.g. $M_y \otimes L_x F$, in (10.56) indicate that the group finite element formulation is producing a nine-point discretisation of $\partial F / \partial x$. Strictly it is only a six-point discretisation since one coefficient in L_x is zero. If the conventional finite element method, is applied to (10.57 and 58) the connectivity is much greater. The connectivity is defined here as the number of nodal groups produced by discretising the convective term.

An increase in the connectivity will increase the operation count and hence the execution time. The connectivity grows with more dimensions, higher-order interpolation or a higher-order nonlinearity in the governing equation. In order to estimate the likely impact on the overall execution time, an operation count comparison is provided in Table 10.10.

The Burgers' equations and compressible Navier-Stokes equations (Sect. 11.6.3) possess quadratically and cubically nonlinear convective terms, respectively. The connectivity for the conventional finite element method with linear interpolation is seen to rise (Table 10.10) with both increased number of dimensions and an increase in the order of the nonlinearity. By contrast, the connectivity of the group formulation is affected by the increase in the number of dimensions but not by the increase in the degree of nonlinearity.

A large connectivity translates into a large operation count to evaluate the steady-state residual. It is assumed that the systems of ordinary differential equations formed by the discretisation are marched in time using split formulations

Table 10.10. Connectivity and operation counts for the conventional and group finite element formulations with linear interpolation

| Equation system | Convective nonlinearity | Conventional F.E.M. | | Group F.E.M. | | Conv. R.O.C. |
		Connectivity (convective nonlinearity)	Residual operation count	Connectivity (convective nonlinearity)	Residual operation count	Group R.O.C.
2-D Burgers'	quadratic	49	828	9	206	4
3-D Burgers'	quadratic	343	12603	27	1308	9
2-D visc comp N-S	cubic	225	6772	9	404	17
3-D visc comp N-S	cubic	3375	217065	27	2349	92

like those developed in Sects. 8.2, 8.3 and 9.5. Typically at each time step the evaluation of the steady-state residual, RHS in (10.56), takes about 50% of the execution time. Consequently the residual operation count for a single time step correlates approximately with the total execution time.

The results shown in Table 10.10 indicate that the group finite element formulation is comparatively more economical as the order of the nonlinearity or the number of dimensions is increased. For three dimensional viscous flow the group formulation will evaluate the steady state residual almost one hundred times more economically than the conventional finite element formulation. This corresponds to approximately a factor of fifty improvement in the overall execution time. If higher-order interpolation than linear is used the relative economy of the group formulation, compared with the conventional finite element formulation, becomes even greater.

The relative execution times for the conventional and group finite element discretisations of the two-dimensional Burgers' equations predicted in Table 10.10 are confirmed in practice (Fletcher 1983b). The fourfold improvement shown in Table 10.10 corresponds to a measured ratio of the execution times of two and a half.

For the steady two-dimensional Burgers' equation with $v = 0.1$ the relative accuracy of the group and conventional finite element formulations can be assessed from the results shown in Fig. 10.11. These results were obtained on a uniform grid with $\Delta x = \Delta y$. The rms errors decrease with grid refinement at about the same rate, and for a given grid, are of about the same magnitude. Both schemes are seen to be more accurate than a three-point finite difference discretisation of (10.57 and 58). The convergence properties are based on a computed solution (Fletcher 1983b) which has a more moderate gradient, but predominantly in the x direction, than that shown in Fig. 10.12.

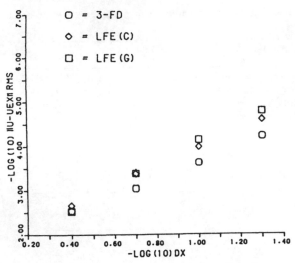

Fig. 10.11. Spatial convergence properties for the discretisations of the steady two-dimensional Burgers' equations. 3-FD = three point finite difference discretisation; LFE(C) = conventional finite element formulation with linear interpolation; LFE(G) = group finite element formulation with linear interpolation

The exact solution is given by (10.63 and 64) with

$$a_1 = a_2 = 110, \quad a_3 = a_4 = 0, \quad a_5 = 1.0, \quad \lambda = 5, \quad x_0 = 1.0 \quad \text{and} \quad v = 0.1 .$$

This solution is tabulated in Fig. 6.14.

The group formulation has also been investigated by Christie *et al.* (1981), who refer to it as the "product approximation". Christie *et al.* have demonstrated that the group formulation is theoretically as accurate as the conventional finite element method for representative model problems.

The broad conclusion from this subsection is that a group finite element formulation must be used in preference to the conventional finite element formulation to achieve acceptable economy when discretising the nonlinear equations governing fluid dynamics (Chap. 11) in more than one dimension. However, the use of a group formulation is not expected to reduce the solution accuracy.

10.4 Two-Dimensional Burgers' Equation

Just as the one-dimensional transport equation has a multidimensional counterpart (Sect. 9.5) so the one-dimensional Burgers' equation can be extended to multidimensions. The two-dimensional Burgers' equations are

$$\frac{\partial \bar{u}}{\partial t} + \bar{u}\frac{\partial \bar{u}}{\partial x} + \bar{v}\frac{\partial \bar{u}}{\partial y} - v\left(\frac{\partial^2 \bar{u}}{\partial x^2} + \frac{\partial^2 \bar{u}}{\partial y^2}\right) = 0 \tag{10.57}$$

and

$$\frac{\partial \bar{v}}{\partial t} + \bar{u}\frac{\partial \bar{v}}{\partial x} + \bar{v}\frac{\partial \bar{v}}{\partial y} - v\left(\frac{\partial^2 \bar{v}}{\partial x^2} + \frac{\partial^2 \bar{v}}{\partial y^2}\right) = 0 . \tag{10.58}$$

The two-dimensional Burgers' equations coincide with the two-dimensional momentum equations for incompressible laminar flow (Sect. 11.5.1) if the pressure terms are neglected.

10.4.1 Exact Solution

Like the one-dimensional Burgers' equation exact solutions can be constructed (Fletcher 1983d) using the Cole–Hopf transformation. In two dimensions the Cole–Hopf transformation introduces a single function $\bar{\Phi}$ to which $\bar{u}$ and $\bar{v}$ in (10.57, 58) are related as

$$\bar{u} = \frac{-2v\dfrac{\partial \bar{\Phi}}{\partial x}}{\bar{\Phi}}, \qquad \bar{v} = \frac{-2v\dfrac{\partial \bar{\Phi}}{\partial y}}{\bar{\Phi}}. \tag{10.59}$$

As a result, (10.57 and 58) are transformed to a single equation

$$\frac{\partial \bar{\Phi}}{\partial t} - \left(\frac{\partial^2 \bar{\Phi}}{\partial x^2} + \frac{\partial^2 \bar{\Phi}}{\partial y^2} \right) = 0 , \tag{10.60}$$

which is the two-dimensional diffusion equation (8.1). For appropriate boundary and initial conditions, (10.60) has an exact solution which, through (10.59) can be used to provide exact solutions of (10.57, 58).

The main interest is in the steady form of (10.57, 58). Therefore exact solutions of the steady part of (10.60), i.e.

$$\frac{\partial^2 \bar{\Phi}}{\partial x^2} + \frac{\partial^2 \bar{\Phi}}{\partial y^2} = 0 \tag{10.61}$$

will be sought. The following exact solution of (10.61) provides considerable control over the corresponding velocity solutions of (10.57 and 58):

$$\bar{\Phi} = a_1 + a_2 x + a_3 y + a_4 xy + a_5 [e^{\lambda(x-x_0)} + e^{-\lambda(x-x_0)}] \cos(\lambda y) , \tag{10.62}$$

where $a_1 - a_5$, λ and x_0 are chosen to give appropriate features to the exact solution. Using (10.59) the corresponding exact solutions of (10.57 and 58) are

$$\bar{u} = \frac{-2v\{a_2 + a_4 y + \lambda a_5 [e^{\lambda(x-x_0)} - e^{-\lambda(x-x_0)}] \cos(\lambda y)\}}{\{a_1 + a_2 x + a_3 y + a_4 xy + a_5 [e^{\lambda(x-x_0)} + e^{-\lambda(x-x_0)}] \cos(\lambda y)\}} \tag{10.63}$$

and

$$\bar{v} = \frac{-2v\{a_3 + a_4 x - \lambda a_5 [e^{\lambda(x-x_0)} + e^{-\lambda(x-x_0)}] \sin(\lambda y)\}}{\{a_1 + a_2 x + a_3 y + a_4 xy + a_5 [e^{\lambda(x-x_0)} + e^{-\lambda(x-x_0)}] \cos(\lambda y)\}} . \tag{10.64}$$

A typical exact solution for a steady 'flow' with a severe internal gradient is shown in Fig. 10.12. This solution is generated by the parameter choice

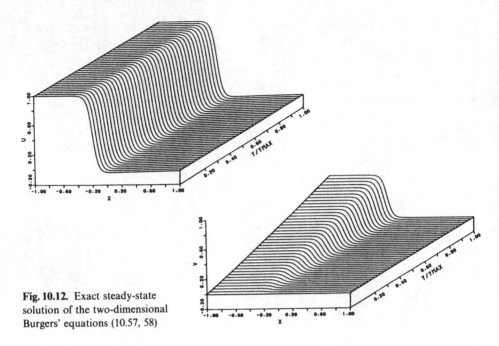

Fig. 10.12. Exact steady-state solution of the two-dimensional Burgers' equations (10.57, 58)

$$a_1 = a_2 = 1.3 \times 10^{13}, \quad a_3 = a_4 = 0, \quad a_5 = 1.0, \quad \lambda = 25, \quad x_0 = 1 \quad \text{and} \quad v = 0.04 \ .$$
$$(10.65)$$

The exact solution (10.63, 64) is particularly useful for assessing the accuracy of various algorithms (Fletcher 1983c), as in Fig. 10.11.

10.4.2 Split Schemes

The development of split schemes for multidimensional equations like the two-dimensional Burgers' equations follows closely the path indicated in Sects. 8.2, 8.3 and 9.5. The nonlinearity introduces an additional complication. How this is handled will be illustrated using the two-dimensional Burgers' equations in conservation form (10.52).

Equations (10.55 and 56), which are the discretised form of (10.52), represent both a group finite element and a finite difference discretisation if the mass operators are given the following interpretation:

$$M_x = (\delta_x, 1 - 2\delta_x, \delta_x) \ , \qquad M_y = (\delta_y, 1 - 2\delta_y, \delta_y)^T \ , \tag{10.66}$$

as in (9.86). The choice $\delta_x = \delta_y = 0$ gives a conventional three-point finite difference discretisation; the choice $\delta_x = \delta_y = \frac{1}{6}$ gives a group Galerkin finite element formulation based on bilinear interpolation.

It would be possible to introduce a three-level time discretisation of (10.55) as was done to generate (9.87). However, the interest here is in using a pseudo-transient formulation (Sect. 6.4) to obtain the steady-state solution. Therefore the following two-level fully implicit ($v=0$, $\beta=1$) time discretisation of (10.55) will be used:

$$M_x \otimes M_y \left[\frac{\varDelta q}{\varDelta t} \right]^{n+1}_{j,k} = \mathrm{RHS}^{n+1} \, , \tag{10.67}$$

which will lead to a split scheme equivalent to the augmented Newton's method (Sect. 6.4.1). To develop a linear system of equations for $\varDelta q^{n+1}$ from (10.67) it is necessary to linearise the nonlinear terms F, G and S in RHS^{n+1} (10.56) about the nth time level. This gives

$$F^{n+1} = F^n + \underline{A}\varDelta q^{n+1} + \cdots \, ,$$

$$G^{n+1} = G^n + \underline{B}\varDelta q^{n+1} + \cdots \, , \tag{10.68}$$

$$S^{n+1} = S^n + \underline{C}\varDelta q^{n+1} + \cdots \, ,$$

where $\underline{A}$, $\underline{B}$ and $\underline{C}$ are evalutated from (10.53) as

$$\underline{A} = \frac{\partial F}{\partial q} = \begin{bmatrix} 2u & 0 \\ v & u \end{bmatrix} \, , \quad \underline{B} = \frac{\partial G}{\partial q} = \begin{bmatrix} v & u \\ 0 & 2v \end{bmatrix} \tag{10.69}$$

and

$$\underline{C} = \frac{\partial S}{\partial q} = (0.5/v) \begin{bmatrix} (3u^2+v^2) & 2uv \\ 2uv & (u^2+3v^2) \end{bmatrix} \, ,$$

Substitution into (10.67) permits the following approximate factorisation to be constructed:

$$[M_x + \varDelta t(L_x\underline{A} - vL_{xx} - 0.5\,M_x\underline{C})]\varDelta q^*_{j,k} = \varDelta t\,\mathrm{RHS}^n \, , \tag{10.70}$$

$$[M_y + \varDelta t(L_y\underline{B} - vL_{yy} - 0.5\,M_y\underline{C})]\varDelta q^{n+1}_{j,k} = \varDelta q^*_{j,k} \, . \tag{10.71}$$

The appearance of the Jacobian matrices $\underline{A}$, etc., makes (10.70 and 71) block-tridiagonal along grid lines in the x and y directions, respectively. Since $\underline{A}$, etc., depend on the solution it is necessary to refactorise the block tridiagonal system at every iteration n. The block tridiagonal system can be solved efficiently using the block Thomas algorithm (Sect. 6.2.5).

The von Neumann stability analysis is applied to (10.70, 71) by temporarily freezing $\underline{A}$, etc. The analysis indicates that the split algorithm (10.70, 71) is unconditionally stable. In practice using too large a time step can introduce a nonlinear instability.

The split scheme has a truncation error of $O(\varDelta t, \varDelta x^2, \varDelta y^2)$. Since only the steady-state solution is of interest it is possible to consider modifications to the left-hand sides of (10.70 and 71) that permit the steady state to be reached with less computational effort. In the steady-state limit $\mathrm{RHS}^n = 0$, so that modifications to the left-hand side have no effect on the steady-state solution, even though they may influence the transient solution. The preferred strategy is to choose left-hand sides that can be solved more efficiently at each step of the iteration without increasing substantially the number of iterations required to reach the steady state.

It was noted in Sect. 6.2.5 that m scalar tridiagonal systems can be solved more efficiently than one $m \times m$ block tridiagonal system of equations. Equations (10.70 and 71) may be each treated as two sequential scalar tridiagonal systems by deleting off-diagonal terms in (10.69). A further economy may be achieved by replacing (10.69) with

$$A \approx u \begin{pmatrix} 1 & 0 \\ 0 & 1 \end{pmatrix}, \quad B \approx v \begin{pmatrix} 1 & 0 \\ 0 & 1 \end{pmatrix} \quad \text{and} \quad C \approx \left(\frac{u^2 + v^2}{v} \right) \begin{pmatrix} 1 & 0 \\ 0 & 1 \end{pmatrix}. \tag{10.72}$$

This means that the same left-hand side is used for each scalar component of (10.70 and 71). That is, the left-hand side need only be factorised (BANFAC) once for two solutions (BANSOL) with different right-hand sides, $\Delta t\,\mathrm{RHS}_1^n$ and $\Delta t\,\mathrm{RHS}_2^n$. Equation (10.72) with (10.70 and 71) is implemented in program TWBURG (Fig. 10.13).

10.4.3 TWBURG: Numerical Solution

The split algorithm (10.70, 71) is implemented in program TWBURG (Fig. 10.13) to obtain steady-state solutions of (10.52) in the computational domain $-1 \le x \le 1$ and $0 \le y \le y_{max}$, where $y_{max} = \pi/6\lambda$. Dirichlet boundary conditions are obtained from the exact solution (10.63, 64). The initial conditions are obtained from a linear interpolation of the boundary conditions in the x direction.

The major parameters used in program TWBURG are described in Table 10.11. Three alternative implementations of (10.70, 71) are available. When $ME = 1$ three-point centred finite difference discretisations are introduced for L_x, L_{xx}, etc., and $\delta_x = \delta_y = 0$. When $ME = 2$ the four-point upwind discretisation, used in Sects. 9.3.2, 9.4.3 and 9.5.2, is used for both L_x and L_y. For $L_x^{(4)}$ the four points $(j-2, k)$ to $(j+1, k)$ are used; for $L_y^{(4)}$ the four points $(j, k-2)$ to $(j, k+1)$ are used. For a typical exact solution, Fig. 10.15, u is negative adjacent to $x = 1$. Strictly in this region the four points $(j-1, k)$–$(j+2, k)$ should be used to evaluate $L_x^{(4)}$. This is not done in TWBURG for coding simplicity. When $ME = 3$ the mass operator parameters δ_x and δ_y are chosen empirically to produce a more accurate solution.

The right-hand side of (10.70) is evaluated in subroutine RHSBU (Fig. 10.14) and the exact solution (10.63, 64) is evaluated in subroutine EXBUR (Fig. 6.11) with the dimensions of UE, VE, etc., changed to 21×21. In addition the value of x_0 (10.65) is set in EXBUR (line 10). Typical output from TWBURG is indicated in Fig. 10.15.

Solutions produced by the three methods corresponding to the parameter choice of (10.65) are shown in Tables 10.12 and 10.13 for a 6×6 grid. The corresponding errors in the solution as a function of x are plotted in Figs. 10.16 and 10.17. Even on such a coarse grid all methods produce comparable accuracy with the largest solution errors coinciding with the location of the severe internal gradient (Fig. 10.12). From the results of Sects. 9.4 and 10.1.4 a greater disparity between the methods might be expected if the gradient were more severe. However, these solutions do indicate that the splitting algorithms can be extended to nonlinear equations without difficulty and that potentially more accurate schemes, AF-4PU and AF-MO, can be implemented reasonably economically.

```
 1
 2  C
 3  C    TWBURG APPLIES APPROXIMATION FACTORISATION TO SOLVE
 4  C    THE STEADY 2-D BURGERS' EQUATIONS FOR U(X,Y) AND V(X,Y)
 5  C
 6       DIMENSION RU(21,21),RV(21,21),U(21,21),V(21,21),UE(21,21),
 7      1VE(21,21),B(5,65),RRU(65),DDU(65),RRV(65),DDV(65),CX(3),CY(3),
 8      2CXQ(4),CYQ(4),EMX(3),EMY(3),DMX(3,3),DMY(3,3),A(5)
 9       COMMON DX,DY,RE,NX,NY,CX,CXQ,CY,CYQ,DMX,DMY,EMX,EMY,RU,RV,U,V
10       OPEN(1,FILE='TWBURG.DAT')
11       OPEN(6,FILE='TWBURG.OUT')
12       READ(1,1)NX,NY,ME,ITMAX,IPR,BET,QX,QY
13       READ(1,2)DTIM,EPS,RE,EM1,EM2
14       READ(1,3)(A(J),J=1,5),AL
15     1 FORMAT(5I5,3F5.2)
16     2 FORMAT(5E10.3)
17     3 FORMAT(2E10.3,4F5.2)
18  C
19       NXP = NX - 1
20       NXPP = NXP - 1
21       NYP = NY - 1
22       NYPP = NYP - 1
23       AN = NXPP*NYPP
24  C
25       CALL EXBUR(UE,VE,A,AL)
26  C
27       CX(1) = -0.5/DX
28       CX(2) = 0.
29       CX(3) =  0.5/DX
30       CY(1) = -0.5/DY
31       CY(2) = 0.
32       CY(3) =  0.5/DY
33       CXQ(1) = QX/DX/3.
34       CXQ(2) = -3.*CXQ(1)
35       CXQ(3) = -CXQ(2)
36       CXQ(4) = -CXQ(1)
37       CYQ(1) = QY/DY/3.
38       CYQ(2) = -3.*CYQ(1)
39       CYQ(3) = -CYQ(2)
40       CYQ(4) = -CYQ(1)
41       CCX =  1./DX/DX/RE
42       CCY =  1./DY/DY/RE
43       IF(ME .NE. 2)QX = 0.
44       IF(ME .NE. 2)QY = 0.
45       EMX(1) = 0.
46       EMY(1) = 0.
47       IF(ME .EQ. 3)EMX(1) = EM1
48       IF(ME .EQ. 3)EMY(1) = EM2
49       EMX(2) = 1. - 2.*EMX(1)
50       EMX(3) = EMX(1)
51       EMY(2) = 1. - 2.*EMY(1)
52       EMY(3) = EMY(1)
53       DO 4 I = 1,3
54       DMX(I,1) = CCX*EMY(I)
55       DMX(I,2) = -2.*DMX(I,1)
56       DMX(I,3) = DMX(I,1)
57       DMY(1,I) = CCY*EMX(I)
58       DMY(2,I) = -2.*DMY(1,I)
59     4 DMY(3,I) = DMY(1,I)
```

Fig. 10.13. Listing of program TWBURG

```
60          IF(ME .EQ. 1)WRITE(6,5)NX,NY,ME,ITMAX
61          IF(ME .EQ. 2)WRITE(6,6)NX,NY,ME,ITMAX
62          IF(ME .EQ. 3)WRITE(6,7)NX,NY,ME,ITMAX
63        5 FORMAT(' 2-D BURGERS EQUATION: AF-FDM,  NX,NY =',2I3,' ME =',
64          1I2,' ITMAX=',I3)
65        6 FORMAT(' 2-D BURGERS EQUATION: AF-4PU,  NX,NY =',2I3,' ME =',
66          1I2,' ITMAX=',I3)
67        7 FORMAT(' 2-D BURGERS EQUATION: AF-MO,  NX,NY =',2I3,' ME =',
68          1I2,' ITMAX=',I3)
69          WRITE(6,8)BET,DTIM,RE,QX,QY,EMX(1),EMY(1)
70        8 FORMAT(' BETA=',F5.2,' DTIM=',F5.3,' RE=',F5.1,' QX=',F4.2
71          1,' QY=',F4.2,' EMX=',F5.3,' EMY=',F5.3)
72          WRITE(6,9)(A(J),J=1,5),AL,DX,DY
73        9 FORMAT(' A= ',2E10.3,3F5.2,' AL=',F5.2,'  DX,DY=',2F8.5,/)
74 C
75 C        GENERATE INITIAL SOLUTION
76 C
77          DO 10 J = 1,NX
78          V(1,J) = VE(1,J)
79          V(NY,J) = VE(NY,J)
80          U(1,J) = UE(1,J)
81       10 U(NY,J) = UE(NY,J)
82          DO 12 K = 2,NYP
83          U(K,1) = UE(K,1)
84          U(K,NX) = UE(K,NX)
85          V(K,1) = VE(K,1)
86          V(K,NX) = VE(K,NX)
87          DO 11 J = 2,NXP
88          AJ = J - 1
89          U(K,J) = U(K,1) + 0.5*AJ*DX*(U(K,NX)-U(K,1))
90       11 V(K,J) = V(K,1) + 0.5*AJ*DX*(V(K,NX)-V(K,1))
91       12 CONTINUE
92          IF(IPR .LE. 1)GOTO 17
93          DO 13 K = 1,NY
94       13 WRITE(6,14)(UE(K,J),J=1,NX)
95       14 FORMAT(4H UE=,7F10.4)
96          DO 15 K = 1,NY
97       15 WRITE(6,16)(VE(K,J),J=1,NX)
98       16 FORMAT(4H VE=,7F10.4)
99       17 DO 20 J = 1,NX
100         DO 18 K = 1,NY
101         RU(K,J) = 0.
102      18 RV(K,J) = 0.
103         DO 19 K = 1,5
104      19 B(K,J) = 0.
105      20 CONTINUE
106         ITER = 0
107 C
108      21 DBT = BET*DTIM
109         CXA = 0.5*DBT/DX
110         CYA = 0.5*DBT/DY
111         CCXA = DBT*CCX
112         CCYA = DBT*CCY
113 C
114         CALL RHSBU(RMSU,RMSV,ME,DTIM)
115 C
116         IF(RMSU .LT. EPS)GOTO 34
117         IF(RMSU .GT. 1.0E+04)GOTO 34
118 C
119 C        TRIDIAGONAL SYSTEMS IN THE X-DIRECTION
120 C
```

Fig. 10.13. (cont.) Listing of program TWBURG

```
121            DO 27 K = 2,NYP
122            DO 23 J = 2,NXP
123            JM = J - 1
124            JP = J + 1
125            B(2,JM) = EMX(1) - CCXA - CXA*U(K,JM) - 0.5*EMX(1)*DBT*RE*
126           1(U(K,JM)*U(K,JM) + V(K,JM)*V(K,JM))
127            B(3,JM) = EMX(2) + 2.*CCXA - 0.5*EMX(2)*DBT*RE*(U(K,J)*U(K,J)
128           1 + V(K,J)*V(K,J))
129            B(4,JM) = EMX(3) - CCXA + CXA*U(K,JP) - 0.5*EMX(3)*DBT*RE*
130           1(U(K,JP)*U(K,JP) + V(K,JP)*V(K,JP))
131            RRU(JM) = RU(K,J)
132            RRV(JM) = RV(K,J)
133            IF(ME .NE. 2)GOTO 23
134            LST = 1
135            IF(J .EQ. 2)LST = 2
136            B(1,JM) = 0.
137            DO 22 L=LST,4
138            LJ = J + L - 3
139         22 B(L,JM) = B(L,JM) + CXQ(L)*U(K,LJ)*DBT
140         23 CONTINUE
141            B(2,1) = 0.
142            B(4,JM) = 0.
143            IF(ME .NE. 2)GOTO 25
144   C
145   C        REDUCE TO TRIDIAGONAL FORM
146   C
147            DO 24 JM = 3,NXPP
148            JMM = JM - 1
149            DUM = B(1,JM)/B(2,JMM)
150            B(2,JM) = B(2,JM) - B(3,JMM)*DUM
151            B(3,JM) = B(3,JM) - B(4,JMM)*DUM
152            B(1,JM) = 0.
153            RRU(JM) = RRU(JM) - RRU(JMM)*DUM
154         24 RRV(JM) = RRV(JM) - RRV(JMM)*DUM
155            B(1,2) = 0.
156   C
157         25 CALL BANFAC(B,NXPP,1)
158            CALL BANSOL(RRU,DDU,B,NXPP,1)
159            CALL BANSOL(RRV,DDV,B,NXPP,1)
160   C
161            DO 26 J = 2,NXP
162            JM = J - 1
163            RU(K,J) = DDU(JM)
164         26 RV(K,J) = DDV(JM)
165         27 CONTINUE
166   C
167   C        TRIDIAGONAL SYSTEMS IN THE Y-DIRECTION
168   C
169            DO 33 J = 2,NXP
170            DO 29 K = 2,NYP
171            KM = K - 1
172            KP = K + 1
173            B(2,KM) = EMY(1) - CCYA - CYA*V(KM,J) - 0.5*EMY(1)*DBT*RE*
174           1(U(KM,J)*U(KM,J) + V(KM,J)*V(KM,J))
175            B(3,KM) = EMY(2) + 2.*CCYA - 0.5*EMY(2)*DBT*RE*(U(K,J)*U(K,J)
176           1 + V(K,J)*V(K,J))
177            B(4,KM) = EMY(3) - CCYA + CYA*V(KP,J) - 0.5*EMY(3)*DBT*RE*
178           1(U(KP,J)*U(KP,J) + V(KP,J)*V(KP,J))
179            RRU(KM) = RU(K,J)
180            RRV(KM) = RV(K,J)
181            IF(ME .NE. 2)GOTO 29
```

Fig. 10.13. (cont.) Listing of program TWBURG

```
182         LST = 1
183         IF(K .EQ. 2)LST = 2
184         B(1,KM) = 0.
185         DO 28 L=LST,4
186         LK = K + L - 3
187      28 B(L,KM) = B(L,KM) + CYQ(L)*V(LK,J)*DBT
188      29 CONTINUE
189         B(2,1) = 0.
190         B(4,KM) = 0.
191         IF(ME .NE. 2)GOTO 31
192   C
193   C     REDUCE TO TRIDIAGONAL FORM
194   C
195         DO 30 KM = 3,NYPP
196         KMM = KM - 1
197         DUM = B(1,KM)/B(2,KMM)
198         B(2,KM) = B(2,KM) - B(3,KMM)*DUM
199         B(3,KM) = B(3,KM) - B(4,KMM)*DUM
200         B(1,KM) = 0.
201         RRU(KM) = RRU(KM) - RRU(KMM)*DUM
202      30 RRV(KM) = RRV(KM) - RRV(KMM)*DUM
203         B(1,2) = 0.
204   C
205      31 CALL BANFAC(B,NYPP,1)
206         CALL BANSOL(RRU,DDU,B,NYPP,1)
207         CALL BANSOL(RRV,DDV,B,NYPP,1)
208   C
209   C     INCREMENT U AND V
210   C
211         DO 32 K = 2,NYP
212         KM = K - 1
213         U(K,J) = U(K,J) + DDU(KM)
214      32 V(K,J) = V(K,J) + DDV(KM)
215      33 CONTINUE
216         ITER = ITER + 1
217         IF(ITER .LE. ITMAX)GOTO 21
218   C
219   C     COMPARE WITH EXACT SOLUTION
220   C
221      34 SUMU = 0.
222         SUMV = 0.
223         DO 36 J = 2,NXP
224         DO 35 K = 2,NYP
225         DU = U(K,J) - UE(K,J)
226         DV = V(K,J) - VE(K,J)
227         SUMU = SUMU + DU*DU
228      35 SUMV = SUMV + DV*DV
229      36 CONTINUE
230         RMSUE = SQRT(SUMU/AN)
231         RMSVE = SQRT(SUMV/AN)
232         WRITE(6,37)ITER,RMSU,RMSV,RMSUE,RMSVE
233      37 FORMAT(' AFTER',I3,' ITERATIONS, RMS-RHS=',2E10.3,
234        1' RMS-ERR=',2E10.3)
235         DO 38 K = 1,NY
236      38 WRITE(6,39)(U(K,J),J=1,NX)
237      39 FORMAT(' U=',7F10.4)
238         DO 40 K = 1,NY
239      40 WRITE(6,41)(V(K,J),J=1,NX)
240      41 FORMAT(' V=',7F10.4)
241         STOP
242         END
```

Fig. 10.13. (cont.) Listing of program TWBURG

Table 10.11. Parameters used in program TWBURG

Parameter	Description
ME	$=1$, approx. fact., three-point f.d. method, AF-FDM
	$=2$, approx. fact., four-point upwind scheme, AF-4PU
	$=3$, approx. fact., mass operator scheme, AF-MO
NX, NY	number of points in the x and y directions
ITMAX	maximum number of iterations
IPR	>1, print the exact solution u and v.
BET	weights the nth and $(n+1)$-th time levels, $=1$ in Tables 10.12 and 10.13
QX, QY	q_x, q_y in the four-point upwind discretisation $L_x^{(4)}$, $L_y^{(4)}$
DTIM, DX, DY	Δt, Δx, Δy
EPS	tolerance on $\|RHS_1^n\|_{rms}$ (10.70) for steady state convergence
RE	Reynolds number, $Re = 1/\nu$
EM1, EM2	δ_x, δ_y in (10.66)
A	a_1 to a_5 in (10.65)
CX, CY	L_x, L_y in (10.70, 71)
CXQ, CYQ	increment to L_x, L_y to produce $L_x^{(4)}$, $L_y^{(4)}$
EMX, EMY	M_x, M_y in (10.70, 71)
DMX, DMY	$M_y \otimes L_{xx}$, $M_x \otimes L_{yy}$ in (10.56)
U, V	dependent variables
UE, VE	exact solution, $(\bar{u}, \bar{v})$, given by (10.63 and 64)
DDU, DDV	components of $\Delta q_{j,k}^*$ and $\Delta q_{j,k}^{n+1}$ on return from BANSOL
RU, RV	components of RHS^n in (10.70); temporary storage for $\Delta q_{j,k}^*$
RRU, RRV	one-dimensional components of RU, RV required by BANSOL
B	quadridiagonal and tridiagonal matrix; left-hand side of (10.70, 71)
RMSU, RMSV	components of $\|RHS^n\|_{rms}$
RMSUE, RMSVE	$\|u - \bar{u}\|_{rms}$; $\|v - \bar{v}\|_{rms}$

Table 10.12. Variation of u with x for $y/y_{max} = 0.4$, NX = 6, NY = 6

x	-1.00	-0.60	-0.20	0.20	0.60	1.00	rms error
Exact	1.0000	1.0000	0.5375	-0.0333	-0.0250	-0.0200	—
AF-FDM	1.0000	1.0022	0.5221	-0.0322	-0.0250	-0.0200	0.0066
AF-4PU, $q_x = 1.0$, $q_y = 0.0$	1.0000	1.0003	0.5247	-0.0326	-0.0256	-0.0200	0.0054
AF-MO $\delta_x = \delta_y = 0.26$	1.0000	1.0004	0.5257	-0.0365	-0.0233	-0.0200	0.0045

```
1
2
3          SUBROUTINE RHSBU(RMSU,RMSV,ME,DTIM)
4  C
5  C       EVALUATES RIGHT-HAND SIDE OF THE 2-D BURGERS' EQUATION
6  C
7          DOUBLE PRECISION RUD,RVD,SUMU,SUMV,AN,DSQRT
8          DIMENSION RU(21,21),RV(21,21),U(21,21),V(21,21),
9         1DMX(3,3),DMY(3,3),EMX(3),EMY(3),CX(3),CY(3),CXQ(4),CYQ(4)
10        2,F(2),G(2),ES(2)
11         COMMON DX,DY,RE,NX,NY,CX,CXQ,CY,CYQ,DMX,DMY,EMX,EMY,RU,RV,U,V
12         NXP = NX - 1
13         NYP = NY - 1
14         SUMU = 0.
15         SUMV = 0.
16         DO 7 J =2,NXP
17         DO 6 K = 2,NYP
18         RUD = 0.
19         RVD = 0.
20         DO 2 N = 1,3
21         NK = K - 2 + N
22         DO 1 M = 1,3
23         MJ = J - 2 + M
24         F(1) = U(NK,MJ)*U(NK,MJ)
25         F(2) = U(NK,MJ)*V(NK,MJ)
26         G(2) = V(NK,MJ)*V(NK,MJ)
27         G(1) = F(2)
28         ES(1) = 0.5*RE*U(NK,MJ)*(F(1) + G(2))
29         ES(2) = 0.5*RE*V(NK,MJ)*(F(1) + G(2))
30         DUM =    CX(M)*EMY(N)*F(1) + CY(N)*EMX(M)*G(1)
31         RUD = RUD+(DMX(N,M)+DMY(N,M))*U(NK,MJ)+EMX(M)*EMY(N)*ES(1) - DUM
32         DUM =    CX(M)*EMY(N)*F(2) + CY(N)*EMX(M)*G(2)
33       1 RVD = RVD+(DMX(N,M)+DMY(N,M))*V(NK,MJ)+EMX(M)*EMY(N)*ES(2) - DUM
34       2 CONTINUE
35         IF(ME .NE. 2)GOTO 5
36         MST = 1
37         IF(J .EQ. 2)MST = 2
38         DO 3 M = MST,4
39         MJ = J - 3 + M
40         RUD = RUD - CXQ(M)*U(K,MJ)*U(K,MJ)
41       3 RVD = RVD - CXQ(M)*U(K,MJ)*V(K,MJ)
42         IF(J .EQ. 2)RUD = RUD - CXQ(1)*U(K,1)*U(K,1)
43         IF(J .EQ. 2)RVD = RVD - CXQ(1)*U(K,1)*V(K,1)
44         NST = 1
45         IF(K .EQ. 2)NST = 2
46         DO 4 N = NST,4
47         NK = K - 3 + N
48         RUD = RUD - CYQ(N)*U(NK,J)*V(NK,J)
49       4 RVD = RVD - CYQ(N)*V(NK,J)*V(NK,J)
50         IF(K .EQ. 2)RUD = RUD - CYQ(1)*U(1,J)*V(1,J)
51         IF(K .EQ. 2)RVD = RVD - CYQ(1)*V(1,J)*V(1,J)
52       5 SUMU = SUMU +  RUD*RUD
53         SUMV = SUMV +  RVD*RVD
54         RU(K,J) = RUD*DTIM
55       6 RV(K,J) = RVD*DTIM
56       7 CONTINUE
57         AN = (NXP-1)*(NYP-1)
58         RMSU = DSQRT(SUMU/AN)
59         RMSV = DSQRT(SUMV/AN)
60         RETURN
61         END
```

Fig. 10.14. Listing of subroutine RHSBU

```
2-D BURGERS EQUATION: AF-MO,  NX,NY =  6  6 ME = 3 ITMAX= 20
BETA= 1.00 DTIM= .010 RE= 50.0 QX= .00 QY= .00 EMX= .260 EMY= .260
A=   .100E+14  .100E+14  .00  .00 1.00 AL=25.00  DX,DY=  .40000  .00419

UE=   1.0000    1.0000    .5505   -.0333    -.0250    -.0200
UE=   1.0000    1.0000    .5491   -.0333    -.0250    -.0200
UE=   1.0000    1.0000    .5448   -.0333    -.0250    -.0200
UE=   1.0000    1.0000    .5375   -.0333    -.0250    -.0200
UE=   1.0000    1.0000    .5271   -.0333    -.0250    -.0200
UE=   1.0000    1.0000    .5132   -.0333    -.0250    -.0200
VE=    .0000     .0000    .0000    .0000     .0000     .0000
VE=    .1051     .1051    .0600    .0000     .0000     .0000
VE=    .2126     .2126    .1204    .0000     .0000     .0000
VE=    .3249     .3249    .1818    .0000     .0000     .0000
VE=    .4452     .4452    .2447    .0000     .0000     .0000
VE=    .5774     .5773    .3097    .0000     .0000     .0000
AFTER 21 ITERATIONS, RMS-RHS=  .624E-04  .801E-05  RMS-ERR=  .449E-02  .120E-02
U=    1.0000    1.0000    .5505   -.0333    -.0250    -.0200
U=    1.0000    1.0002    .5426   -.0354    -.0239    -.0200
U=    1.0000    1.0003    .5349   -.0364    -.0233    -.0200
U=    1.0000    1.0004    .5275   -.0365    -.0233    -.0200
U=    1.0000    1.0003    .5203   -.0354    -.0238    -.0200
U=    1.0000    1.0000    .5132   -.0333    -.0250    -.0200
V=     .0000     .0000    .0000    .0000     .0000     .0000
V=     .1051     .1051    .0586   -.0004     .0002     .0000
V=     .2126     .2126    .1180   -.0008     .0004     .0000
V=     .3249     .3250    .1790   -.0009     .0005     .0000
V=     .4452     .4453    .2425   -.0007     .0004     .0000
V=     .5774     .5773    .3097    .0000     .0000     .0000
```

Fig. 10.15. Typical output from program TWBURG

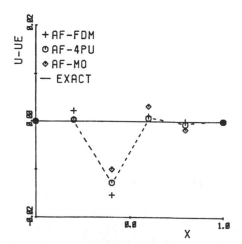

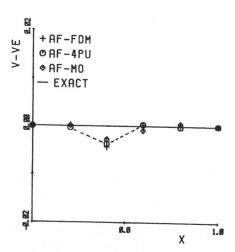

Fig. 10.16. Error distribution for the u solution of the two-dimensional Burgers' equation at $y/y_{max} = 0.4$

Fig. 10.17. Error distribution for the v solution of the two dimensional Burgers' equation at $y/y_{max} = 0.4$

Table 10.13. Variation of v with x for $y/y_{max} = 0.4$, $NX = 6$, $NY = 6$

x	-1.00	-0.60	-0.20	0.20	0.60	1.00	Error
Exact	0.3249	0.3249	0.1818	0.0000	0.0000	0.0000	—
AF-FDM	0.3249	0.3251	0.1774	0.0003	0.0000	0.0000	0.0018
AF-4PU $q_x = 1.0$, $q_y = 0.0$	0.3249	0.3246	0.1781	0.0003	-0.0002	0.0000	0.0015
AF-MO $\delta_x = \delta_y = 0.26$	0.3249	0.3250	0.1790	-0.0009	0.0005	0.0000	0.0012

10.5 Closure

Computational techniques, introduced in conjunction with the diffusion equation in Chaps. 7 and 8, and the linear convection and transport equations in Chap. 9 have been extended to a family of nonlinear equations, Burgers' equations, in one and two dimensions. Burgers' equations contain the same form of convective nonlinearity as in many of the fluid dynamic governing equations.

Burgers' equations possess readily computable exact solutions for many combinations of initial and boundary conditions. For this reason they are appropriate model equations on which to test various computational techniques. This feature has been exploited, for the one-dimensional Burgers' equation, in Sects. 10.1.4 and 10.3.1 and, for the two-dimensional Burgers' equations, in Sects. 10.3.2 and 10.4.3.

To take advantage of the Thomas algorithm it is necessary to linearise the convective nonlinearity. This is required for both implicit schemes in one dimension (Sect. 10.1.3) and for split schemes in multidimensions (Sect. 10.4.2). This linearisation extends to systems of equations (Sect. 10.2) without difficulty.

The Fourier analysis which was directly applicable to linear equations (Sects. 9.2.1 and 9.4.3) is still made use of in the von Neumann stability analysis *after freezing the nonlinearity*. The modified equation approach (Sect. 9.2.2) is applicable to nonlinear equations. However, products of higher-order derivatives appear in sufficient number and magnitude as to make the identification of dissipative and dispersive properties less precise. It is usually helpful to analyse the equivalent linear equation as an intermediate step, since this equation often demonstrates qualitatively equivalent dissipative and dispersive behaviour. The construction of more accurate schemes often requires a certain amount of empiricism imposed on the guidelines provided by the corresponding linear equations.

Most of the computational techniques developed for linear equations extend to nonlinear equations. However convective nonlinearities are often handled more effectively in conserved variables, e.g. F in (10.3). Conserved variables will be used, where appropriate, in Chaps. 14–18.

The nonlinear nature of the governing equation makes the application of the conventional Galerkin finite element method uneconomical, particularly in multi-dimensions and if higher-order interpolation is used. This problem is overcome by casting the equations in conservation form and in introducing approximate solutions for the conserved groups. The group formulation is often more accurate particularly where the conservation form of the governing equations is preferred on physical grounds.

10.6 Problems

1-D Burgers' Equation (Sect. 10.1)

10.1 Determine the truncation error of the scheme, (10.20), with the nonlinear coefficients frozen, for the two cases: (i) $q = 0$, (ii) $\delta = 0$. Show that the optimum values of δ and q are given by (10.26) and (10.27), respectively.

10.2 Modify program BURG to implement a Crank-Nicolson finite difference discretisation of (10.1) and compare the solution with that of the Crank-Nicolson finite difference discretisation of (10.3) for the conditions associated with Table 10.5. How does the comparison change when the integration is made to a larger time?

10.3 Modify program BURG to implement (10.26 and 27) and compare solutions for the intermediate cell Reynolds number conditions appropriate to Table 10.4.

10.4 Program BURG can be used to obtain steady-state solutions of (10.37) with $\bar{F} = 0.5(\bar{u}^2 - \bar{u})$. Implement these changes for the general three-level scheme, equivalent to (8.26), and obtain solutions with i) $\gamma = 0$, $\beta = 1$, ii) $\gamma = 0$, $\beta = 0.5$, iii) $\gamma = 0.5$, $\beta = 1.0$. Compare the number of iterations to reach the steady state, for the conditions corresponding to Table 10.9, but with a uniform grid.

10.5 For the convection diffusion equation (10.28) code the nonuniform discretisation (10.30–32) and confirm the results shown in Tables 10.6 and 10.7.

10.6 Develop a four-point upwind discretisation on a nonuniform grid with one free parameter (equivalent to q). Assume that r_x is constant but not equal to unity. Apply this discretisation to the modified Burgers' equation (10.37) with the two-level fully implicit marching scheme (10.38). Compare the accuracy of the solutions with those shown in Tables 10.8 and 10.9 for various values of the free parameter.

Systems of Equations (Sect. 10.2)

10.7 Modify program SHOCK (Vol. 2, Fig. 14.17) to implement the explicit four-point upwind scheme introduced in Sect. 9.4.3. Obtain solutions corresponding to those shown in Fig. 10.9 with sufficient artificial dissipation to suppress excessive oscillations. It may be necessary to use a small value of Δt

(small Courant number, C) to obtain stable solutions Compare the sharpness of the shock profile with that shown in Fig. 10.9.

10.8 Extend the Crank-Nicolson finite difference scheme (10.44) to generate a Crank-Nicolson mass operator scheme with variable δ for the system of equations (10.40). Modify SHOCK (Fig. 14.17) to execute this scheme and compare the quality of the solution with that shown in Fig. 10.9, for δ chosen empirically.

Group Finite Element Method (Sect. 10.3)

10.9 Implement the conventional finite element method with Crank-Nicolson time differencing in program BURG, for the one-dimensional Burgers' equation, and compare the solution accuracy with that produced by the group finite element method and the Crank-Nicolson mass operator scheme for the conditions corresponding to Table 10.5.

10.10 Estimate the connectivity and residual operation count for the conventional and group finite element formulations applied to the two-dimensional Burgers' equations for quadratic interpolation.

10.11 Modify program TWBURG to implement the conventional finite element method applied to (10.57, 58) and compare the accuracy and economy with that of the group finite element method applied to (10.52) for the conditions corresponding to Tables 10.12 and 10.13. How does the comparison change on a finer grid?

2-D Burgers' Equation (Sect. 10.4)

10.12 Write a program, based on subroutine EXBUR, to generate exact solutions of the two-dimensional Burgers' equations that produce more severe internal gradients.

10.13 Obtain solutions using program TWBURG on 6×6, 11×11 and 21×21 grids for the conditions corresponding to Tables 10.12 and 10.13, with $0 \leq \delta_x$, $\delta_y \leq 0.30$, $0 \leq q_x$, $q_y \leq 1.0$. Decide on optimal choices for δ_x, δ_y and q_x, q_y for each grid.

10.14 Execute program TWBURG with the following choice of parameters: $a_1 = a_2 = 0.01101$, $a_3 = a_4 = 0$, $a_5 = 1.0$, $\lambda = 5$, $x_0 = 1.0$ and $v = 0.1$. This choice produces a moderate boundary gradient adjacent to $x = 1.0$. What are optimal choices for δ_x, δ_y and q_x, q_y on an 11×11 grid for this case?

Appendix

A.1 Empirical Determination of the Execution Time of Basic Operations

A computer program, COUNT, is provided in Fig. A.1 which allows the execution time of various types of operation to be determined approximately. The type of operation corresponding to different values of ICT are given in Table A.1.

For each operation a nested pair of DO loops is executed. The inner loop repeats the given operation 10 000 times. The outer loop repeats the inner loop N times. For N sufficiently large the time from the input of ICT and N (line 15) to the output of ICT and N (line 48) can be determined manually. Alternatively a timing subroutine, if available, can be used to measure the CPU time between lines 15 and 48.

Table A.1. Operations considered in Program COUNT

ICT	Operation	Execution time (secs) for $N = 1000$
1	Empty DO loops	12.3
2	Replacement	15.8
3	Addition and replacement	19.1
4	Subtraction and replacement	19.1
5	Multiplication and replacement	19.0
6	Division and replacement	34.8
7	Integer IF statement	15.3
8	Power and replacement	81.9
9	Square root and replacement	111
10	Sin and replacement	112
11	Exponential and replacement	123
12	Array addition and replacement	25.1

The execution times shown in Table A.1 are appropriate to the Supermicrocomputer (SUN Sparc St1) in Table 4.4. The execution time for ICT = 1 is subtracted from cases ICT = 2 and 7. For all other cases the execution time for ICT = 2 is subtracted. For the fixed point operations (FX) shown in Table 4.4 the integer statement (line 6) is activated and the library function evaluations (lines 38, 40 and 42) are replaced with labelled CONTINUE statements to satisfy the GOTO statement (line 21). It is recommended that program COUNT be compiled in an unoptimised mode to obtain consistent results.

```
 1  C    COUNT FACILITATES THE DETERMINATION OF THE EXECUTION
 2  C    TIME FOR SPECIFIC OPERATIONS, DEPENDING ON THE VALUE OF ICT.
 3  C    N IS THE NUMBER OF OUTER LOOPS.
 4  C
 5       DIMENSION E(10000),F(10000),G(10000)
 6  C    INTEGER A,B,C,E,F,G
 7       B = 3.0
 8       C = 2.0
 9       IA = 8
10       DO 21 J = 1,10000
11       F(J) = 2.0
12    21 G(J) = 1.0
13       WRITE(*,22)
14    22 FORMAT(5X,'VALUE FOR ICT AND N',/)
15       READ(*,*)ICT,N
16  C
17       DO 14 I = 1,N
18       DO 13 J = 1,10000
19  C
20       GOTO(1,2,3,4,5,6,7,8,9,10,11,12),ICT
21     1 CONTINUE
22       GOTO 13
23     2 A = C
24       GOTO 13
25     3 A = B + C
26       GOTO 13
27     4 A = B - C
28       GOTO 13
29     5 A = B*C
30       GOTO 13
31     6 A = B/C
32       GOTO 13
33     7 IF(I .GT. 0)GOTO 13
34       GOTO 13
35     8 A = B**IA
36       GOTO 13
37     9 A = SQRT(B)
38       GOTO 13
39    10 A = SIN(B)
40       GOTO 13
41    11 A = EXP(B)
42       GOTO 13
43    12 E(J) = F(J) + G(J)
44    13 CONTINUE
45    14 CONTINUE
46  C
47       WRITE(*,15)ICT,N
48    15 FORMAT(5X,'  ICT=',I7,5X,'  N=',I7)
49       STOP
50       END
```

Fig. A.1. Listing of Program COUNT

A.2 Mass and Difference Operators

It is indicated in Sects. 5.5.1 and 5.5.2 that the finite element method can be interpreted as providing a term-by-term discretisation if directional mass and difference operators are identified explicitly. The origin of these operators within the Galerkin finite element formulation, Sects. 5.1 and 5.3, will be indicated here in relation to the two-dimensional transport equation (9.81)

$$\frac{\partial \bar{T}}{\partial t} + u \frac{\partial \bar{T}}{\partial x} + v \frac{\partial \bar{T}}{\partial y} - \alpha_x \frac{\partial^2 \bar{T}}{\partial x^2} - \alpha_y \frac{\partial^2 \bar{T}}{\partial y^2} = 0 \ . \tag{A.1}$$

An approximate solution is introduced for $\bar{T}$ as

$$T = \sum_{i=1}^{I} T_i \phi_i(x, y) , \tag{A.2}$$

where $\phi_i(x, y)$ are two-dimensional bilinear Lagrange interpolating functions, equivalent to (5.60). Equation (A.2) is substituted into (A.1), the Galerkin weighted residual integral (5.5 and 10) is evaluated and the result can be written

$$M_x \otimes M_y \dot{T} + u M_y \otimes L_x T + v M_x \otimes L_y T - \alpha_x M_y \otimes L_{xx} T - \alpha_y M_x \otimes L_{yy} T = 0 , \tag{A.3}$$

where $\otimes$ is the tensor product and $\dot{T} \equiv dT/dt$. The terms M_x and M_y are directional mass operators and are defined by

$$M_x = A \left\{ \frac{1}{6}, \frac{1+r_x}{3}, \frac{r_x}{6} \right\} , \qquad M_y = B \left\{ \frac{r_y}{6}, \frac{1+r_y}{3}, \frac{1}{6} \right\}^T \tag{A.4}$$

where

$$A = 2/(1+r_x) , \qquad B = 2/(1+r_y) .$$

The directional difference operators are given by

$$L_x = A \left\{ -\frac{0.5}{\Delta x}, 0, \frac{0.5}{\Delta x} \right\} , \qquad L_{xx} = A \left\{ \frac{1}{\Delta x^2}, -\frac{(1+r_x)}{r_x \Delta x^2}, \frac{1}{r_x \Delta x^2} \right\}$$

$$L_y = B \left\{ \frac{0.5}{\Delta y}, 0, -\frac{0.5}{\Delta y} \right\}^T , \qquad L_{yy} = B \left\{ \frac{1}{r_y \Delta y^2}, -\frac{(1+r_y)}{r_y \Delta y^2}, \frac{1}{\Delta y^2} \right\}^T , \tag{A.5}$$

where r_x and r_y are grid growth ratios

$$r_x = \frac{x_{j+1} - x_j}{x_j - x_{j-1}} , \qquad r_y = \frac{y_{k+1} - y_k}{y_k - y_{k-1}} , \tag{A.6}$$

so that on a uniform grid $r_x = r_y = 1$.

It can be seen that there is a term-by-term correspondence between the original equation (A.1) and the discretised equation (A.3). The source of this correspondence can be found by considering a single term $\partial\bar{T}/\partial x$ in (A.1). Application of the finite element method to (A.1) includes the following contribution from $\partial\bar{T}/\partial x$:

$$I = \sum_i \left(\sum_e \iint \phi_m \frac{\partial \phi_i}{\partial x} \, dx \, dy \right) T_i , \tag{A.7}$$

where $\sum_e$ denotes the contributions from the elements adjacent to node m. The Lagrange interpolating functions ϕ_m can be written as the product of one-dimensional interpolating functions

$$\phi_m = \phi_m^{(x)} \phi_m^{(y)} , \tag{A.8}$$

where $\phi_m^{(x)}$, $\phi_m^{(y)}$ are given by (5.45 and 46). Consequently the contributions to the integral in (A.7) can be split into directional components

$$I = \sum_i \left(\sum_e \int \phi_m^{(y)} \phi_i^{(y)} \, dy \int \phi_m^{(x)} \frac{d\phi_i^{(x)}}{dx} \, dx \right) T_i \ . \tag{A.9}$$

As a result it is convenient to introduce the operators

$$M_{y_i} = \frac{B}{\Delta y_{k-1/2}} \sum_e \int \phi_m^{(y)} \phi_i^{(y)} \, dy \tag{A.10}$$

and

$$L_{x_i} = \frac{A}{\Delta x_{j-1/2}} \sum_e \int \phi_m^{(x)} \frac{d\phi_i^{(x)}}{dx} \, dx \ , \tag{A.11}$$

where it is assumed that the mth Galerkin node coincides with the global grid point (j, k), so that

$$\Delta x_{j-1/2} = x_j - x_{j-1} \quad \text{and} \quad \Delta y_{k-1/2} = y_k - y_{k-1} \ . \tag{A.12}$$

Thus the term $\partial \bar{T} / \partial x$ in (A.1) is discretised as $M_y \otimes L_x T$. On the global grid, this has the form

$$M_y \otimes L_x T = AB \left[\frac{r_y}{6} \left\{ \frac{T_{j+1,k+1} - T_{j-1,k+1}}{2\Delta x} \right\} + \frac{1+r_y}{3} \left\{ \frac{T_{j+1,k} - T_{j-1,k}}{2\Delta x} \right\} \right.$$

$$\left. \times \frac{1}{6} \left\{ \frac{T_{j+1,k-1} - T_{j-1,k-1}}{2\Delta x} \right\} \right] \ . \tag{A.13}$$

Where second derivatives occur in the governing equation (A.1), application of integration by parts leads to the following definition of the operator L_{xx}, at an interior point:

$$L_{xx_i} = \frac{-A}{\Delta x_{j-1/2}} \sum_e \int \left(\frac{\partial \phi_m^{(x)}}{\partial x} \right) \left(\frac{d\phi_i^{(x)}}{dx} \right) dx \ . \tag{A.14}$$

Comparable definitions to those given in (A.10, 11 and 14) are available for M_x, L_y and L_{yy}, respectively. In practice the integrals in (A.10), etc., are evaluated by introducing element-based coordinates (ξ, η) as in (5.58–60). For linear elements,

$$\phi_m^{(x)} = 0.5(1 + \xi\xi_m) \quad \text{with} \quad \xi_m = \pm 1 \quad \text{and} \quad -1 \leq \xi \leq 1 \ . \tag{A.15}$$

Equations (A.10, 11, 14) are applicable for any order of Lagrange interpolation. For quadratic interpolation the operators in (A.4, 5) would have five components associated with corner nodes and three components associated with mid-side nodes.

It is apparent from (A.5) that the form of the L_x and L_{xx} operators is similar to finite difference discretisation. In contrast, the mass operators (A.4) arise from the integral nature of the Galerkin method. Therefore, in a three-dimensional problem $\partial \bar{T}/\partial x$ would be discretised as

$$\partial \bar{T}/\partial x \rightarrow M_y \otimes M_z \otimes L_x T . \tag{A.16}$$

For brick elements with trilinear interpolation this implies that $M_y \otimes M_z \otimes L_x$ would be at most a 27-point operator.

On a uniform grid there is a connection between the role of the mass operators and Padé differencing. A fourth-order evaluation of $\partial \bar{T}/\partial x$ using Padé differencing is obtained by solving the tridiagonal system

$$\frac{1}{6}\left(\frac{\partial T}{\partial x}\right)_{j-1} + \frac{2}{3}\left(\frac{\partial T}{\partial x}\right)_j + \frac{1}{6}\left(\frac{\partial T}{\partial x}\right)_{j+1} = \frac{T_{j+1} - T_{j-1}}{2\Delta x} , \tag{A.17}$$

or

$$M_x \frac{\partial T}{\partial x} = L_x T . \tag{A.18}$$

It may be noted that here the mass operator is applied in the same direction as the derivative is taken. In contrast in the Galerkin finite element method,

$$\partial T/\partial x \rightarrow M_y \otimes L_x T , \tag{A.19}$$

and an explicit formula involving the transverse direction appears. However, on a uniform grid (A.19) is also a fourth-order accurate discretisation.

To obtain fourth-order accuracy in discretising $\partial^2 \bar{T}/\partial x^2$ using the equivalent of (A.18) or (A.19), it is necessary to generalise the form of the mass operator, as in (8.44) to $M_x = M_y^T = \{\delta, 1 - 2\delta, \delta\}$. The choice $\delta = 1/12$ gives fourth-order accuracy, Table 8.2.

References

Chapter 1

Arlinger, B.G. (1986): "Computation of Supersonic Flow including Leading-Edge Vortex Flows using Marching Euler Techniques", in Proc. Int. Symp. Comp. Fluid Dynamics, ed. by K. Oshima (Japan Computational Fluid Dynamics Society, Tokyo) Vol. 2, pp. 1–12

Bailey, F.R. (1986): "Overview of NASA's Numerical Aerodynamic Simulation Program", in Proc. Int. Symp. Comp. Fluid Dynamics, ed. by K. Oshima (Japan Computational Fluid Dynamics Society, Tokyo) Vol. 1, pp. 21–32

Baker, A.J. (1983): *Finite Element Computational Fluid Mechanics* (McGraw-Hill, New York)

Book, D.L. (ed.) (1981): *Finite-Difference Techniques for Vectorised Fluid Dynamics Calculations*, Springer Ser. Comput. Phys. (Springer, New York, Berlin, Heidelberg)

Bourke, W., McAveney, B., Puri, K., Thurling, R. (1977): Methods Comput. Phys. **17**, 267–325

Chapman, D.R., Mark, H., Pirtle, M.W. (1975): Astronaut. Aeronaut. 22–35

Chapman, D.R. (1979): AIAA J. **17**, 1293–1313

Chapman, D.R. (1981): In 7th Int. Conf. Numer. Methods in Fluid Dynamics, ed. by W.C. Reynolds, R.W. MacCormack, Lecture Notes in Physics, Vol. 141 (Springer, Berlin, Heidelberg) pp. 1–11

Chervin, R. (1989): Cray Channels **11**, 6–9

Cullen, M.J.P. (1983): J. Comput. Phys. **50**, 1–37

Cullen, M.J.P. (1990): Private communication

Dongarra, J. (1989): "Performance of Various Computers Using Standard Linear Equations Software in a Fortran Environment", Tech. Memo 23, Argonne National Laboratory, Argonne, Ill.

Fletcher, C.A.J. (1984): *Computational Galerkin Methods*, Springer Ser. Comput. Phys. (Springer, Berlin, Heidelberg)

Gentzsch, W., Neves, K.W. (1988): "Computational Fluid Dynamics: Algorithm & Supercomputers", Agardograph 311, NATO

Glowinski, R. (1984): *Numerical Methods for Nonlinear Variational Problems*, Springer Ser. Comput. Phys. (Springer, Berlin, Heidelberg)

Grassl, C., Schwarzmeier, J. (1990): Cray Channels **12**, 18–21

Green, J.E. (1982): In *Numerical Methods in Aeronautical Fluid Dynamics*, ed. by P.L. Roe (Academic Press, London) pp. 1–32

Haltiner, G.J., Williams, R.T. (1980): *Numerical Prediction and Dynamic Meteorology*, 2nd ed. (Wiley, New York)

Hockney, R.W., Jesshope, C.R. (1981): *Parallel Computers* (Adam Hilger, Bristol)

Holst, T.L., Thomas, S.D., Kaynak, U., Grundy, K.L., Flores, J., Chaderjian, N.M. (1986): "Computational Aspects of Zonal Algorithms for Solving the Compressible Navier-Stokes Equations in Three Dimensions", in Proc. Int. Sym. Comp. Fluid Dynamics, ed. by K. Oshima (Japan Computational Fluid Dynamics Society, Tokyo) Vol. 1, pp. 113–122

Holt, M. (1984): *Numerical Methods in Fluid Dynamics*, 2nd ed., Springer Ser. Comput. Phys. (Springer, Berlin, Heidelberg)

Hoshino, T. (1989): PAX Computer, Highly Parallel Processing in Scientific Calculation (Addison-Wesley, Reading, Mass.)

Jameson, A. (1989): Science **245**, 361–371

Krause, E. (1985): Comput. Fluids **13**, 239–269

Kutler, P. (1985): AIAA J. **23**, 328–338

Levine, R.D. (1982): Sci. Am. **246**, 112–125

Ortega, J.M., Voigt, R.G. (1985): SIAM Rev. **27**, 147–240

Peyret, R., Taylor, T.D. (1983): *Computational Methods for Fluid Flow*, Springer Ser. Comput. Phys. (Springer, Berlin, Heidelberg)

Richtmyer, R.D., Morton, K.W. (1967): *Difference Methods for Initial Value Problems*, 2nd ed. (Wiley, New York)

Roache, P.J. (1976): *Computational Fluid Dynamics* (Hermosa, Albuquerque, N.M.)

Rubbert, P.E. (1986): "The Emergence of Advanced Computational Methods in the Aerodynamic Design of Commercial Transport Aircraft", in Proc. Int. Symp. Comp. Fluid Dynamics, ed. by K. Oshima (Japan Computational Fluid Dynamics Society, Tokyo) Vol. 1, pp. 42–48

Simon, H. (1989): Supercomputer Course Notes, Canberra, Australia

Thomasset, F. (1981): *Implementation of Finite Element Methods for Navier-Stokes Equations*, Springer Ser. Comput. Phys. (Springer, Berlin, Heidelberg)

Tobak, M., Peake, D. (1982): Annu. Rev. Fluid Mech. **14**, 61–85

Turkel, E. (1982): Comput. Fluids **11**, 121–144

Chapter 2

Ames, W.F. (1969): *Numerical Methods for Partial Differential Equations* (Barnes and Noble, New York)

Belotserkovskii, O.M., Chushkin, P.I. (1965): In *Basic Developments in Fluid Dynamics*, ed. by M. Holt (Academic, New York) pp. 1–126

Chester, C.R. (1971): *Techniques in Partial Differential Equations*, (McGraw-Hill, New York)

Courant, R., Friedrichs, K.O. (1948): *Supersonic Flow and Shock Waves* (Interscience, New York)

Courant, R., Hilbert, D. (1962): *Methods of Mathematical Physics, Vol II* (Interscience, New York)

Fletcher, C.A.J. (1983): "The Galerkin Method and Burgers' Equation", in *Numerical Solution of Differential Equations*, ed. by J. Noye (North-Holland, Amsterdam) pp. 355–475

Garabedian, P. (1964): *Partial Differential Equations* (Wiley, New York)

Gelfand, I.M., Shilov, G.E. (1967): *Generalised Functions, Vol. 3. Theory of Differential Equations* (Academic, New York)

Gustafson, K.E. (1980): *Partial Differential Equations and Hilbert Space Methods* (Wiley, New York)

Hellwig, G. (1964): *Partial Differential Equations, An Introduction* (Blaisdell, New York)

Jeffrey, A., Taniuti, T. (1964): *Nonlinear Wave Propagation with Applications to Physics and Magneto-hydrodynamics* (Academic, New York)

Lighthill, M.J. (1958): *Fourier Analysis and Generalised Functions* (Cambridge University Press, Cambridge)

Whitham, G. (1974): *Linear and Nonlinear Waves* (Wiley, New York)

Chapter 3

Gear, C.W. (1971): *Numerical Initial Value Problems in Ordinary Differential Equations* (Prentice-Hall, Englewood Cliffs, N.J.)

Hamming, R.W. (1986): *Numerical Methods for Scientists and Engineers* (Dover, New York)

Lambert, J.D. (1973): *Computational Methods in Ordinary Differential Equations* (Wiley, Chichester)

Liepmann, H., Roshko, A. (1957): *Elements of Gas Dynamics* (Wiley, New York)

Chapter 4

Carnahan, B., Luther, H.A., Wilkes, J.O. (1969): *Applied Numerical Analysis* Wiley, New York)

Dahlquist, G., Bjorck, A. (1974): *Numerical Methods*, trans. by N. Anderson (Prentice-Hall, Englewood Cliffs, N.J.)

Finlayson, B.A. (1980): *Nonlinear Analysis in Chemical Engineering* (McGraw-Hill, New York)

Fletcher, C.A.J. (1983): J. Comput. Phys. **51**, 159–188

Gentzsch, W., Neves, K.W. (1988): "Computational Fluid Dynamics: Algorithms & Supercomputers", Agardograph 311, NATO

Hindmarsh, A.C., Gresho, P.M., Griffiths, D.F. (1984): Int. J. Numer. Methods Fluids, **4**, 853–897

Isaacson, E., Keller, H.B. (1966): *Analysis of Numerical Methods* (Wiley, New York)

Mitchell, A.R., Griffiths, D.F. (1980): *The Finite Difference Method in Partial Differential Equations* (Wiley, Chichester)

Morton, K.W. (1980): Int. J. Numer. Methods Eng. **15**, 677–683

Noye, J. (1984): Chap. 2 in *Numerical Solution of Differential Equations*, ed. by J. Noye (North-Holland, Amsterdam)

Richtmyer, R.D., Morton, K.W. (1967): *Difference Methods for Initial-Value Problems* (Interscience, New York)

Smith, G.D. (1985): *Nonlinear Solution of Partial Differential Equations*, 3rd ed. (Oxford University Press, Oxford)

Sod, G.A. (1985): *Numerical Methods in Fluid Dynamics* (Cambridge University Press, Cambridge)

Trapp, J.A., Ramshaw, J.D. (1976): J. Comput. Phys. **20**, 238–242

Chapter 5

Baker, A.J. (1983): *Finite Element Computational Fluid Mechanics* (McGraw-Hill, New York)

Bathe, K.J., Wilson, E.L. (1976): *Numerical Methods in Finite Element Analysis* (Prentice-Hall, Englewood Cliffs, N.J.)

Bourke, W., McAvaney, B., Puri, K., Thurling, R. (1977): Methods Comput. Phys. **17**, 267–325

Brigham, E.O. (1974): *The Fast Fourier Transform* (Prentice-Hall, Englewood Cliffs, N.J.)

Canuto, C., Hussaini, M.Y., Quateroni, A., Zang, T.A. (1987): *Spectral Methods in Fluid Dynamics*, Springer Ser. Comput. Phys. (Springer, Berlin, Heidelberg)

Cooley, J.W., Tukey, J.W. (1965): Math. Comp. **19**, 297–301.

Finlayson, B.A. (1972): *The Method of Weighted Residuals and Variational Principles* (Academic, New York)

Finlayson, B.A. (1980): *Nonlinear Analysis in Chemical Engineering* (McGraw-Hill, New York)

Fletcher, C.A.J. (1984): *Computational Galerkin Methods*, Springer Ser. Comput. Phys. (Springer, Berlin, Heidelberg)

Fox, L., Parker, I.B. (1968): *Chebyshev Polynomials in Numerical Analysis* (Oxford University Press, Oxford)

Gottlieb, D., Orszag, S.A. (1977): *Numerical Analysis of Spectral Methods: Theory and Applications* (SIAM, Philadelphia)

Haltiner, G.J., Williams, R.T. (1980): *Numerical Prediction and Dynamic Meteorology*, 2nd ed. (Wiley, New York)

Hussaini, M.Y., Zang, T.A. (1987): Ann. Rev. Fluid Mech **19**, 339–367

Isaacson, E., Keller, H.B. (1966): *An Analysis of Numerical Methods* (Wiley, New York)

Jameson, A. (1989): Science **245**, 361–371

Jennings, A. (1977): *Matrix Computation for Engineers and Scientists* (Wiley, Chichester)

Ku, H.C., Hatziavramidis, D. (1984): J. Comput. Phys. **56**, 495–512

Ku, H.C., Taylor, T.D., Hirsch, R.S. (1987): Comput. Fluids, **15**, 195–214

Lax, P., Wendroff, B. (1960): Commun. Pure Appl. Math. **13**, 217–237

Mitchell, A.R., Wait, R. (1977): *The Finite Element Method in Partial Differential Equations* (Wiley, Chichester)

Oden, J.T., Reddy, J.N. (1976): *An Introduction to the Mathematical Theory of Finite Elements* (Wiley, New York)

Orszag, S.A. (1971): Stud. Appl. Math. **50**, 293–327

Peyret, R. (1986): "Introduction to Spectral Methods", Lect. Ser. 1986–04, Von Karman Inst. for Fluid Dynamics, Belgium

Peyret, R., Taylor, T.D. (1983): *Computational Methods for Fluid Flow*, Springer Ser. Comput. Phys. (Springer, New York, Berlin, Heidelberg)

Strang, G., Fix, G.J. (1973): *An Analysis of the Finite Element Method* (Prentice-Hall, Englewood Cliffs, N.J.)

Thomasset, F. (1981): *Implementation of Finite Element Methods for Navier-Stokes Equations*, Springer Ser. Comput. Phys. (Springer, Berlin, Heidelberg)

Voigt, R.G., Gottlieb, D., Hussaini, M.Y. (1984): *Spectral Methods for Partial Differential Equations* (SIAM, Philadelphia)

Zienkiewicz, O.C. (1977): *The Finite Element Method*, 3rd ed. (McGraw-Hill, London)

Chapter 6

Axelsson, O., Gustafsson, I. (1979): J. Inst. Math. Its Appl. **23**, 321–337

Brandt, A. (1977): Math. Comput. **31**, 333–390

Briggs, W.L. (1987a): In *Multigrid Methods*, ed. by S. McCormick, SIAM Frontier Series, Vol. 3 (SIAM, Philadelphia)

Briggs, W.L. (1987b): *A Multigrid Tutorial* (SIAM, Philadelphia)

Brigham, E.O. (1974): *The Fast Fourier Transform* (Prentice-Hall, Englewood Cliffs, N.J.)

Broyden, C. (1965): Math. Comput. **19**, 577–593

Cooley, J.W., Lewis, P.A.W., Welch, P.D. (1970): J. Sound and Vibrations, **12**, 315–337

Dahlquist, G., Bjorck, A. (1974): *Numerical Methods*, trans. by N. Anderson (Prentice-Hall, Englewood Cliffs, N.J.)

Dorr, F.W. (1970): SIAM Rev. **12**, 248–263

Duff, I.S. (1981): "A Sparse Future", in *Sparse Matrices and Their Uses*, ed. by I.S. Duff (Academic, London) pp. 1–29

Duff, I.S., Erisman, A., Reid, J.K. (1986): *Direct Methods for Sparse Matrices* (Clarendon, Oxford)

Engelman, M.S., Strang, G., Bathe, K.J. (1981): Int. J. Numer. Methods Eng. **17**, 707–718

Fletcher, C.A.J. (1983): Int. J. Numer. Methods Fluids **3**, 213–216

Fletcher, R. (1976): In *Numerical Analysis*, ed. by G.A. Watson, Lecture Notes in Mathematics, Vol. 506 (Springer, Berlin, Heidelberg)

Fletcher, R. (1980): *Practical Methods of Optimization, Vol 1, Unconstrained Optimization* (Wiley, Chichester)

Forsythe, G.E., Malcolm, M.A., Moler, C.B. (1977): *Computer Methods for Mathematical Computations* (Prentice-Hall, Englewood Cliffs, N.J.)

George, A., Liu, J. (1981): *Computer Solution of Large Sparse Positive Definite Systems* (Prentice-Hall, Englewood Cliffs)

Gerald, C. (1978): *Applied Numerical Analysis*, 2nd ed. (Addison-Wesley, Reading, Mass.)

Hackbusch, W. (1985): *Multigrid Methods and Applications*, Springer Ser. Comput. Math. (Springer, Berlin, Heidelberg)

Hageman, L.A., Young, D.M. (1981): *Applied Iterative Methods* (Academic, New York)

Hestenes, M.R., Stiefel, E.L. (1952): Nat. Bur. Stand. J. Res. **49**, 409–436

Hockney, R.W. (1970): Meth. Comp. Phys., **9**, 135–211

Hockney, R.W., Jesshope, C.R. (1981): *Parallel Computers* (Adam Hilger, Bristol)

Isaacson, E., Keller, H.B. (1966): *Analysis of Numerical Methods* (Wiley, New York)

Jackson, C.P., Robinson, P.C. (1985): Int. J. Numer. Methods Eng. **21**, 1315–1338

Jennings, A. (1977a): *Matrix Computation for Engineers and Scientists* (Wiley, Chichester)

Jennings, A. (1977b): J. Inst. Math. Its Appl. **20**, 61–72

Khosla, P.K., Rubin, S.G. (1981): Comput. Fluids **9**, 109–121

Lin, A. (1985): Int. J. Numer. Methods Fluids **5**, 381–391

Markham, G. (1984): "A Survey of Preconditioned Conjugate-Gradient-type Methods Applied to Asymmetric Systems", CERL Report TPRD/L/AP 137/M83

Ortega, J.M., Rheinboldt, W.C. (1970): *Iterative Solution of Nonlinear Equations in Several Variables* (Academic, New York)

Powell, M.J.D. (1976): "A View of Unconstrained Optimization", in *Optimization in Action*, ed. by L.C.W. Dixon (Academic, London) pp. 117–152

Rheinboldt, W.C. (1974): "On the Solution of Large, Sparse Sets of Nonlinear Equations", TR-324, University of Maryland

Rubin, S.G., Khosla, P.K. (1981): Comp. Fluids 9, 163–180

Schneider, G.E., Zedan, M. (1981): Numer. Heat Transfer 4, 1–19

Shanno, D.F. (1983): Comput. Chem. Eng. 7, 569–574

Sonneveld, P., Wesseling, P., de Zeeuw, P.M. (1985): In *Multigrid Methods for Integral and Differential Equations*, ed. by D.J. Paddon, H. Holstein (Oxford Science Publications, Oxford) pp. 117–167

Stone, H.L. (1968): SIAM J. Numer. Anal. 5, 530–558

Stuben, K., Trottenberg, U. (1982): In *Multigrid Methods*, ed. by W. Hackbusch, U. Trottenberg, Lecture Notes in Mathematics, Vol. 960 (Springer, Berlin, Heidelberg)

Swartztrauber, P.N. (1977): SIAM Rev. 19, 490–501

Varga, R. (1962): *Matrix Iterative Analysis* (Prentice-Hall, Englewood Cliffs, N.J.)

Vinsome, P.K.W. (1976): "ORTHOMIN, An Iterative Method for Solving Sparse Sets of Simultaneous Linear Equations", in Proc. 4th Symp. on Reservoir Engineering (SPE of AIME, Los Angeles) pp. 150–159

Wachpress, E.L. (1966): *Iterative Solution of Elliptic Systems and Applications to the Neutron Diffusion Equations of Reactor Physics* (Prentice-Hall, Englewood Cliffs, N.J.)

Zedan, M., Schneider, G.E. (1985): Numer. Heat Transfer 8, 537–557

Chapter 7

Beam, R.M., Warming, R.F. (1979): "An Implicit Factored Scheme for the Compressible Navier-Stokes Equations II: The Numerical ODE Connection," AIAA Paper 79–1446

Dahlquist, G. (1963): BIT 3, 27–43

Gear, C.W. (1971): *Numerical Initial Value Problems in Ordinary Differential Equations* (Prentice-Hall, Englewood Cliffs, N.J.)

Gentzsch, W., Neves, K.W. (1988): "Computational Fluid Dynamics: Algorithms & Supercomputers", Agadograph 311, NATO

Holt, M. (1984): *Numerical Methods in Fluid Dynamics*, 2nd ed., Springer Ser. Comput. Phys. (Springer, Berlin, Heidelberg)

Lambert, J.D. (1973): *Computational Methods in Ordinary Differential Equations* (Wiley, Chichester)

Mitchell, A.R., Griffiths, D.F. (1980): *The Finite Difference Method in Partial Differential Equations* (Wiley-Interscience, New York)

Noye, B.J. (1983): In *Numerical Solution of Differential Equations*, ed. by J. Noye (North-Holland, Amsterdam)

Ortega, J.M., Voigt, R.G. (1985): SIAM Rev. 27, 149–240

Richtmyer, R.D., Morton, K.W. (1967): *Difference Methods for Initial-Value Problems* (Interscience, New York)

Seward, W.L., Fairweather, G., Johnston, R.L. (1984): IMA J. Numer. Anal. 4, 375–425

Trapp, J.A., Ramshaw, J.D. (1976): J. Comput. Phys. 20, 238–242

Widlund, O.B. (1967): BIT 7, 65–70

Chapter 8

Douglas, J., Gunn, J.E. (1964): Numer. Math. 6, 428–453

Dwoyer, D., Thames, F. (1981): "Accuracy and Stability of the Time-Split Finite Difference Schemes", AIAA Paper 81–1005.

Fletcher, C.A.J. (1984): *Computational Galerkin Methods*, Springer Ser. Comput. Phys. (Springer, Berlin, Heidelberg)

Fletcher, C.A.J., Srinivas, K. (1984): Comp. Meth. Appl. Mech. Eng. 46, 313–327

Gourlay, A.R. (1970): J. Inst. Math. Its Appl. 6, 375–390

Gourlay, A.R. (1977): "Splitting Methods for Time Dependent Partial Differential Equations", in *The State of the Art in Numerical Analysis*, ed. by D. Jacobs (Academic, London)

Marchuk, G.I. (1974): *Numerical Methods in Weather Prediction* (Academic, New York)

Mase, G.E. (1971): *Continuum Mechanics* (McGraw-Hill, New York)

Mitchell, A.R., Fairweather, G. (1964): Numer. Math. **6**, 285–292

Mitchell, A.R., Griffiths, D.F. (1980): *The Finite Difference Method in Partial Differential Equations* (Wiley-Interscience, New York)

Peaceman, D.W., Rachford, H.H. (1955): SIAM J. **3**, 28–41

Yanenko, N. N. (1971): *The Method of Fractional Steps*, trans. by M. Holt (Springer; New York, Berlin, Heidelberg)

Chapter 9

Baker, A.J., Kim, J.W. (1987): Int. J. Numer. Methods Fluids, **7**, 489–520

Brooks, A.N., Hughes, T.J.R. (1982): Comput. Methods Appl. Mech. Eng. **32**, 199–259

Brown, G.M. (1960): AIChE J. **6**, 179–183

Carey, G.F., Sepehrnoori, K. (1980): Comput. Methods Appl. Mech. Eng. **22**, 23–48

de Vahl Davis, G., Mallinson, G.D. (1976): Comput. Fluids **4**, 29–43

Donea, J. (1984): Int. J. Numer. Methods Eng. **20**, 101–119

Donea, J., Giuliani, S., Laval, H., Quartapelle, L. (1984): Comput. Methods Appl. Mech. Eng. **45**, 123–145

Finlayson, B.A. (1980): *Nonlinear Analysis in Chemical Engineering* (McGraw-Hill, New York)

Fletcher, C.A.J. (1983): Comput. Methods Appl. Mech. Eng. **37**, 225–243

Griffiths, D.F., Mitchell, A.R. (1979): In *Finite Elements for Convection Dominated Flows*, ed. by T.J.R. Hughes (ASME, New York), pp. 91–104

Haltiner, G.J., Williams, R.T. (1980): *Numerical Prediction and Dynamic Meteorology*, 2nd ed. (Wiley, New York)

Hindmarsh, A.C., Gresho, P.M., Griffiths, D.F. (1984): Int J. Numer. Methods Fluids **4**, 853–897

Jennings, A. (1977): *Matrix Computation for Engineers and Scientists* (Wiley, Chichester)

Klopfer, G.H., McRae, D.S. (1983): AIAA J. **21**, 487–494

Leonard, B.P. (1979): Comput. Methods Appl. Mech. Eng. **19**, 59–98

Morton, K.W. (1971): Proc. R. Soc. London, Ser. A **323**, 237–253

Noye, J. (1983): Chap. 2 in *Numerical Solution of Differential Equations*, ed. by J. Noye (North-Holland, Amsterdam)

Pinder, G.F., Gray, W.G. (1977): *Finite Element Simulation in Surface and Subsurface Hydrology* (Academic, New York)

Raithby, G.D. (1976): Comput. Methods Appl. Mech. Eng. **9**, 75–94

Roache, P.J. (1976): *Computational Fluid Dynamics* (Hermosa, Albuquerque, N.M.)

Steinberg, S., Roache, P.J. (1985): J. Comput. Phys. **57**, 251–284

Warming, R.F., Hyett, B.J. (1974): J. Comput. Phys. **14**, 159–179

Chapter 10

Basdevant, C., Deville, M., Haldenwang, P., Lacroix, J.M., Onazzoni, Peyret, R., Orlandi, P., Patera, A.T. (1986): Comput. Fluids **14**, 23–41

Burgers, J.M. (1948): Adv. Appl. Mech. **1**, 171–19

Christie, I., Griffiths, D.F., Mitchell, A.R., Sanz-Serna, J.M. (1981): Inst. Appl. J. Numer. Anal. **1**, 253–266

Cole, J.D. (1951): *Q. Appl. Math.* **9**, 225–236

Fletcher, C.A.J. (1979): J. Comput. Phys. **33**, 301–312

Fletcher, C.A.J. (1983a): Chap. 3 in *Numerical Solution of Differential Equations*, ed. by J. Noye (North-Holland, Amsterdam) pp. 355–475

Fletcher, C.A.J. (1983b): Comput. Methods Appl. Mech. Eng. **37**, 225–243

Fletcher, C.A.J. (1983c) J. Comput. Phys. **51**, 159–188

Fletcher, C.A.J. (1983d) Int. J. Numer. Methods Fluids **3**, 213–216

Fletcher, C.A.J. (1984): *Computational Galerkin Methods*, Springer Ser. Comput. Phys. (Springer, Berlin, Heidelberg)

Hamming, R.W. (1973): *Numerical Methods for Scientists and Engineers*, 2nd ed. (McGraw-Hill, New York)

Hopf, E. (1950): Commun. Pure Appl. Math. **3**, 201–230

Kim, H.J., Thompson, J.F. (1990): AIAA J. **28**, 470–477

Klopfer, G.H. McRae, D.S. (1983): AIAA J. **21**, 487–494

Lerat, A., Peyret, R. (1975): Rech. Aerosp. 1975–2, 61–79

Noye, J. (1983): Chap. 2 in *Numerical Solution of Differential Equations*, ed. by J. Noye (North-Holland, Amsterdam) pp. 130–353

Peyret, R., Taylor, T.D. (1983): *Computational Methods for Fluid Flow*, Springer Ser. Comput. Phys. (Springer, Berlin, Heidelberg)

Richtmyer, R.D., Morton, K.W. (1967): *Difference Methods for Initial-Value Problems* (Interscience, New York)

Thommen, H.U. (1966): Z. Angew. Math. Phys. **17**, 369–384

Thompson, J.F., Warsi, Z.U.A., Martin, C.W. (1985): *Numerical Grid Generation, Foundations and Applications* (North-Holland, New York)

Whitham, G. (1974): *Linear and Nonlinear Waves* (Wiley, New York)

Yanenko, N.N, Fedotova, Z.I., Tusheva, L.A., Shokin, Y.I. (1983): Comput. Fluids **11**, 187–206

Subject Index

Contents of

Computational Techniques for Fluid Dynamics 2

Specific Techniques for Different Flow Categories

Contents of **Computational Techniques for Fluid Dynamics 1**
Fundamental and General Techniques

Printing: COLOR-DRUCK DORFI GmbH, Berlin
Binding: Buchbinderei Lüderitz & Bauer, Berlin